LIQUID CRYSTALS & PLASTIC CRYSTALS

Volume 1

Physico-Chemical Properties and Methods of Investigation

ELLIS HORWOOD SERIES IN PHYSICAL CHEMISTRY

Series Editor: T. MORRIS SUGDEN, F.R.S.

Director 'Shell' Research Limited, Thornton Research Centre; Associate Professor of Molecular Biology, University of Warwick

Founded as a library of fundamental books on important or growing areas in physical chemistry, this series will serve chemists in industrial research, and in teaching or advanced study

Published or in active preparation

Liquid Crystals and Plastic Crystals

Vol. 1 Preparation, Constitution and Applications
Vol. 2 Physico-Chemical Properties and Methods of Investigation

GRAY, G. W., *University of Hull*
WINSOR, P. A., *'Shell' Research Ltd* } *Editors*

Kinetics and Mechanisms of Polymerization Reactions

ALLEN, P. E. M., *University of Adelaide*
PATRICK, C. R., *University of Birmingham*

Polymer Physics

HEARLE, J. W. S., *University of Manchester Institute of Science and Technology*

Mechanisms of Reactions in Solution

KOHNSTAM, G., *University of Durham*

Electrochemical Phase Formation

THIRSK, H. R.
ARMSTRONG, R. D.
HARRISON, J. A. } *University of Newcastle-upon-Tyne*

Surface Chemistry

PARFITT, G. D., *Tioxide International*
JAYCOCK, M. J., *Loughborough University of Technology*

LIQUID CRYSTALS
&
PLASTIC CRYSTALS

Volume 1

PHYSICO-CHEMICAL PROPERTIES and METHODS OF INVESTIGATION

Editors

G. W. GRAY
Department of Chemistry, University of Hull

P. A. WINSOR
'Shell' Research Limited

ELLIS HORWOOD LIMITED
Chichester

Halsted Press : a division of
JOHN WILEY & SONS, Inc.
New York - London - Sydney - Toronto

First published in 1974 by

ELLIS HORWOOD LIMITED
Coll House, Westergate, Chichester, Sussex, England

this book bears his colophon reproduced from
James Gillison's drawing of the
ancient Market Cross, Chichester.

Distributed in:

Australia, New Zealand, South-east Asia by
JOHN WILEY & SONS AUSTRALASIA PTY LIMITED
110 Alexander Street, Crow's Nest, N.S.W. Australia

Europe, Africa by
JOHN WILEY & SONS LIMITED
Baffins Lane, Chichester, Sussex, England

N. & S. America and the rest of the world by
HALSTED PRESS a division of
JOHN WILEY & SONS INC.
605 Third Avenue, New York, N.Y. 10016, U.S.A.

Library of Congress Catalog No. 73-11504
ISBN (Ellis Horwood) 8-5312003 X
ISBN (Halsted Press) 0-470-32339-6

Printed in Great Britain by
Butler & Tanner Ltd., Frome and London.

Contents

Preface

"Liquid crystals" represent a number of different states of matter in which the degrees of molecular order lie intermediate between the almost perfect long-range positional and orientational order found in solid crystals and the statistical long-range disorder found in ordinary isotropic amorphous liquids and gases.

These intermediate states of matter were first termed "liquid crystals" by Lehmann (1889) because they frequently possess low rigidity in association with optical anisotropy. They were later termed "les états mesomorphes" by Friedel (1922) to avoid the ambiguities and controversies inherent in a term such as "liquid crystals" or "crystalline liquids" and to indicate explicitly their character as quite distinct states intermediate between the "perfectly ordered periodic structure" of solid crystals and the "perfectly disordered structure" of the amorphous type. The nature, properties and relationships of these mesomorphic states to one another and to the "perfectly ordered" solid crystals and the "perfectly disordered" amorphous gases, liquids and solids form the main theme of this book.

In current usage the terms liquid crystal, mesomorphic (or mesomorphous) phase, mesophase and, more rarely, paracrystal or paracrystalline phase are employed almost synonymously. The term "plastic crystal" or "plastic mesophase" is used to describe a particular class of mesophase in which the molecules though orientationally completely disordered reside, with minor fluctuations in position, at the points of a cubic lattice.

Compounds which under suitable conditions give rise to mesophases, we will, for convenience, refer to as "mesogens". Mesogens fall naturally into two main classes which may be further subdivided:

Class 1

Certain non-polar or moderately polar organic compounds. Examples are:

(*a*) numerous compounds with fairly rigid "rod"- or "lath"-like molecules, e.g.,

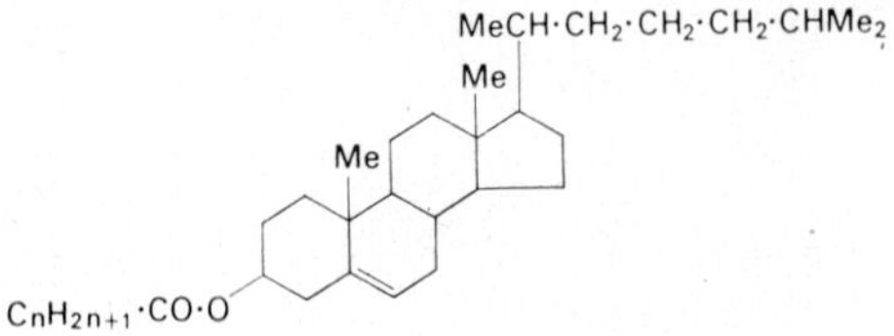

$MeCH \cdot CH_2 \cdot CH_2 \cdot CH_2 \cdot CHMe_2$

Me

Me

$C_nH_{2n+1} \cdot CO \cdot O$

(*b*) numerous compounds with globular molecules (p. 12), e.g., CMe_4, $Me_3C \cdot CMe_3$, cyclohexane, CCl_4, C_2Cl_6, camphor.

Class 2

Certain amphiphilic organic compounds, that is compounds that contain within the same molecule localized lipophilic and hydrophilic groups which show, in marked degree, tendencies to confer solubility in organic solvents and in water respectively. Examples are:

(*a*) anionic amphiphiles, e.g.,

$$Me \cdot (CH_2)_n \cdot CO_2^- Na^+, K^+, NH_4^+ \text{ or } Ca_{\frac{1}{2}}^+$$

soaps of the fatty acids

$$CH_2 \cdot CO \cdot OCH_2 \cdot CH(C_2H_5)CH_2 \cdot CH_2 \cdot CH_2Me$$
$$|$$
$$CH(SO_3^- Na^+)CO \cdot O \cdot CH_2 \cdot CH(C_2H_5)CH_2 \cdot CH_2 \cdot CH_2Me$$

Aerosol OT

(*b*) cationic amphiphiles, e.g.,

$$C_{16}H_{33} \cdot NMe_3^+ Cl^-, Br^- \text{ or } I^-$$

(*c*) non-ionic amphiphiles, e.g.,

$$n\text{-}C_8H_{17}NH_2,\ n\text{-}C_{12}H_{25} \cdot O \cdot (CH_2 \cdot CH_2O)_{6 \text{ to } 12}H \text{ etc.}$$

The mesogens of Class 1 we shall term non-amphiphilic mesogens to distinguish them from the amphiphilic mesogens of Class 2.

The amphiphilic mesogens frequently form mesophases, either at room temperature or at higher temperatures, which can incorporate into their structure considerable amounts of water and/or organic compounds. On this account the mesophases formed by amphiphilic mesogens have commonly been called "lyotropic", i.e., solvent-induced mesophases. It is important to remember, however, that the majority of the anionic amphiphilic mesogens also form mesophases when in the pure condition. Thus, the branched-chain amphiphiles, such as Aerosol OT, are mesomorphous at room temperature and the straight-chain amphiphiles, which are crystalline at room temperature, form mesophases at temperatures above that at which the solid crystal lattice breaks down. This temperature is, of course, usually depressed by the presence of a second component (water etc). Further, the nature of the mesophases formed by the solvent-containing systems is itself highly dependent on temperature and on composition.

The non-amphiphilic mesogens have mainly been studied as pure compounds with emphasis on the transition temperatures for the mesomorphic and non-mesomorphic states. The non-amphiphilic mesophases are therefore commonly called "thermotropic", i.e., temperature-induced, but in this case it is important to remember that, like the amphiphilic mesophases, they can also incorporate considerable amounts of organic solvents, although usually not of water, into their structure with attendent modification of their character and properties and with large alterations of the transition temperatures. Thus, both lyotropic and thermotropic behaviour are characteristic of all mesophases without distinction and the merit of continuing to use these terms for classification is questionable since they may suggest the existence of differences which may not be real. However, the terms have in some cases been employed in the present work according to the choice of the individual authors.

The different classes of mesogens have often been studied by independent groups of investigators with some lack of intercommunication between the groups.

The non-amphiphilic mesophases formed by lath-like molecules were initially, and to a considerable extent also latterly, studied largely by organic chemists particularly interested in the relationships between the chemical constitutions of the mesogens and the types and temperature ranges of the pure mesophases produced.

The non-amphiphilic cubic, optically isotropic mesophases formed by globular molecules, the so-called plastic crystals, were discovered by Timmermanns (1935) when collecting data for his book *Les Constantes Physiques des Composés Organiques Cristallisés*. These mesophases have consequently mainly been regarded as plastic solid crystals rather than as liquid crystals or mesophases and information concerning them has been largely confined to the literature of the solid state. This classification has usually been justified by the fact that their constituent molecules occupy mean positions at the points of a regular three-dimensional cubic lattice and to this extent show order similar to that in solid crystals. However, the molecules undergo relatively free thermal rotatory motions about the lattice points so that there is complete long-range orientational disorder between them. In this respect, of all the mesophases, plastic crystals most closely resemble the amorphous liquid.

Although the mesophases derived from ammonium and potassium oleates (typical amphiphilic mesogens) were among the earliest liquid crystalline phases to be recognized, the extensive study of amphiphilic mesophases came later than that of non-amphiphilic series and virtually commenced about 1920 with the phase-rule studies of MacLennan and of McBain and his school. Studies in this field have received much industrial support on account of their relevance to soap technology and to the synthetic detergent industry and its ancillaries (laundering, preparation of emulsifiable oils, emulsion polymerization, etc.)

Because of the degree of isolation of these different fields of mesophase study, it happens that no recent treatment has appeared which gives a balanced account of all aspects of the subject. Thus, the articles by Friedel [1], Chatelain [2], Brown and Shaw [3], Saupe [4], Sackmann and Demus [5] and the monograph by Gray [6] are heavily weighted on the side of the non-amphiphilic mesophases formed by molecules of lath-like character.

Articles [7–11] which consider the cubic mesophases or "plastic crystals" formed by the non-amphiphilic globular molecules, are largely confined to the literature of the solid state.

The reviews by McBain [12, 13, 14] and the more recent reviews by Skoulios [15], Winsor [16, 17], Luzzati [18] and Ekwall, Mandell and Fontell [19] discuss only the amphiphilic liquid crystalline phases, though including within these those of the cubic type.

It is hoped that the two volumes of this work, by collecting together under multiple authorship material relating to all the above aspects of the subject, will produce a comprehensive account so that the mutual relevance of studies in the different branches may be evident and so that investigators in each branch may benefit from the researchs of those engaged in the others.

The articles provided by the different contributors each represent the individual views and terminology of the writers and each article forms a self-contained unit. This necessarily results in duplication but avoids the inconvenience of frequent cross reference. Further, since on some questions different writers may hold contrary opinions, there may be apparent inconsistencies between the views expressed in their various exposés. It is hoped, however, that this may not prove disconcerting to the reader but give the impression of actively developing research in which, though much has been achieved, many questions remain yet to be answered.

It must be admitted that, for reasons of time and space, certain omissions exist in the text. Thus, some mesophases formed by less familiar organic compounds and by some inorganic compounds [20] have not been explicitly included. Further, only mesophases in thermodynamic equilibrium (metastable equilibrium in the case of the monotropic non-amphiphilic mesophases) have received systematic treatment. Certain non-equilibrium systems of liquid crystalline character, e.g., virus preparations and the tactosols formed by dispersions of inorganic compounds have been mentioned only incidentally.

When this work was first planned, detailed consideration of the plastic crystalline mesophases formed by globular molecules was not envisaged. However, during compilation the development of knowledge concerning the amphiphilic cubic mesophases and the recent characterization by X-ray diffraction of cubic mesophases among the mesophases formed by the lath-like non-amphiphilic mesogens has made the relationship of the plastic crystalline mesophases to the general field of liquid crystals outstandingly evident. The general significance of the plastic crystalline mesophases is therefore considered in broad perspective in the Introduction to Volume 1, and a more detailed account of their properties and structures is given in Chapter 2.2. The amphiphilic cubic mesophases have been fully considered in the general context of their relationships to the other amphiphilic mesophases.

The format adopted in the present work is as follows.

Volume 1, *Preparation, Constitution and Applications*, deals broadly with the formation and constitution of liquid crystalline and plastic crystalline mesophases and provides an account of the recently developed and still fast developing technological applications of non-amphiphilic liquid crystals.

Chapter 1 presents a short introduction which attempts to place the broad field of mesophase studies in chronological and structural perspective.

Chapter 2 deals with the classification and organization of mesophases formed by non-amphiphilic and by amphiphilic compounds.

Chapter 3 considers the microstructures of liquid crystalline phases, including disclinations, fluctuations, light scattering and dynamical properties, in terms of the Continuum Theory.

Chapter 4 discusses the influence of composition and structure on the formation of liquid crystals by non-amphiphilic mesogens.

Chapter 5 deals with the influence of composition and temperature on the formation of mesophases in amphiphilic systems. In the case of the "fused" mesophases, particular attention is given to the interpretation of the observed behaviour afforded by the R-Theory.

Chapter 6 discusses the significance of liquid crystals in relation to cell membranes and life processes.

Chapter 7 describes some scientific and technological applications of non-amphiphilic liquid crystals. It is of interest that although amphiphilic mesophases are of great technological importance and their study has received much industrial support on this account, their importance lies more in their avoidance than in their use. Thus formation of the highly viscous "middle" mesophase may give rise to difficulties in pumping operations in soap manufacture. Further, in the formulation of the very numerous technical water-dispersible preparations containing surface-active agents (cleaning liquids, dispersible metal-working oils, insecticidal and fungicidal preparations, etc.) the formation of mesophases may restrict the range of temperature available for stable storage.

In contrast, although the non-amphiphilic liquid crystals were originally studied almost wholly for their academic interest, they have in recent years been found to possess many interesting scientific and technological applications.

Volume 2, *Physico-Chemical Properties and Methods of Investigation*, discusses in more detail certain of the physico-chemical properties of the various mesophases and shows how studies of these particular properties have been used in elucidating the structures of the mesophases. The topics covered in the ten chapters include electron microscopy, optical properties, X-ray diffraction, electrical properties, infrared, Raman, visible and ultraviolet spectroscopy, magnetic resonance spectroscopy and thermal properties. Rheological properties are discussed elsewhere in relation to particular mesophases.

REFERENCES

[1] Friedel, G. *Annls. Phys.* **18,** 273 (1922).
[2] Chatelain, P. *Bull. Soc. fr. Minér. Cristallogr.* **77,** 323 (1954).
[3] Brown, G. H. and Shaw, W. G. *Chem. Rev.* **57,** 1097 (1957).
[4] Saupe, A. *Angew. Chem. Int. Edn.* **7,** 97 (1968).
[5] Sackmann, H. and Demus, D. *Fortsch. d. chem. Forsch.* **12,** 349 (1969).
[6] Gray, G. W. *Molecular Structure and the Properties of Liquid Crystals*, Academic Press, London and New York (1962).
[7] Timmermanns, J. *J. Phys. Chem., Solids* **18,** 1 (1961).
[8] Staveley, L. A. K. *Ann. Rev. Phys. Chemistry* **13,** 351 (1962).
[9] Westrum, E. F., Jr. and McCullough, J. P. *Physics and Chemistry of the Organic Solid State*, Interscience/Wiley (1963), p. 89.
[10] Aston, J. G. *Physics and Chemistry of the Organic Solid State*, Interscience/Wiley (1963), p. 543.
[11] Smith, G. W. *International Science and Technology* (January 1967), p. 72.
[12] McBain, J. W. *Advances in Colloid Science* **1,** 99 (1942) (Interscience/Wiley (1942)).
[13] McBain, J. W. *Colloid Chemistry* **5,** 102 (1944) (edited by J. Alexander), Reinhold, New York (1944).
[14] McBain, J. W. *Colloid Science*, D.C. Heath & Co., Boston (1950), p. 240.
[15] Skoulios, A. *Advances in Colloid and Interface Science* **1,** 79 (1967).
[16] Winsor, P. A. *Chem. Rev.* **68,** 1 (1968).
[17] Winsor, P. A. *Molec. Crystals Liqu. Crystals* **12,** 141 (1971).
[18] Luzzati, V. *Biological Membranes*, Academic Press, London (1968), Chap. 3, p. 71.
[19] Ekwall, P., Mandell, L. and Fontell, K. *Molec. Crystals Liqu. Crystals* **8,** 157 (1969).
[20] Cf. Ubbelohde, A. R. *Melting and Crystal Structure*, Oxford University Press (1965).

1

Introduction

G. W. Gray and P. A. Winsor

In this chapter we shall place ideas now held concerning the nature of mesophases in chronological perspective and indicate broad relationships and differences between the various types of mesophase both non-amphiphilic and amphiphilic.

The earliest clear recognition of "Liquid Crystals" is usually attributed to Reinitzer [1], who in 1888 noted the colour phenomena, now known to be characteristic of many cholesteric mesophases, which arise when melts of cholesteryl acetate or benzoate are cooled. He recorded that similar colour phenomena had been found earlier by other workers with cholesteryl chloride, cholesterylamine and the silver salt of cholesteric acid. In addition he observed, using a carefully purified sample of cholesteryl benzoate, that this "melted" at 145·5°C to give a cloudy fluid which, on further heating, suddenly clarified at 178·5°C; subsequent cooling gave colour effects similar to those observed with cholesteryl acetate. "It was principally this remarkable phenomenon of the presence of two melting points, if one may call them such, and of the occurrence of the colour effect which led me to the opinion that here, and with the acetate, physical isomerism must be present. On this account I asked Professor Lehmann at Aachen to study these relationships more closely."

Reinitzer also observed that on cooling the clear melt of cholesteryl benzoate, "there appeared at a certain point a deep blue colour which spread out rapidly through the whole mass and almost as quickly disappeared again, leaving in its place a uniform turbidity. The mass then remained turbid for some time but fluid; on further cooling a similar colour effect appeared for the second time to be followed by crystallization of the mass and a simultaneous disappearance of the colour effect."

Lehmann [2], using the polarizing microscope, confirmed Reinitzer's observations and recorded similarities in the melting behaviour and plasticity of the "plastic crystalline" phase of cholesteryl benzoate (probably the cholesteric phase) and the cubic higher temperature phase of silver iodide which he also regarded as within the realm of "Fliessende Krystalle".

Later [3] he observed similar phenomena on cooling the amorphous melts of cholesteryl esters of fatty acids. Cholesteryl benzoate and these esters, except the formate, give, besides birefringent mesophases, modifications which are apparently mesomorphous but optically isotropic [4] or nearly so [2, 3, 5]. These somewhat confusing observations are recorded here on account of their possible relevance to recently discovered cubic mesophases to be considered below (cf. p. 12).

Following and concurrently with Lehmann's studies, synthetic work was done by organic chemists in Germany to ascertain whether effects such as those observed by Reinitzer and Lehmann were due to impurities and, later, to unravel the relationships between chemical constitution and the capacity to form liquid crystals.

This work established the formation of liquid crystals by a wide range of organic compounds with "lath-like" molecules (cf. Chap. 4.1). A critique and bibliography of this work and of Lehmann's publications was issued by Friedel in 1922 [6]. In this account, he proposed a clarified systematization and nomenclature which has since been widely adopted. This was based mainly on observations with the polarizing microscope, including those of Lehmann, Friedel himself and other French investigators. His own studies were illustrated by excellent photomicrographs.

Although Lehmann's term "liquid crystal" remains in general use it was vehemently opposed by Friedel, on the thesis that "liquid crystals" were neither true liquids nor true crystals but represented two new states of matter intermediate between the solid crystal and the amorphous liquid.

Seeking an unambiguous adjective to describe these two new forms he wrote "Since without exception, the ranges of stability of the two new types are interposed, for a particular compound, between that of the crystalline type which prevails at low temperatures and that of the amorphous type, which is found at higher temperatures; since also their molecular structures form, in a way, two echelons between the periodic perfectly ordered structure which is that of the crystal type and the perfectly disordered structure of the amorphous type, the term mesomorphic would seem suitable" . . . "This term would seem moreover sufficiently vague not to be able to raise difficulties" . . . in the event of the subsequent discovery of new facts.

Further, seeking terms to distinguish individually each of the two types and to avoid the use of the words "liquid" and "crystal", he wrote "what is needed are two adjectives, not having any too specific meaning, which would serve to distinguish the two forms simply and without ambiguity. I will call smectic (σμῆγμα, soap) materials, particles, phases, etc., of the first type" (Lehmann's Fliessende Kr., Schleimig Flüssige Kr., liquides à coniques) "because . . . the oleates of ammonium and potassium in particular were the first materials of this class to be characterized. I will call nematic (νῆμα, thread) the forms, phases, etc., of the second type (Lehmann's Flüssige Kr., Tropfbar Flüssige Kr., liquides à fils) on account of the linear discontinuities, twisted like threads—which are one of their salient characteristics".

In enumerating only two mesomorphic types, smectic and nematic,* Friedel rejected Lehmann's classification of cubic silver iodide as liquid crystalline [2]. According to Friedel "this compound, in the cubic form which it adopts above 146°C, is a soft plastic solid, otherwise with all the properties of a crystalline substance and none of those of mesomorphic substances with which it shows nothing in common".

It is probably partly on account of this dictum, which in the light of facts since discovered must be regarded as mistaken [7], that the mesomorphic character of the large group of cubic "plastic" crystals first characterized by Timmermanns [8] in 1935 has been widely overlooked. It is rarely that the mesomorphic character of the cubic plastic crystal phase has been explicitly noted [9]. Yet this phase, since it is definitely not of the amorphous type and equally definitely not of the "perfectly ordered" crystal type (since there is no long-range orientational order between individual constituent molecules and even the periodic order is far from perfect (cf. p. 15)), falls clearly within Friedel's definition of mesomorphic and thus constitutes a well-defined third mesomorphic type. Had the term "plastic mesophase" rather than "plastic crystal" been adopted by Timmermanns, who actually indicated clearly the mesomorphic character of his newly discovered phase, (cf. Chap. 2.2, p. 49) the impact of his discovery on liquid crystals might have been greater. One may legitimately wonder whether the term "plastic crystals" might also be included in Friedel's objection that "l'example des 'cristaux liquides' nous donne l'occasion de constater l'action fâcheuse d'un mot mal choisi". None the less, both the terms "liquid crystal" and "plastic crystal" are now established. To avoid confusion, however, it must be realized that these terms represent neither true liquids nor true crystals but additional and distinct phases of matter.

Amphiphilic Mesophases

In addition to the three types of non-amphiphilic mesophase, smectic, nematic and plastic (= cubic), several related types of amphiphilic mesophase are known. The chronology of the discovery of these mesophases will not be detailed since it is relatively recent [10, 11, 12] and covered later in this book. In order however that the characters of the non-amphiphilic and the amphiphilic mesophases may be discussed side by side, we will anticipate here certain matters which are considered more fully later.

The mesophases formed by typical amphiphiles may be regarded as based, not on structural arrangements of individual molecules, as in most non-amphiphilic mesophases, but on the arrangement of multimolecular units, termed "aggregates" or "micelles", which are themselves commonly "liquid" or "fused" in character (Chap. 5). Although many amphiphiles form mesophases in the pure state, many amphiphilic mesophases also incorporate one or more additional components, frequently including

* A further distinction has often been made between nematic and cholesteric mesophases, but Friedel considered the cholesteric form to be a modification (nématique cholestérique) of the ordinary nematic mesophase (nématique proprement dit).

water, which may be located within, at the surface of and/or between the micellar units, in the last case being termed "inter-micellar liquid". The micellar units and inter-micellar liquid constitute jointly the amphiphilic mesophases, which, like the non-amphiphilic mesophases (either uni- or multi-component) represent single thermodynamically stable (or sometimes thermodynamically meta-stable or "monotropic") phases.

In the amphiphilic mesophases the shapes of the micelles are determined by intra-micellar forces, i.e., by intermolecular forces operative both within and at the surface of the micelles. These forces will depend on the compositions of the micelles and of the inter-micellar liquid (if any). Both these compositions will depend on the bulk composition and on the temperature,

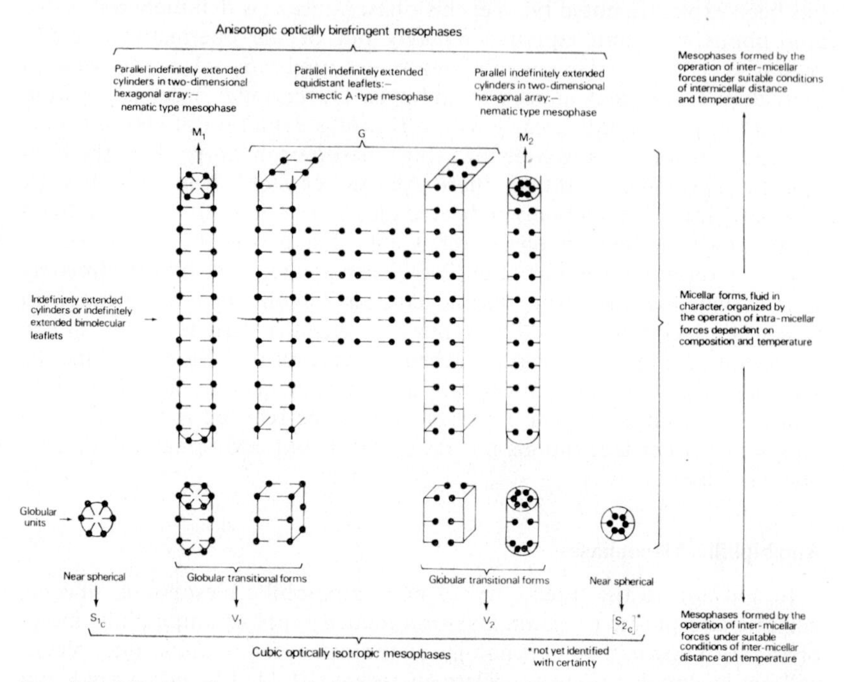

FIG. 1.1. Micellar forms and their interactions which give rise to amphiphilic mesophases

on account of its influence on molecular distribution and orientation. The shapes of the micelles may, therefore, vary with temperature and, in multi-component systems, with the proportions of the components (cf. Chap. 5). According to conditions, micelles of a variety of shapes and sizes may arise. However, although in the amorphous liquid phases formed at higher temperatures or/and higher solvent content a variety of micellar shapes and sizes is often present in thermal equilibrium, in the individual mesophases themselves the micellar forms are more definite, each form (or in some instances possibly a particular combination of forms) giving a specific mesophase. These more definite micellar forms may be classified (Fig. 1.1) as indefinitely extended rods, indefinitely extended sheets, or globular units, i.e., spheres or not too greatly oblate or prolate spheroids.

If these micelles are in sufficiently close proximity (i.e. the amount of inter-micellar liquid is not sufficient to cause too great a separation of the micelles), and the temperature is not too high (i.e. the disorganizing effect of thermal motion is not too great) the operation of inter-micellar forces causes the micelles to adopt a thermodynamically stable arrangement showing a minimum potential energy. This stable arrangement constitutes the mesophase. Thus, the formation of the amphiphilic mesophases depends on two broadly separable sets of interactions each affected by thermal motion, i.e., the intra-micellar interactions (short-range) determining the micellar shape and size, and the inter-micellar interactions (often of longer range) determining the mutual orientation of the micelles. The structures of the numerous types of amphiphilic mesophases are discussed in Chapter 5. At present they will be considered briefly in correlation with the chronological development of ideas concerning the structures of the non-amphiphilic mesophases.

Structures of Individual Mesophases

General Considerations

The molecular structures of the non-amphiphilic mesophases were considered by Friedel [6]; his conclusions, largely based on studies with the polarizing microscope, were supported by citations of work in other fields (e.g. X-ray diffraction, behaviour in electric and magnetic fields).

Although his views have required some revision in the light of more recent work (cf. Chap. 2.1), his account still represents a clear exposé of the subject. His general conclusions are recounted here in the words of his own summary.

"The molecular structures to which we have been led for smectic and nematic bodies are in remarkable accord with the order of succession of the four types of matter in the scale of temperatures. To low temperatures belongs the crystalline type in which the spatial periodic repetition of the units is perfectly defined and in which thermal agitation plays only a very insignificant rôle (un rôle très effacé). At higher temperatures the smectic type still exhibits a certain degree of periodic repetition but of the perfectly periodic structure of the crystal there persists only the parallelism of a common direction of the molecules and their disposition on equidistant surfaces. Thermal agitation has broken other liaisons. At still higher temeratures the nematic phase exhibits thermal agitation as having broken all periodicity, all regular distribution of corresponding points of the molecules, and as permitting only the parallelism of a common direction of the molecules to persist, a parallelism moreover that is only mean parallelism and disturbed in detail by a visible motion" (Brownian motion). "Finally at still higher temperatures all detailed order disappears and only mean properties remain; here is the amorphous phase."

THE SMECTIC PHASE

Friedel considered, on account of the optically uniaxial character of all smectic phases which he studied, that on the individual planes the molecules are arranged at random and turned in all ways except in having one direction in common oriented normal to the plane. It should be remarked, however, that a biaxially ordered arrangement of the molecules within each individual layer would still be compatible with the uniaxial character of the smectic phase provided that the successive planar layers were themselves arranged with a random succession of the axial directions peculiar to each individual layer.

As will be discussed in Chapter 5 such a state of affairs must be envisaged for at least some of the smectic phases of the amphiphilic series [13]. Further it is now known (Chap. 2.1) that with certain non-amphiphilic mesogens a number of polymorphic smectic phases [14a] arises in which a higher degree of order occurs than in the smectic phases studied by Friedel (now termed "smectic A" [14a]). Although Friedel noted the possibility of smectic polymorphism, he was sceptical of its actual occurrence.

The characterization and classification of smectic phases have been particularly advanced by the work of Arnold, Demus, Sackmann and associates [14a]. In all these phases (with the qualified exception of the "smectic D" phase—cf. below), the common feature is the parallel mean orientation of the molecules, a feature also common to the solid and to the nematic mesophase considered in the next section. The different smectic mesophases, and ultimately the solid, arise by the superimposition on this parallel arrangement (the sole feature in the nematic phase) of successive additional degrees of order with falling temperature. A schematic representation of the possible nature of the processes is given in Fig. 1.2.

In the amphiphilic series of smectic mesophases the planar layers of the non-amphiphilic series (smectic A types (Fig. 1.2)) are typically replaced

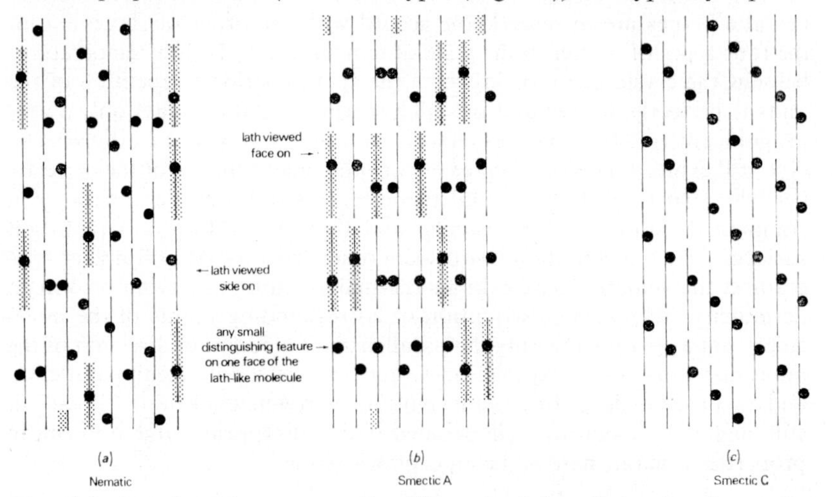

FIG. 1.2. For legend see opposite. The extended polar order drawn for simplicity in (c) does not occur in practice.

FIG. 1.2. Types of arrangement of the lath-like molecules in certain non-amphiphilic mesophases

Each column, (a), (b) and (c), represents diagrammatically a cross-section through an idealized arrangement with the molecules longitudinally aligned in the plane of the paper parallel to a plane surface that lies at right angles to the plane of the paper and parallel to the left-hand margin. This cross-section will be repeated indefinitely above and below the plane of the paper.

(a) The only regular feature of the arrangement is the mean longitudinal parallelism of the molecules (cf. Chap. 2.1). The molecules are otherwise disordered both in the plane of the paper and in the plane at right angles to the direction of mean parallelism. In the latter plane the cross-sections of the laths are in complete two-dimensional isotropic disorder. This represents the classical uniaxial structure for the nematic phase. The lateral equidistance and parallelism of the molecules as drawn in the figure represent statistical mean equidistance and parallelism. There are considerable deviations on either side of the mean (cf. Chap. 2.1).

(b) In addition to possessing mean longitudinal parallelism each molecule conditions the mean position of its neighbour so that the mean distance between their centres (or ends) is minimized. This results in planes of structural discontinuity which are perpendicular to the direction of parallelism. The layers of molecules so constituted can slide with equal facility in all directions over the planes of discontinuity. There are no additional regular features of molecular arrangement. This represents the classical uniaxial structure for Friedel's smectic phase, now classified as "smectic A".

(c) Additional regularities of molecular arrangement result in the mean relative positions of minimum energy for neighbours lying not with the mid-points of the molecules at a minimized mean separation distance, but with certain non-corresponding points at this minimized distance. This results in planes of structural discontinuity inclined to the direction of mean parallelism of the molecules. The layers of molecules so constituted are regarded as free to glide over the planes of discontinuity. The details of the additional regularities of molecular arrangement which give rise to this "smectic C" structure have not yet been clearly characterized. One hypothetical possibility is indicated in Fig. 1.2(c). Although the molecules are here represented as showing mean orientational order both in the plane of the paper and in the plane at right angles to the direction of mean parallelism, they do not show regular positional order in the latter plane. In this plane the lath cross-sections, on this model, show two-dimensional nematic order.

(d) (not illustrated in figure).
With some compounds the acquisition on fall of temperature of further regularities of arrangement, e.g., a two-dimensional hexagonal arrangement of the parallel molecules in each individual layer or a restriction in the freedom of gliding at the planes of discontinuity, can give rise to additional forms of smectic mesophase (smectic B, etc.). Such forms may apparently be derived from smectic A or from smectic C.

(e) (not illustrated).
The final step in the ordering process gives rise to the solid crystal in which the molecules retain their parallelism and in addition acquire virtually complete orientation and fixed positions at the points of a perfectly periodic three-dimensional crystal lattice.

Note: When a compound forms two or more mesophases they always follow the succession (a), (b), (c), (d) above with falling temperature, although intermediate phases may be absent, i.e. one may observe (a), (c), (d); (a), (d); (c), (d); etc. In a few instances, a further mesophase of "cubic" type (not represented on Fig. 1.2) may be found in the higher temperature range of the series (cf. Chap. 2.2).

by a succession of parallel and equidistant amphiphilic "bimolecular leaflets" of indefinite extension (Fig. 1.1). These mesophases closely resemble the non-amphiphilic smectic phases of Friedel (smectic A) and show analogous uniaxial optical behaviour.

Polymorphism found among the smectic mesophases in the amphiphilic series is apparently often due to different preferred mutual orientations of the polar ends of the molecules on the face of the individual leaflets under different conditions of composition and/or temperature. The uniaxial optical behaviour found with these phases would be accounted for by random orientational stacking of the individually biaxial layers. Amphiphilic smectic polymorphy due to the hydrocarbon chains or polar groups adopting a degree of order more closely approaching the crystalline type (cf. Chap. 5) is also known.

The Nematic Mesophase and its Cholesteric Modification

Friedel classified as a single mesomorphic type the ordinary nematic phase ("nématique proprement dit") and the cholesteric phase ("nematique cholestérique") which arises with optically active substances which otherwise would yield an ordinary nematic phase. Thus, if an unresolved dl mixture yields a nematic phase, then even partial resolution transforms the nematic phase, without discontinuity, into a cholesteric phase (Chap. 2.1). Further, with a nematic phase formed by a compound which is not resolvable, incorporation of any optically active material results in conversion, without discontinuity, to a cholesteric phase (Chap. 2.1). In spite of their essential structural identity, the nematic mesophase "proprement dit" and the cholesteric phase show apparently widely divergent properties so that the validity of their grouping together by Friedel has sometimes been questioned. However, work in recent years has fully supported Friedel's reasoning.

(a) *The ordinary nematic phase*

The structure of the nematic phase is much more mobile than that of the smectic phase. This is shown by the high fluidity of the nematic phase, which is sometimes more fluid than the amorphous liquid to which it gives rise when heated. It is also manifest in spontaneous Brownian movements of the thread-like discontinuities which give the nematic phase its name. According to Friedel, the periodicity of molecular arrangement in the smectic phase is absent in the nematic phase and the centres of the molecules are distributed at random as in the amorphous liquid. However, on account of forces arising from their mutual interaction, the molecules adopt a common mean parallel orientation. The uniaxial character of the phase makes it necessary to consider the molecules as rotated in all ways possible about their common direction so that the assembly possesses a symmetry of revolution about their common direction which corresponds to the optic axis.* Similar considerations apply in the case of the smectic

* This does not mean that each individual molecule can rotate freely [15, 16]. Rather, as the result of many small thermal rotatory displacements, the average orientational order of the molecular cross-sections, in planes normal to the optic axis, is random.

A phase. Friedel continues "one cannot refrain from remarking that for the one as for the other of the two mesomorphic forms, the existence of a direction common to all the molecules is in accord with the observation of Vorländer on the chemical formulae of these mesomorphous compounds. As the chemists see it, the molecule is always linear without bifurcations of any importance, and remarkably long. Probably rectilinear in the mesomorphic state, it has on this account the form of a needle, so to speak. Indeed one can hardly see, particularly with a very mobile liquid such as nematic phases often are, how molecules of this shape, constituting as it were a packet of needles, would be able to arrange themselves in a manner such that the assembly has a normal density other than by arranging themselves parallel to one another or at least close to parallel. It is true that the same compound at higher temperatures, shows the amorphous form. But it remains legitimate to imagine that at the point of transition to the amorphous phase the molecule, though remaining linear, ceases to be rigid and becomes capable of bending back on itself and thus entangled with its neighbours. Measurements by Schenck and Eichwald on the viscosity would be in good accord with this idea for nematic compounds for which the coefficient of internal friction increases sharply on passing from the nematic to the liquid state." These views agree with present day ideas except that, in contrast to the almost complete disorder proposed by Friedel, de Vries [17] has concluded that with some nematic phases studied by X-ray diffraction, more ordered groups (= cybotactic groups) arise locally, with fall of temperature, as precursors of the smectic state.

An interesting fact recorded by Sandquist [18, 19] and discussed by Friedel is that on adding a small quantity of water to crystalline 9-bromophenanthrene-3-sulphonic acid one obtains a thick pasty smectic mesophase; addition of further water gives a cloudy, much more mobile, nematic phase, while addition of still more water finally gives an amorphous solution. Thus, the crystal on addition of solvent has passed through the same sequence of phases as is observed in other cases on rise of temperature. It would appear that the units in these aqueous solution phases are monomolecular and not, as with the more typical aliphatic amphiphiles, multimolecular.

In micellar solutions of the latter type, containing rod-shaped micelles of indefinite length, these micelles, under appropriate conditions of temperature and concentration, arrange themselves parallel to one another in two-dimensional hexagonal array, an arrangement first found by Bernal and Fankuchen [20] for the virus particles in liquid crystalline solutions of tobacco mosaic virus. The "middle" phases M_1 and M_2, so constituted (Chap. 5), unlike the non-amphiphilic nematic phases formed by molecules of finite length, are highly viscous immobile phases in which the local micellar arrangement, in contrast to the local arrangement in the non-amphiphilic nematic phases, is highly persistent. This difference between what we might term the amphiphilic micellar nematic phases and the non-amphiphilic molecular nematic phases, is not surprising in view of the great difference in the dimensions of the parallel units in the two types.

The amphiphilic micellar nematic phases M_1 and M_2, like the non-amphiphilic molecular nematic phases, are optically uniaxial but, in accord with the different structures of their constituent units, usually optically negative. Although the M_1 phase is recorded as occurring only in the presence of water, numerous examples of the M_2 phase are known in which the amphiphile is present in the pure state. With certain "branched chain" amphiphiles, the pure amphiphile forms the M_2 mesophase at room temperature. With other amphiphiles which are crystalline at room temperature, the M_2 mesophase is formed on heating to a temperature above the limit of stability of the crystal lattice (Chap. 5).

(b) *The cholesteric nematic phase*

The cholesteric nematic phase (usually more simply termed the cholesteric phase) possesses striking and unique optical properties which were very fully described by Friedel. Cholesteric phases also exhibit certain properties which resemble those of the nematic phase and others which are similar to those of the smectic phase. Friedel therefore asked "Do these phases belong to one or the other of these two types, of which they are only a particular mode, or do they actually constitute a third form of mesomorphic phase?" In attempting to elucidate this point he cited, among others, the following facts.

Most cholesteric materials (i.e. as known at that time) are derivatives of cholesterol. But optically active amyl cyanobenzalaminocinnamate shows the same properties which are therefore not specific to cholesterol derivatives. Also certain cholesterol derivatives show only a smectic phase between the crystal and amorphous melt and no cholesteric phase.

Many compounds form the cholesteric phase as their sole mesomorphous phase. Others form successively a smectic phase at lower temperatures and a cholesteric phase at higher temperatures, always with a discontinuous transition between these phases. But no compound known to Friedel (or since discovered) forms both a nematic phase and a cholesteric phase. Thus, although the smectic phase shows an independent existence in all cases, the nematic phase and the cholesteric phase replace one another. A given material in the temperature interval between the crystalline phase and the amorphous phase may show a nematic phase or a cholesteric phase, but never the two. Further, in this respect mixtures behave similarly to pure substances. Friedel considered that these facts presented a very cogent reason for thinking that the cholesteric phase is a particular form of the nematic phase.

Further supporting evidence lay in the observation that cholesteric phases and nematic phases are miscible in all proportions, the mixtures, being homogeneous and behaving like a pure cholesteric mesophase.

Other evidence was drawn from a very careful consideration of the optical properties. This cannot be considered here other than extremely briefly, but the following important conclusion reached by Friedel from his study of the "plane texture" may be quoted. "The plane texture, and the screw-characteristics ("torsions") of the sense and pitch which it implies, are necessarily linked to a molecular dissymmetry" (d- or l-) "which is not, however, the cause of the optical rotatory power" (of the plane texture) "but determines this indirectly by giving rise to the arrangement. Cholesteric phases thus seem to be the form taken by the nematic phases when they are endowed with molecular holoaxial asymmetry."* This conclusion agrees entirely with modern results (Chap. 2.1) establishing that the incorporation of any effectual excess either of d- or of l- molecules into any ordinary nematic phase converts it, without discontinuity, into a cholesteric phase. Conversely, when the preponderance of one or other optical isomer is eliminated, the cholesteric phase is converted without discontinuity to an ordinary nematic phase.

In this connection the as yet unanswered question arises whether one would expect to generate cholesteric properties on incorporation of optically active molecules into the amphiphilic M_1 and M_2 phases which as already discussed are essentially of nematic type.

The cholesteric properties, discovered by Conmar Robinson (cf. Chap. 4.3), for certain solutions of poly-γ-benzyl-L-glutamate, may be of significance in this connection. In solutions above a certain concentration, the polymer molecules possess a helical conformation comparable to that of a rod of barley sugar and these helical rod-like molecules play a rôle in constituting the mesophase which might be considered as analogous to that of the cylindrical rod-like micelles in M_1 and M_2 amphiphilic phases. In agreement with this, the helical rods are locally arranged with respect

* This conclusion may be further elucidated by the following remarks. When confined between slide and coverslip a nematic phase may assume the "homogeneous texture" in which the molecules lie parallel to one another and to the plane of the slide and coverslip. This may be regarded as due to the layer of molecules attached to the slide being anchored thereto and fixed in a common direction. The nematogenic molecules in the layer above this anchored layer will be oriented by the fixed molecules in a mean direction parallel to these but with angular thermal fluctuations to right and left of parallel of equal mean amplitude. This mean parallelism will be transmitted to succeeding layers. If, however, the non-asymmetric nematogenic molecules are replaced by asymmetric molecules then right- and left-handed displacements (of equal angular amplitude) of molecules of the second layer from parallelism with the neighbouring fixed molecules of the surface layer, are configurationally inequivalent and would not be of equal energetic amplitude. Conversely, displacements of equal energetic amplitude, actually the determining factor, will not be of equal angular amplitude and will be displaced either slightly to the right or to the left of mean parallelism according to the optic sign of the asymmetric molecules. Thus the mean direction of the molecules in the second layer will be twisted from parallelism to that in the first. A similar twist will be transmitted in turn by the second to the third layer and so on until, when a cumulative twist of 180° has been obtained, a layer will be reached in which the mean molecular direction will once more be the same as that in the first layer. This process will be repeated indefinitely. It is the helical arrangement and the resulting periodic repetition of similarly aligned molecular layers which gives rise to the optical properties (colour effects, optical rotatory power, etc.) of the plane texture and from these properties the repeat distance may be calculated. It may vary widely according to the mean asymmetry of the molecules or mixture of molecules constituting the phase and also with the temperature. With a continuous series of mixtures of molecules of d- and l- types, at the point where the mean asymmetry passes through zero the plane texture is replaced by the corresponding ordinary nematic homogeneous texture. The effect of molecular asymmetry, which produces a twist in the plane texture of the cholesteric phase, should likewise give a twist between the successive layers of molecules arranged as in smectic C in Figure 1.2. This effect has recently been observed (cf. Chap. 2.1) [22].

to one another in parallel two-dimensional hexagonal array [21] just as with the micellar units of M_1 and M_2 mesophases.

Optically Isotropic Cubic Mesophases

Friedel did not record the existence of any cubic mesophase, possibly because his conclusions were mainly based on the use of the polarizing microscope to study the birefringent patterns ("textures") adopted by smectic, nematic or cholesteric preparations confined between slide and coverslip. Cubic mesophases, being optically isotropic, could then easily remain unnoticed. It had, much earlier, been noted [2, 3] that a coloured optically isotropic or near optically isotropic "modification" is produced as a first stage in cooling certain cholesterol compounds below the stability limit of the amorphous melt. On further cooling, the substance adopts the ordinary cholesteric condition. Lehmann was of the opinion that these two forms represent distinct phases. Recent observations with cholesteryl myristate and nonanoate [5a, 5b] apparently support this conclusion.

In a careful study of the cholesteryl esters ranging from the formate to the stearate, Gray [4] observed the production of the optically isotropic condition with all the esters, except the formate, but inclined to regard this condition as representing a homeotropic texture of the ordinary cholesteric phase rather than an additional individual mesophase.

However, that the isotropic condition represents an additional phase is made more probable by the recent identification by Sackmann and associates [14a] of an optically isotropic mesophase whose X-ray diffraction pattern reveals cubic symmetry. This mesophase, capable of separating as polyhedral particles, occurs at higher temperatures in the series of mesophases formed by the 4′-n-hexadecyloxy- and 4′-n-octadecyloxy-3′-nitrobiphenyl-4-carboxylic acids (cf. [14b]).

A very extensive series of cubic mesophases was recognized in 1935 by Timmermanns [8] who termed this series "plastic crystals". In the course of a later review [23] he said:

"Some twenty-five years ago, while collecting data for my book, *Les Constantes Physiques des Composés Organiques Cristallisés*, I saw directly, when I came to the study of the chapter on the heat of melting, that the molecular heat of fusion and freezing temperature increase in ascending a homologous series. I tried to nullify this effect by dividing the values of the molecular heat by the absolute temperature to obtain the molecular entropy of melting. There was, however, still no constancy of the melting entropy in a homologous series. However, some peculiar compounds, of usually very simple constitution had a very low entropy of melting. These compounds were 'globular', i.e. they were either symmetrical around their centre (CH_4, CCl_4, pentaerythritol, etc.); or would give a sphere by rotation around an axis (cyclohexane, camphor, etc.). All these compounds had an entropy of melting lower than 5 e.u. (cal/deg/mol), a quantity I represent by the symbol σ, and so far this arbitrary limit has not been exceeded. Thus, cyclohexane, with σ equal to 2·27, is globular, while methylcyclohexane, with σ equal to 11·00, is non-globular; the reason seems to be that in methylcyclohexane the lateral CH_3 group would stop the rotation. The comparative study of many other properties of organic crystals revealed that such compounds had many other remarkable values of their constants. Usually their melting point is relatively high. While n-pentane melts at −141°C, tetramethylmethane melts at −16°C. However, they have a transition point lower down with a high value of the entropy of transition (for $C(CH_3)_4$ this is at −133°C with $\sigma = 4{\cdot}40$). With a high melting point goes a high saturated vapour pressure of the plastic crystals. The triple point (V + L + S) may be higher than atmospheric pressure so that sometimes, under ordinary pressure, the crystals sublime directly—as C_2Cl_6 does—instead of melting first.

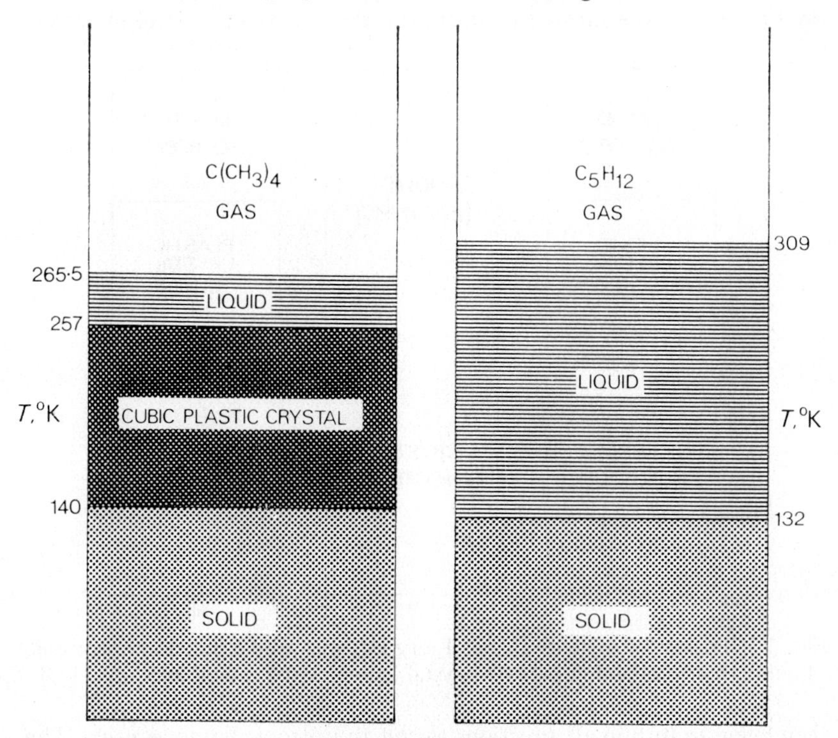

FIG. 1.3. Phases for neopentane (globular) compared with those for normal pentane (non-globular) [23]

"Between the melting and the transition points they crystallize almost all in the cubic system, the system with the highest symmetry, and take another form below the transition points. (Cyclopentane though globular is hexagonal; in some cases the transition takes place in two steps as with hexachloroethane, and hydrogen chloride.) Figure 1 (= Fig. 1.3 of the present text) illustrates this for a globular molecule, $C(CH_3)_4$ (one transition), and a non-globular molecule, normal-$C_5\,H_{12}$, which has no transition in the crystal."

It is to be noted that Timmermanns classified his "plastic crystal" as a "solid crystal" even though he was aware of its close relationship to the more familiar "liquid crystals" as is made very clear by Fig. 1.4 (from his review) and from his concluding remarks as follow:

"In considering this large domain, I think that we are able to make certain general conclusions. The phenomenon of melting is not very easy to understand because our theoretical knowledge of the inner states of liquids is much more superficial than for crystals and gases. At the end of the last century, the discovery of liquid crystals had proved that the melting process is a super-position of two different effects, although both result from the thermal movements of molecules. When the temperature is high enough the thermal energy becomes so great that the crystal boundaries are broken and the substance liquefies; if there is room, the free molecules

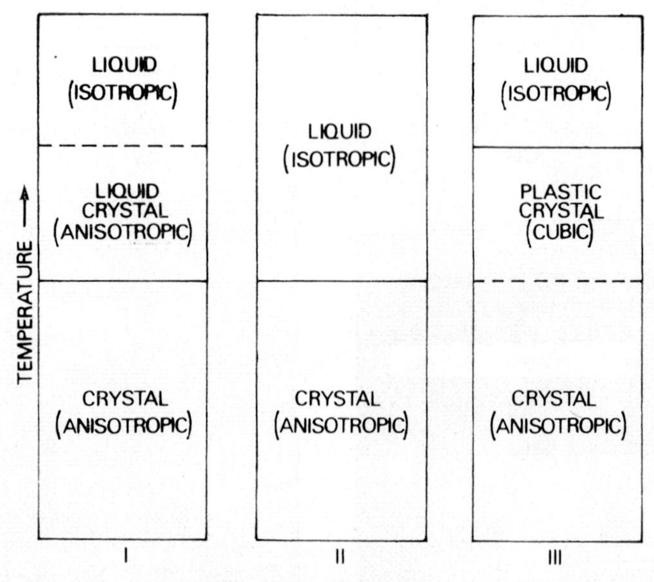

FIG. 1.4. Phases for systems giving: I. crystal, anisotropic liquid and isotropic liquid; II. crystal and liquid; III. crystal, cubic plastic crystal and liquid [23]

then begin to turn in all directions giving an isotropic liquid as usual. The compounds that give anisotropic liquids are generally very long molecules which, because of double bonds are unable to bend, having the shape of a 'lath', and which need much space to rotate freely. Therefore, some directions in the liquid are privileged and it remains anisotropic till the higher transition point where, by expansion, it becomes isotropic.*

"In the plastic crystals, on the contrary, the freedom to rotate exists already in the crystal state, because the molecules are globular in shape, but the coherence of the crystal is broken only at a higher temperature, the melting point. In the two cases the liquefaction and the isotropy do not occur at the same temperature. In liquid crystals, by heating, the fluidity comes first but in plastic crystals the isotropy comes first. This is illustrated

* Cf., however, Friedel's explanation on p. 9.

in Figure 7. [Fig. 1.4 of this text.] They are complementary from this point of view as pointed out in my first paper on the subject, in forging the nomenclature of the globular compounds or plastic crystals."

If however we recall Friedel's definition of a solid crystal as the type "in which the spatial periodic repetition of the units is perfectly defined and in which thermal agitation plays only a very insignificant rôle" we see that plastic crystals should not be included. In plastic crystals the units undergo thermal rotatory or tumbling motions fairly freely, in some cases even more freely (Chap. 2.2) than in the amorphous liquid [27]. That even the "periodic repetition" of the units is not perfectly defined but accompanied by appreciable translational displacements of the tumbling units seems to be shown by the blurred X-ray diffraction patterns (Chap. 2.2) given by plastic crystals. Thus, "plastic crystals" like "liquid crystals" do not belong either to the crystalline solid state or to the liquid state of matter but form a further mesomorphic phase of comparable status to the nematic and cholesteric mesophases. However, the classification of plastic crystals as solid crystals by Timmermanns has resulted in studies of "plastic crystals" and "liquid crystals" being largely divorced from one another because of an artificial distinction which might have been avoided had both been classified initially as mesomorphic states.

Although in the non-amphiphilic series studies of cubic and anisotropic mesophases have thus been largely divorced, this has not happened in the amphiphilic series even though, as before, the two groups of mesophases apparently arise from globular and highly anisometric units (in this case, micelles) respectively (Fig. 1.1 and Chap. 5).

Cubic mesophases in the amphiphilic series were first characterized by Luzzati and collaborators [28, 29] who, indeed, remarked in earlier accounts that these phases are actually "true crystals" in that the units are located at the points of a cubic lattice. In conformity with this, these mesophases, under certain conditions, can separate from a precursive phase in well defined cubic forms [30] and the cubic forms observed in aqueous preparations of tomato bushy stunt virus by Bowden and Pirie [31] may also represent a related type of mesophase. However on account of the occurrence of the cubic mesophases of the amphiphilic series in close relationship with other "liquid crystalline" mesophases, the essential structural affinity of all these mesophases was from the first clearly evident.

The rotation and globular character of the micelles in the cubic amphiphilic mesophases, in comparison with the lack of rotation of the highly extended micelles of the anisotropic mesophases, has been indicated by nuclear magnetic resonance (NMR) studies [32, 33], although various alternative structures containing rigid networks of small, almost isometric but rod-like units have been proposed [29]. It is difficult to see, however, how such immobile structures could be reconciled with the NMR findings. How globular micelles and the related cubic phases can arise in the general sequence of amphiphilic mesophases is discussed in Chapter 5. It should be mentioned here that, although many of the cubic mesophases in amphiphilic systems contain water, such mesophases, like the amphiphilic anisotropic mesophases, can also arise with pure anhydrous amphiphiles.

Thus calcium ω-phenylundecanoate forms a cubic mesophase at room temperature (cf. Chap. 5).

Finally, to return to the non-amphiphilic cubic mesophases; in those formed by the globular molecules it appears that the units are usually single molecules although, occasionally, three or more molecules can rotate together [24, 26]. In the case of the cubic mesophases formed by 4′-n-hexadecyloxy- and 4′-n-octadecyloxy-3′-nitro-biphenyl-4-carboxylic acids [14a, 14b] it would appear, however, that globular assemblies of parallel molecules must be involved. The isotropy of the phase would then result from the absence of long-range orientational order between the individually anisotropic units. A similar conclusion might again apply to the coloured optically isotropic high-temperature "Modifications" formed by cholesteryl esters. In this case the assemblies would presumably be small units each individually possessing the screw-like arrangement of molecules found in the plane texture of the cholesteric phase. Those units suitably oriented with respect to the direction of vision could then give rise to the colours characteristic of this "Modification".

REFERENCES

[1] Reinitzer, F. *Mh. Chem.* **9,** 421 (1888).
[2] Lehmann, O. *Z. phys. Chem.* **4,** 462 (1889).
[3] Lehmann, O. *Z. phys. Chem.* **56,** 750 (1906).
[4] Gray, G. W. *J. chem. Soc.*, 1956, 3733.
[5] Price, F. P. and Wendorff, J. H. (a) *J. phys. Chem.* **75,** 2839 (1971). (b) *J. phys. Chem.* **76,** 276 (1972).
[6] Friedel, G. *Annls Phys.* **18,** 273 (1922).
[7] cf. Ubbelohde, A. R. *Melting and Crystal Stucture*, Clarendon Press, Oxford (1965), p. 66.
[8] Timmermanns, J. *Bull. Soc. Chem. Belg.* **44,** 17 (1935).
[9] Smith, G. W. *International Science and Technology*, January 1967, p. 72.
[10] cf. Winsor, P. A. *Chem. Rev.* **68,** 1 (1968) and bibliography.
[11] MacLennan, K. *J. Soc. Chem. Ind.* **42,** 392 (1923).
[12] McBain, J. W. and Elford, W. J. *J. chem. Soc.* 1926, 421.
[13] Winsor, P. A. *Molec. Crystals Liqu. Crystals*, **12,** 141 (1971).
[14] (a) See Review by Sackmann, H. and Demus, D. *Fortsch. d. chem. Forsch.* **12,** 349 (1969).
[14] (b) Gray, G. W. *Symposium of the Faraday Soc.* **5,** 94 (1971).
[15] Gray, G. W. *Molecular Structure and the Properties of Liquid Crystals*, Academic Press, London (1962), p. 258.
[16] de Vries, A. *Molec. Crystals Liqu. Crystals* **11,** 361 (1970).
[17] de Vries, A. *Molec. Crystals Liqu. Crystals* **10,** 219 (1970).
[18] Sandquist, H. *Ber.* **18,** 2054 (1915), *Kolloidzeitschrift* **19,** 113 (1916).
[19] See also Balaban, I. E. and King, H. *J. chem. Soc.*, 1927, 3068.

[20] Bernal, J. D. and Fankuchen, I. *J. gen. Physiol.* **25,** 111 (1941).
[21] Robinson, C., Ward, J. C. and Beevers, R. D. *Discuss. Faraday Soc.* **25,** 29 (1958).
[22] Leclercq, M., Billard, J. and Jacques, J. *Molec. Crystals Liqu. Crystals* **8,** 367 (1969).
[23] Timmermanns, J. *J. Phys. Chem. Solids* **18,** 1 (1961).
[24] Staveley, L. A. K. *Ann. Rev. Physical Chem.* **13,** 351 (1962).
[25] Westrum, E. F., Jr. and McCullough, J. F. *Physics and Chemistry of the Organic Solid State*, Interscience Publishers, London (1963), p. 89.
[26] Aston, J. G. *ibid.*, p. 543.
[27] Powles, J. G., Williams, D. E. and Smyth, C. P. *J. chem. Phys.* **21,** 136 (1953).
[28] Luzzati, V., Mustacchi, H. and Skoulios, A. E. *Discuss. Faraday Soc.* **25,** 43 (1958).
[29] For review and bibliography see Luzzati, V. *Biological Membranes*, Academic Press, London, (1968), Chap. 3, p. 71.
[30] Rogers, J. and Winsor, P.A. *5th International Congress on Surface Activity*, Barcelona (1968) p. 933.
[31] Bowden, F. L. and Pirie, N. W. *Brit. J. Exp. Pathology* **19,** 251 (1938).
[32] Gilchrist, C. A., Rogers, J. Steel, G., Vaal, E. G. and Winsor, P. A. *J. Colloid Interface Sci.* **25,** 409 (1967).
[33] Lawson, K. D. and Flautt, T. J. *J. phys. Chem.* **72,** 2066 (1968).

2

Classification and Organization of Mesomorphous Phases formed by Non-amphiphilic and Amphiphilic Compounds

2.1 SMECTIC, NEMATIC AND CHOLESTERIC MESOPHASES FORMED BY NON-AMPHIPHILIC COMPOUNDS

A. Saupe

Classification of Thermotropic* Liquid Crystals

Thermotropic liquid crystals are conventionally classified into nematics, cholesterics or smectics. The term smectic liquid crystal is not particularly specific and is used for all thermotropic liquid crystals which are neither nematic nor cholesteric.

Nematic liquid crystals can be formed by compounds that are optically inactive or by racemic modifications, and consist of molecules that have a more or less elongated shape†. Characteristic for the nematic molecular order is the partial parallel orientation of the molecules with an axis that coincides or nearly coincides with the mean position of the long geometrical axis of the molecules (Fig. 2.1.1(*a*)). The distribution of the molecular centres of gravity, however, is without long range order as in normal amorphous isotropic liquids.

Admitting all kinds of orientational order, one can generalize the term nematic so that it includes all liquids with spontaneous orientational order. So far only uniaxial order without polarization has been observed and

* By thermotropic liquid crystals are meant those liquid crystals formed by the action of heat alone on pure solids or mixtures of solids—in this section, the solids under consideration are non-amphiphilic.

† Nematic liquid crystals formed by disc-shaped molecules occur commonly during carbonization of graphitizable materials [31].

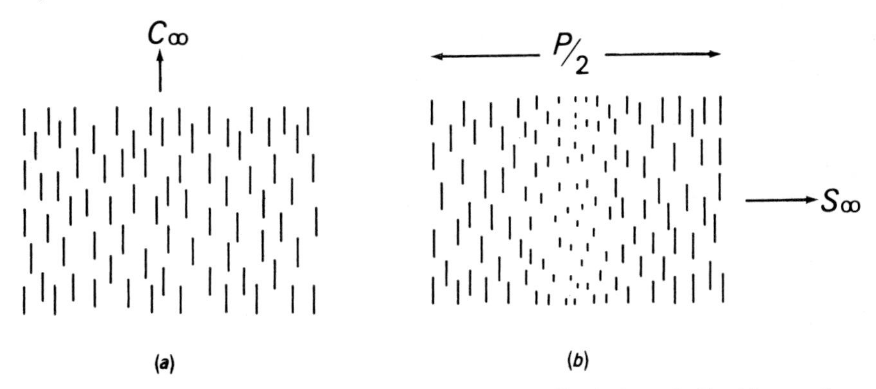

FIG. 2.1.1 Schematic structure of (*a*) nematic and (*b*) cholesteric liquid crystals. The lines represent the molecules in their preferred orientations; their varying lengths in (*b*) are a measure of the out of plane inclination of the molecules.

we will use the term nematic only in its original narrow sense, if not otherwise indicated. A uniaxial long range orientational order without polarization accounts for the macroscopic properties observed for nematics, e.g., the strongly anisotropic uniaxial optical properties, the dielectric properties and the curvature elastic properties connected with a continuous variation in space of the preferred orientation or, in other words, of the direction of the local optic axis.

Cholesteric liquid crystals are formed by optically active compounds or optically active mixtures. They are isomorphous and miscible with nematic liquid crystals, but while in nematics the equilibrium structure corresponds to the uniform parallel alignment, in cholesterics it corresponds to the planar uniformly twisted structure. In this structure the direction of the local optic axis is constant in planes that lie perpendicular to the twist axis and turns regularly along this axis (Fig. 2.1.1(*b*)). The distance for a 360° turn is usually referred to as the pitch (*P*). The spontaneously twisted structures of cholesterics have peculiar optical properties and may also give rise to optical patterns in thin films that resemble those for smectic phases.

It is generally assumed that smectic phases have a laminar structure which is developed in addition to an orientational long range order similar to that in nematics. A laminar structure means that the density of distribution of the molecular centres has its highest value along equidistant surfaces. The molecular centres have to be suitably defined for this purpose in relation to the molecular shape and molecular interactions. They coincide only for symmetrical molecules with the centres of mass.

It is useful to make a subdivision of smectics into two groups. We follow the proposal by Herrmann [1] and differentiate between smectic phases with unstructured layers ("*ohne Ordnung*") and smectic phases with structured layers ("*mit Ordnung*"). In the first group the distribution of the molecular centres within the layers corresponds to that of a two-dimensional liquid. It is without long range order. In the second group the layers are built up regularly so that the positions of the molecular centres lie on a two-dimensional lattice. The most common smectics are in the first group which includes the phases classified as smectic *A* and smectic *C*. A representative of the second group is classified as smectic *B*.

Recently, much work on a classification of smectics has been done in Halle by Sackmann and Demus [2] and their co-workers who defined the smectic classes mentioned above. The work started with the dissertation by Arnold [3] which contains the major guidelines underlying their classification. It consists essentially in the attempt to determine experimentally the number of different groups of isomorphous smectic phases* and to characterize them by their textures. Isomorphous liquid crystals are considered as equivalent and characterized by the same symbol.

Isomorphism was established by studying miscibility. A liquid crystal formed by compound I is isomorphous with the liquid crystal formed by compound II if the two compounds can be mixed in the liquid crystalline state under consideration without the occurrence of a concentration range where the two phases coexist separated by phase boundaries. While uninterrupted miscibility establishes isomorphism, the converse is not necessarily true. It is therefore difficult to establish non-isomorphism by considering miscibility only and other properties have to be taken into account. An additional plausible assumption is useful here:

Two liquid crystal phases formed by the same compound but separated by a well defined phase transition are considered as qualitatively different in their structures and therefore not isomorphous.

The two assumptions have been found to be self consistent and have not led to contradictory results, although a large number of binary mixtures has been studied. The phase diagrams of binary mixtures are often rather complex. An example is shown in Fig. 2.1.2. It has been chosen because of a remarkable feature not commonly observed. In a certain concentration range the mixtures form a nematic phase in contrast to the pure compounds which form only smectic phases. This reminds one of those amphiphilic systems in which the liquid crystal phases incorporate solute and solvent, but in our case a stable nematic phase can be observed in the mixtures because the thermal stability of the smectic phases is reduced. In each of the pure compounds the nematic phase is present only as a latent phase because of the higher thermal stability of the smectic phase.

For a practical and relatively fast classification of liquid crystals, the optical patterns or textures observed in thin layers ($<0{\cdot}3$ mm) with a polarizing microscope are most useful. This optical method has been used since the discovery of liquid crystals and led to their classification as nematics, cholesterics and smectics. It has also been used by Arnold, Sackmann and Demus for an additional test of their classification of smectic phases and has been further developed by them. In general, we can expect that different liquid crystals will show different textures and that isomorphous liquid crystals will show textures with characteristic

* The term phase is used in its wider sense. Non-isomorphous states are considered to belong to different phases.

features in common. This is true for the most common liquid crystal phases and supports the classification by miscibility. There are, however, limitations and a complete classification of smectic phases by textures is not always possible. It can happen that similar textures are observed with two liquid crystal states separated by a phase transition.

Non-isomorphous liquid crystals should differ in the molecular long range order characteristic for these phases. Such differences in structure should be recognizable by X-ray studies and have indeed been established

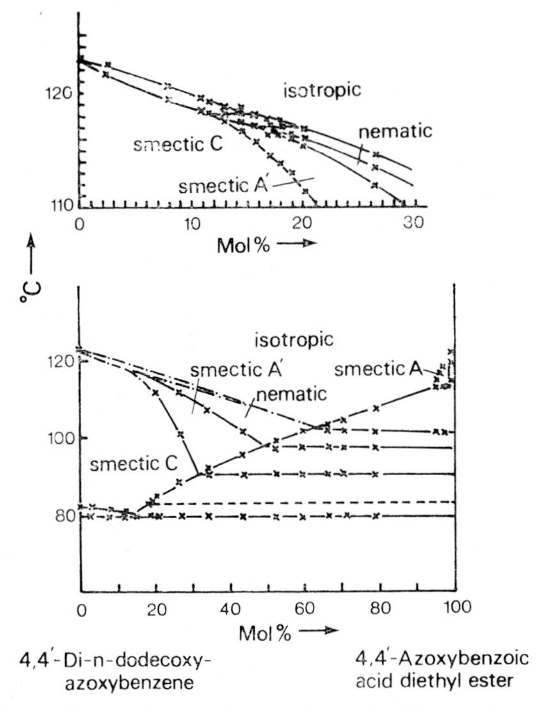

FIG. 2.1.2. Phase diagram for a mixture of two liquid crystal compounds. The upper part shows an enlarged section (from reference 4).

in a number of cases, but much work remains to be done. X-ray studies may also reveal structural differences between liquid crystals that are isomorphous because of differences in short range order. This is especially true for nematics where considerable differences in the packing of the molecules have been found and a subclassification of nematics based on these results has been proposed [5]. This may be useful since the macroscopic properties depend on short range order. The long range orientational order is, however, of the same kind in all nematics that have been studied and they are all isomorphous and miscible.

TABLE 2.1.1. *Transition temperatures*[a] *and enthalpies*[b]

Compound	Solid → Liquid Crystal		Smectic $B \to A, C$		Smectic $C \to A$		Smectic → Nematic		Liquid Crystal → Normal Liquid		Ref.
	T	H	T	H	T	H	T	H	T	H	
C_1A	118 N	7·06							N 135	0·137	7
C_2A	137 N	6·41							N 168	·326	7
C_3A	115 N	6·42							N 124	·161	7
C_4A	102 N	5·00							N 137	·247	7
C_5A	76 N	3·48							N 123	·173	7
C_6A	81 N	9·88							N 129	·250	7
C_7A	74 N	9·77					*C* 95	0·380	N 124	·243	7
C_8A	80 *C*	10·07					*C* 108	·282	N 126	·343	7
C_9A	76 *C*	9·11					*C* 113	·394	N 122	·421	7
$C_{10}A$	78 *C*	9·21					*C* 121	·553	N 123	·751	7
$C_{11}A$	81 *C*	9·81							*C* 121	2·40	7
$C_{12}A$	82 *C*	10·04							*C* 123	2·86	7
$C_{18}A$	94 *B*	18·88	99 *C*	2·52					*C* 115	5·40	8
$C_{10}B$	65 *C*	10·88			73	0·024			*A* 88	1·77	9
$C_{12}B$	79 *B*	17·97			82	·024			*A* 88	2·10	9
C_2D	114 *A*	4·74							*A* 123	1·21	10
C_2EC_2	82 *B*	6·53	118 *A*	0·502			*A* 156	1·22	N 159	0·12	10
$C_{12}EC_5$	75 *B*	6·69	96 *C*	1·309	107	0·146			*A* 133	2·02	8

C_nA = $CH_3(CH_2)_{n-1}$—O—C_6H_4—N(→O)=N—C_6H_4—O—$(CH_2)_{n-1}CH_3$

C_nB = $CH_3(CH_2)_{n-1}$—O—C(=O)—C(CH_3)=CH—C_6H_4—N(→O)=N—C_6H_4—CH=C(CH_3)—C(=O)—O—$(CH_2)_{n-1}CH_3$

C_2D = C_2H_5—O—C(=O)—C_6H_4—N(→O)=N—C_6H_4—C(=O)—O—C_2H_5

C_nEC_m = $CH_3(CH_2)_{n-1}$—O—C_6H_4—CH=N—C_6H_4—CH=CH—C(=O)—$O(CH_2)_{m-1}CH_3$

[a] T in °C; required liquid crystal states are specified (N = nematic, A, B, C = smectic A, B, C); lower temperature states left, higher temperature states right of T,

[b] H in kcal mol^{-1}.

Nematic Liquid Crystals

(1) General Properties

Thick nematic layers (>1 mm) appear cloudy. On heating, they transform at the "clearing" point to the normal or amorphous isotropic liquid, and on cooling, at the lower transition point to a smectic liquid crystal or the solid crystal. The transition to the solid crystal from the normal liquid or from a liquid crystalline state is usually subject to the effects of supercooling. Sometimes, indeed, liquid crystalline states can be obtained only by supercooling. These metastable states are called monotropic to differentiate them from the stable liquid crystal phases, the enantiotropic liquid crystals. The transition from the normal liquid phase to a liquid

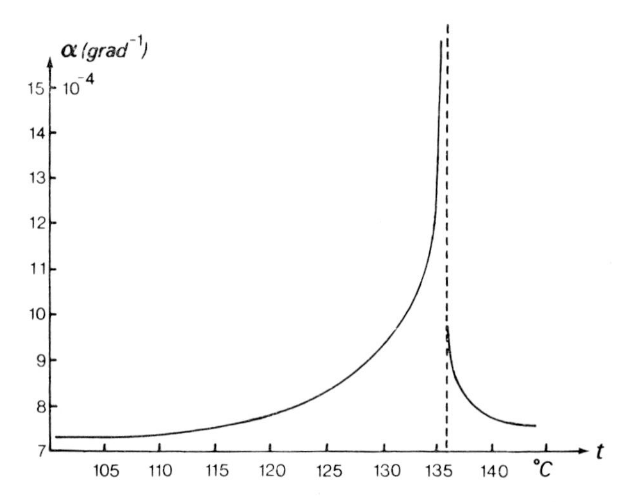

Fig. 2.1.3. Thermal expansion coefficient of 4,4′-dimethoxyazoxybenzene in the nematic and normal liquid state (from reference 11).

crystal phase and the transitions between liquid crystal phases are usually completely reversible and show no detectable supercooling. Exceptions have, however, been reported for liquids of high viscosity [6].

The nematic liquid to amorphous isotropic liquid transition is a first order transition. The transition energies are generally small, but vary considerably from 0·1 to about 1 kcal mol^{-1}. In homologous series the transition energies increase with increasing molecular length (see Table 2.1.1). No apparent change in viscosity is noticeable at the isotropic-nematic transitions; nematics can be poured just as easily as normal liquids.

The smectic phase to which a nematic phase may transform on cooling is often one with unstructured layers, e.g., smectic *A* or *C*. Transitions to a smectic with structured layers, e.g., smectic *B* have been found only in a few cases [2]. These transitions are also in general of first order. The transition enthalpies to smectics with unstructured layers are usually in about the same range as the enthalpies of nematic to isotropic liquid transitions (see Table 2.1.1). For the transition to a smectic with structured layers we expect a higher energy. No experimental results for these transitions are available. The thermal volume expansion in nematics is not very different from that of normal liquids. Specific heats and compressibilities are also similar. Near to the isotropic-nematic transitions there are, however, pretransitional effects which lead to an increase of these quantities. Figure 2.1.3 shows the thermal expansion coefficient of *p*-azoxyanisole. At the isotropic-nematic transition there is a small discontinuous decrease in volume; for *p*-azoxyanisole it is approximately 0·3%.

The parallel orientation of the molecules is far from being as complete as the idealized structure in Fig. 2.1.1(*a*) suggests. Thermal fluctuations

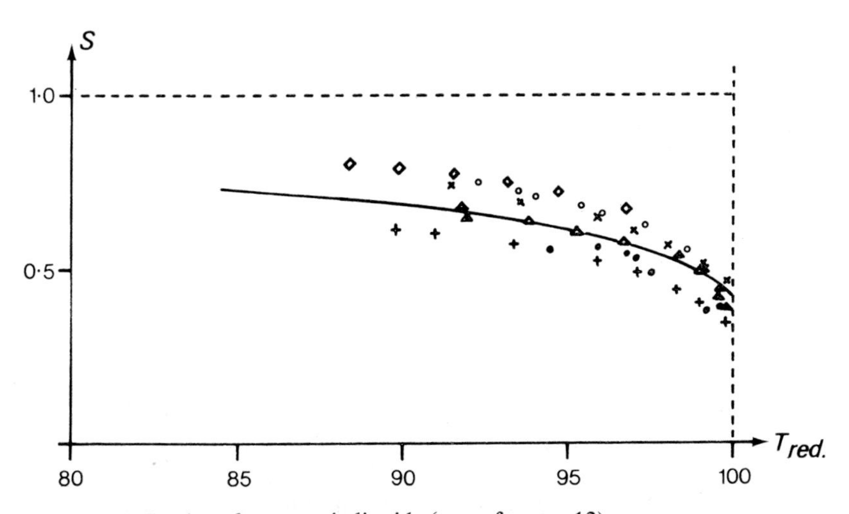

FIG. 2.1.4. *S*-values for nematic liquids (see reference 12).
△ 2,4-nonadienoic acid
● and + 4,4′-dimethoxyazoxybenzene
○ and × 4,4′-di-ethoxyazoxybenzene
◇ 4-n-hexyloxybenzoic acid

are very strong and large deviations from the perfect parallel orientation exist. A measure for the fluctuations gives the degree of order defined as the average value

$$S = \frac{1}{2}\langle 3 \cos^2 \theta - 1\rangle$$

where θ is the momentary angle between the long axis of the molecule and the axis of preferred orientation or local symmetry axis. For ideal nematic order all molecules are perfectly parallel and $S = 1$. In amorphous isotropic liquids $S = 0$, corresponding to random distribution of the axes.

Experimentally observed S values for nematics are given in Fig. 2.1.4 together with a theoretical curve derived in using a simple inner field theory for the nematic state. A reduced temperature scale is used in Fig. 2.1.4 defined by $T_{red.} = 100T/T_c$, where T_c is the isotropic to nematic transition temperature and T the actual temperature, all measured in °K. S is strongly temperature dependent in nematics. There is also a considerable difference between S values for different compounds at the same reduced temperature. The S values at the clearing point are generally between 0·3 and 0·4, but at the lower temperature limit values near to 0·8 may be reached.

(2) NEMATIC TEXTURES

When liquid crystals are studied with a polarizing microscope, they are usually observed as thin films between glass plates. On cooling from the amorphous isotropic liquid, circular birefringent areas appear at first. The appearance of the textures that form after the completed transition often depend considerably on the layer thickness. Thicker nematic layers may show the typical threaded *Schlieren* texture. The well defined thread-like structures may move and float around in the nematic. The term nematic refers to these features, being derived from the Greek word "*nēma*", a thread.

The threads have elastic properties. In Fig. 2.1.5 a nematic thread is shown one part of which is attached to the upper surface of an uncovered droplet and the other part to the lower glass surface. The two parts are connected by a smooth thread running through the bulk of the liquid. In Fig. 2.1.5(*a*) the thread is extended and curved due to flow of the liquid. In Fig. 2.1.5(*b*) the liquid is at rest and the thread has contracted to a straight line.

In thinner layers the threaded texture changes to the *Schlieren* texture with point-like disclinations (Fig. 2.1.6). The appearance between crossed polarizers is characterized by dark brushes which start from points, in which the direction of extinction is not defined. Usually points with two or four dark brushes can be observed. On turning the polarizer and analyzer simultaneously, the direction of the brushes at the disclinations turn either in the same direction (positive disclinations) or in the opposite direction (negative disclinations). Along a circle around a disclination, the directions of extinction therefore change continuously. They turn by plus or minus 2π for a completed circle around a disclination with four brushes and by plus or minus π around a disclination with two brushes. The existence of undisturbed disclinations with two brushes proves the absence of polarization in the nematic molecular structure, because a physically equivalent point is reached after a change in direction of the optic axis by 180°. By a suitable surface treatment it is possible to obtain films with uniform molecular alignment. Figure 2.1.7 shows a film that has been aligned between rubbed surfaces. The optic axis is parallel to the direction of rubbing. Thread-like double lines are visible which correspond to inversion walls. The direction of the optic axis turns across these walls by 180° [26]. With untreated surfaces, the surface effect may produce an irregular texture reminding one of polished marble (the marble texture).

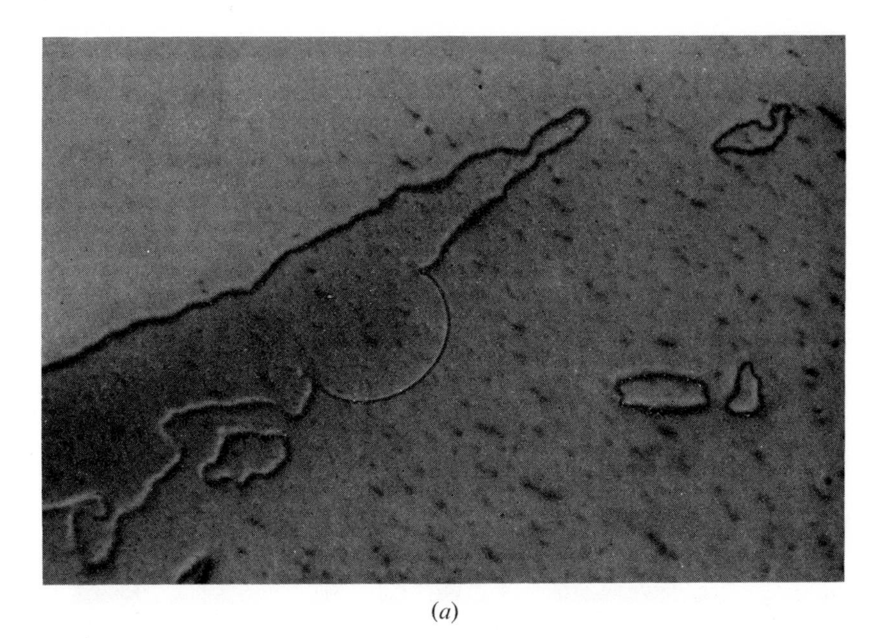

(*a*)

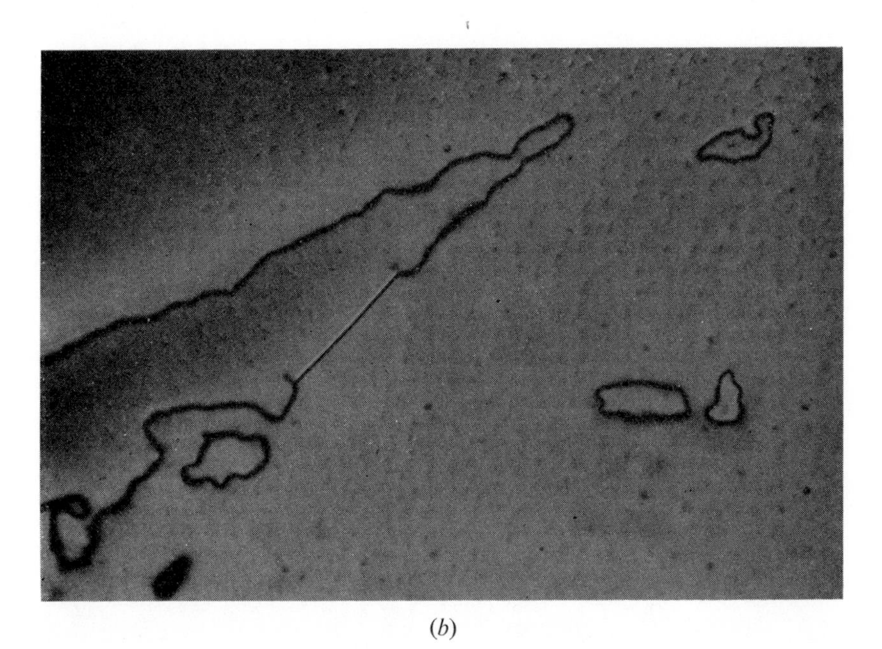

(*b*)

FIG. 2.1.5. Threaded *Schlieren* texture of nematic 4-methoxybenzylidene-4′-n-butylaniline as an uncovered film on a glass slide at 25°C; thickness ≈ 100 μm; (*a*) disturbed film, (*b*) film at rest; crossed polarizers, magnification ×90.

FIG. 2.1.6. *Schlieren* texture with disclination 'points' of nematic 4-n-caproyloxy-4′-ethoxyazoxybenzene; film between glass slides at 130°C; thickness ≈ 5 μm; crossed polarizers; magnification ×200; n-caproyl = $CH_3(CH_2)_4CO$.

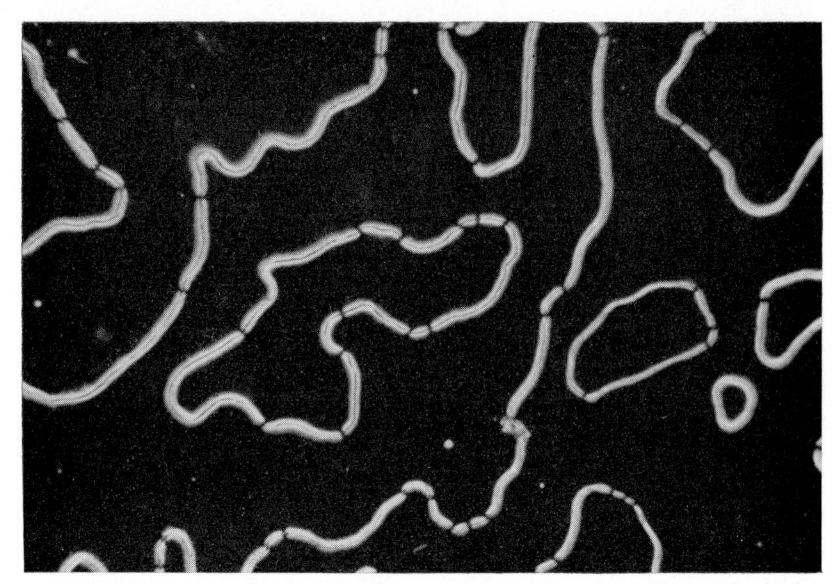

FIG. 2.1.7. Parallel aligned film of nematic 4-methoxybenzylidene-4′-n-butyl-aniline between rubbed glass slides; thickness ≈ 5 μm; crossed polarizers perpendicular and parallel to direction of rubbing; magnification ×60.

It is also possible to obtain uniformly aligned films with the optic axis normal to the surface. Such films with a pseudo-isotropic texture show no birefringence in orthoscopic observation when viewed vertically. Their existence proves the uniaxial properties of nematics. The perpendicular alignment can easily be disturbed and by slightly touching the cover glass of a sample viewed between crossed polarizers, a flash-like brightness can be produced. If the sample is tilted, so that the axis of observation makes a small angle with the optic axis one observes between crossed polarizers a blueish picture of a grainy appearance which is restlessly changing due to thermal fluctuations (Brownian motion). Such fluctuations are observed under other conditions and cause the turbidity of nematic liquids in their macroscopic appearance. Pseudo-isotropic textures are often obtained with compounds that form, in addition to the nematic, a smectic *A* phase at lower temperatures (see for instance the description of the textures in reference [13]). Special surface treatments can be of help in producing these textures in other cases. A surface coating of lecithin for instance produces such a texture with MBBA (4-methoxybenzylidene-4′-n-butylaniline) [14] and cleaning the surface thoroughly with an acid promotes it with PAA (4,4′-dimethoxyazoxybenzene, *p*-azoxyanisole). The results of surface treatments are, however, often not very reproducible. Traces of impurities in the compound and on the surfaces can be of importance.

In addition to a molecular alignment parallel or perpendicular to the surface, an inclined alignment can also be obtained, preferably with compounds that form at lower temperatures a smectic *C* phase in the *Schlieren* texture. 4,4′-Di-n-octyloxyazoxybenzene for instance may give an alignment in the nematic phase that is temperature dependent. Near the clearing point the nematic may show a uniform parallel alignment to the surfaces. At lower temperatures, the alignment becomes less uniform as parts of the sample assume a tilted alignment. Near the transition to smectic *C*, the alignment can be fairly uniform again, but now this is tilted with a tilt angle that corresponds to that of the smectic structure. The explanation for the influence of the adjacent smectic phase on the alignment is that the molecular layers covering the glass surfaces can develop earlier and at a higher temperature in the nematic (*a*) structure corresponding to that of the lower temperature smectic. The molecular orientation thus given at the surfaces determines then, in turn, the orientation in the bulk of the nematic film. It is accordingly inclined or perpendicular depending on whether a tilted or perpendicular alignment of the molecular axis occurs in the smectic layers.

Cholesteric Liquid Crystals

(1) General Properties

As the name indicates cholesteric liquid crystals can be observed with cholesterol derivatives, e.g., cholesteryl benzoate and cholesteryl acetate. Non-sterol derivatives can, however, also exist in a cholesteric liquid crystal state, the only necessary condition being optical activity of the compounds. Accordingly, a nematic liquid crystal can be transformed to a cholesteric by dissolving an optically active compound in it. The optical activity of the molecules reduces the symmetry so that the uniform parallel alignment no longer corresponds to an energy extreme. The structure with the lowest energy, therefore, is now in general uniformly twisted as indicated in Fig. 2.1.1(*b*).

TABLE 2.1.2. *Transition temperatures*[a] *and enthalpies*[b] *of cholesteric*[c] *compounds* [15]

	Solid → Liquid or Liquid Crystal		Smectic *A* → Cholesteric		Cholesteric → Normal Liquid	
Cholesteryl	*T*	*H*	*T*	*H*	*T*	*H*
Propionate	99·6 Ch	5·21			114·1	0·16
Laurate	91·3	7·91	81·4	0·34	88·6	0·32
Myristate	71·0 *A*	11·35	79·1	0·42	84·6	0·36
Ethylcarbonate	83·7 Ch	5·04			105·7	0·16

[a] *T* in °C. [b] *H* in kcal mol^{-1}. [c] The liquid crystal state involved is indicated (Ch = cholesteric; *A* = smectic *A*) beside the temperature.

Cholesteric liquid crystals are miscible (isomorphous) with nematics and their thermodynamic properties are correspondingly very similar. Table 2.1.2 shows a selection of transition enthalpies for phase transitions between smectic *A* and cholesterics and between cholesterics and normal amorphous isotropic liquids. No experimental values for the degree of order in cholesterics are available, but it can be assumed that the range of *S* values is about the same as that for nematics.

Hydrodynamic and optical properties of cholesterics depend on textures and may differ strongly from those of nematics. The most striking feature is the strong optical activity and the selective light reflection observable with cholesteric films of a planar texture. The optical activity and the selective light reflection are due to the twisted structure and are directly related to the pitch of the twisting. The pitch can be strongly temperature dependent in particular near to the transition point to a smectic *A* [16]. It increases rapidly when the transition temperature is approached due to a build-up of a short range order that anticipates the layered structure of the adjacent smectic *A* phase [17]. Since the optical properties are directly related to the pitch, they are also temperature dependent. This temperature dependence can be extremely strong so that cholesteric films are very suitable for thermo-graphing and for temperature measurements. They have found useful applications as a diagnostic tool in medicine (see Chap. 7.1).

(2) TEXTURES

Cholesteric liquid crystals can occur in three textures. On cooling, an isotropic texture or modification can be obtained first at the liquid to cholesteric liquid crystal transition, which at a lower temperature changes to a birefringent texture. This behaviour is observed with cholesteryl nonanoate or those homologues which form strongly twisted cholesteric liquid crystals. The transition point is then not readily observed optically and one can get the impression that supercooling is occurring. In other cases when the birefringent texture is formed directly the transition can be easily observed. There is not always a formation of circular droplets as with nematics; elongated banana shaped forms may appear first, usually referred to as *bâtonnets*.

The optically isotropic texture has a higher optical rotatory power than the amorphous isotropic liquid. Its light scattering is due to thermal fluctuations and is also much increased relative to the amorphous liquid, but not to a degree comparable with that of nematics. Even several centimeter thick samples can appear nicely transparent. Under mechanical strain or disturbances the isotropic modification becomes noticeably birefringent; on release of the stress, the sample relaxes and loses its birefringence. Its viscosity is relatively high and it can assume a gel-like character. It is not clear what molecular arrangement is present in this modification. It has been suggested that it is based on a regular lattice of alignment singularities [18], but experimental evidence is still lacking.

Figure 2.1.8 shows the birefringent "focal-conic" texture of a film of cholesteryl nonanoate that has developed out of the isotropic texture. The "focal-conic" texture nucleates in discrete points from which it grows in all directions with the same speed, forming circular areas, until finally the whole film is covered, developing a pattern as shown in Fig. 2.1.8.

A reversal of the change in texture is not in general possible. The isotropic modification can only be readily regained by heating above the transition point and renewed cooling.

The "focal-conic" texture derives its name from the conic sections that are sometimes readily visible (see under the section on smectics).

The "focal-conic" texture of cholesterics changes to the "planar" texture (Fig. 2.1.9) when the cover glass is shifted. In this texture the sample is uniformly aligned with the twist axis perpendicular to the plane of the film. There are, however, often alignment discontinuities which form pattern-like cracks (Fig. 2.1.9). In the planar texture a cholesteric can show reflection colours. The wavelength of the light at the centre of the reflection band is, for perpendicular incidence, equal to the length of the pitch multiplied by a mean refractive index.

Wedge-shaped samples, when the surfaces are prepared for a parallel alignment, show a particular feature called Grandjean steps or threads (Fig. 2.1.10). The threads follow lines of equal thickness and are connected with a discontinuous change of the pitch. In the wedge-shaped sample the pitch decreases with increasing thickness because the orientation at

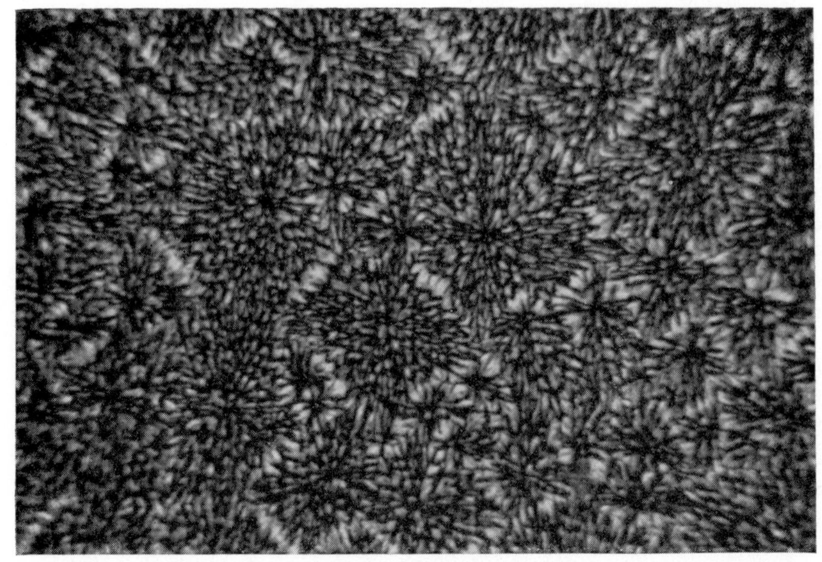

FIG. 2.1.8. "Focal-conic" texture of cholesteric cholesteryl nonanoate; film between glass slides at 80°C; crossed polarizers; magnification ×96.

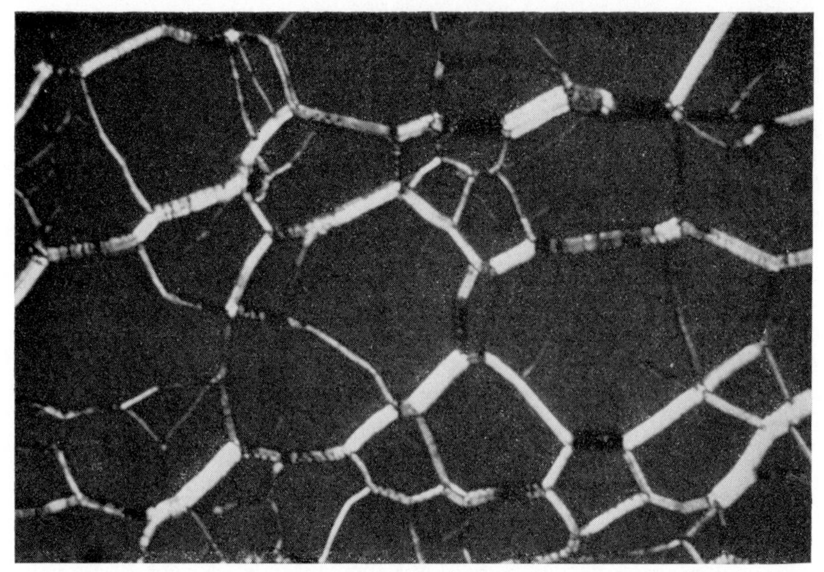

FIG. 2.1.9. Planar texture of cholesteric cholesteryl nonanoate; film between glass slides at 80°C; crossed polarizers; magnification ×300.

the surfaces is fixed. When the deformation energy becomes too high, a disclination line forms that allows a discontinuous change of pitch.

In samples that show reflection colours, the variation of the pitch is directly observable by the colour changes. In Fig. 2.1.10 the colour changes from yellow to red with increasing thickness (bright to dark in

FIG. 2.1.10. Wedge shaped layer of a cholesteric liquid crystal showing Grandjean steps (reference [16]); film between glass slides; crossed polarizers; magnification ×30.

the black and white reproduction) and at each step back to yellow. These changes in colour are more pronounced in the thinner part of the sample, because of the stronger relative change of pitch.

Uniaxial Smectic Liquid Crystals with Unstructured Layers

(1) General Properties and Structures

Smectic phases of this class have an infinite-fold symmetry axis perpendicular to the layers. They include the most common smectic phase, type *A*, which has the molecular structure shown in Fig. 2.1.11(*a*). Its orientational order is of the same kind as in a nematic. The degree of order is higher and ranges probably between 0·8 and 0·9 and is less temperature dependent. The smectic *A* phase was originally described by Friedel.

The symmetry class of smectic *A* is $D_{\infty h}$ when formed by optically inactive compounds or racemic modifications; with optically active compounds its symmetry reduces to D_{∞}. In the equilibrium structure, the layers are parallel with the molecules uniformly aligned for both symmetry classes; no differences in the characteristic macroscopic properties are noticeable. When additional smectic phases are formed by a compound, smectic *A* is always the highest temperature smectic. On heating it transforms accordingly either to a nematic or cholesteric or into the normal amorphous liquid phase. The enthalpy for the transition to the nematic or cholesteric is generally $<1{\cdot}3$ kcal mol^{-1} (see Tables 2.1.1 and 2.1.2). If the transition leads directly to the normal amorphous liquid, the enthalpy is considerably higher (>2 kcal mol^{-1}) as the layered structure disappears together with the orientational order.

There is another smectic with the same symmetry as smectic *A*. The molecules in this phase build double layers (Fig. 2.1.11(*b*)). It can only be formed by compounds with molecules that have no centre of symmetry. The orientational order in contrast to smectic *A* is polar. In each layer the molecules are oriented in the same direction and the layers, therefore,

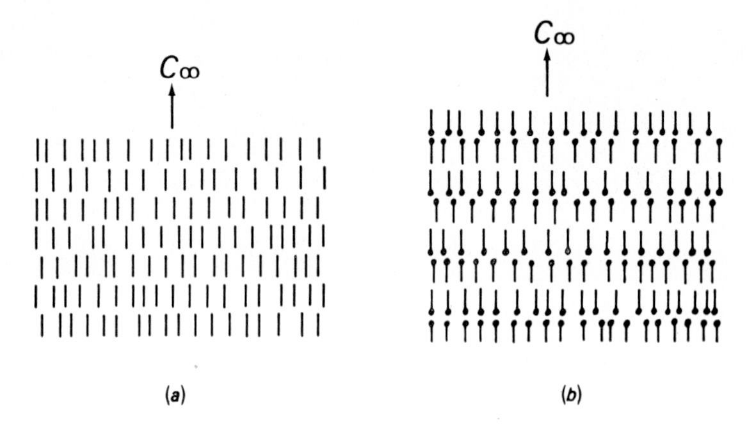

FIG. 2.1.11. Schematic structures of smectic liquid crystals with $D_{\infty h}$ symmetry. The lines represent molecules in their preferred orientation.
(*a*) smectic *A*, unstructured layers without polarity;
(*b*) unstructured layers with polar orientational order, the polarity alternating in successive layers.

have a polarity. But the polarization alternates in successive layers so that the completed structure has no polarization. A double layer formation will directly affect only the dielectric properties. Its influence on other macroscopic properties will be less important and it can be expected that textures and mechanical properties show the same features in both phases.

A formation of double layers is often found with molecules that are amphiphilic. The smectics formed by thallium soaps, for which the double layer structure was established by Herrmann [19], provide an example. There are also indications that double layers can be formed by non-amphiphilic molecules and that a transition from a single to a double layer structure can take place. Some 4-n-alkoxybenzylidene-4′-amino-propiophenones

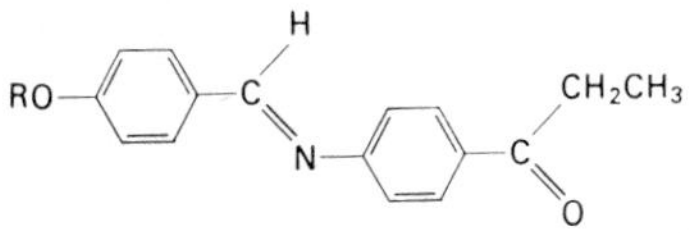

show in addition to a smectic *A* a second smectic which is also optically uniaxial, has the same textures as smectic *A* and shows additional features indicating that it consists of unstructured layers [20]. It has been suggested though that the second phase forms no compensating double layers, but is ferroelectric. There are, however, no observations to confirm this suggestion; indeed no liquid crystal is known for which ferroelectric properties have been established.

(2) Textures

The typical textures obtained with smectic *A* liquid crystals are focal-conic textures. Examples are shown in Fig. 2.1.12–2.1.14. A characteristic feature is the elliptical disclination lines at the liquid crystal-glass interface. In isolated circular areas they tend to group in a regular arrangement as in Fig. 2.1.14.

To each elliptical disclination line belongs a second line that follows the branch of a hyperbola. It lies in the plane perpendicular to the ellipse which passes through the long axis of the ellipse. The vertices of the ellipse pass through the foci of the hyperbola and *vice versa*. The appearance of this focal-conic texture is directly related to the layered structure of the smectic. Due to the limited depth of sharpness of focus, the disclination lines tracing the hyperbolae are only visible in Fig. 2.1.12 as short diffuse lines marking one focus of each ellipse.

In thinner layers, the focal-conic texture changes its appearance and gives the fan texture or more accurately the fan-shaped focal-conic texture (Fig. 2.1.15). The ellipses are no longer recognizable as such. They lie in planes perpendicular to the film and are arranged along the edges of the fan-like areas. The lines following the hyperbolae are now visible. They appear as essentially straight lines along which the direction of extinction changes discontinuously. This change becomes smaller with the distance from the accompanying ellipse and the lines gradually disappear.

A shift of the cover glass often changes the fan texture suddenly to a pseudo-isotropic texture. The molecular layers are then parallel to the plane of the film and the shear force needed to shift the cover glass decreases noticeably. In contrast to nematics, a slight shift of the cover glass has no optical effects on a pseudo-isotropic smectic *A* film.

A pseudo-isotropic texture can also form spontaneously at the phase transition, particularly when this takes place on cooling from a pseudo-isotropic nematic. The transition is then difficult to observe except by the disappearance of Brownian motion.

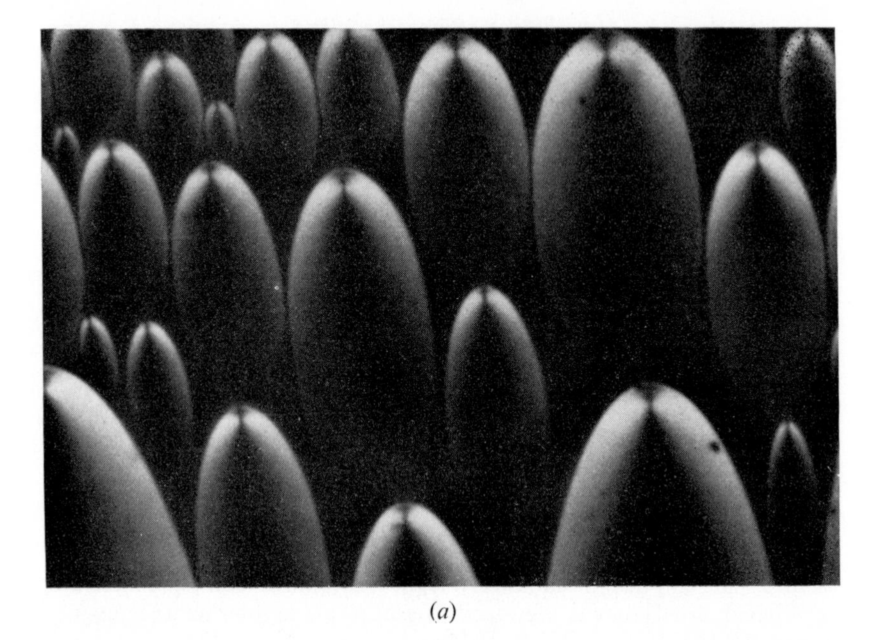

(*a*)

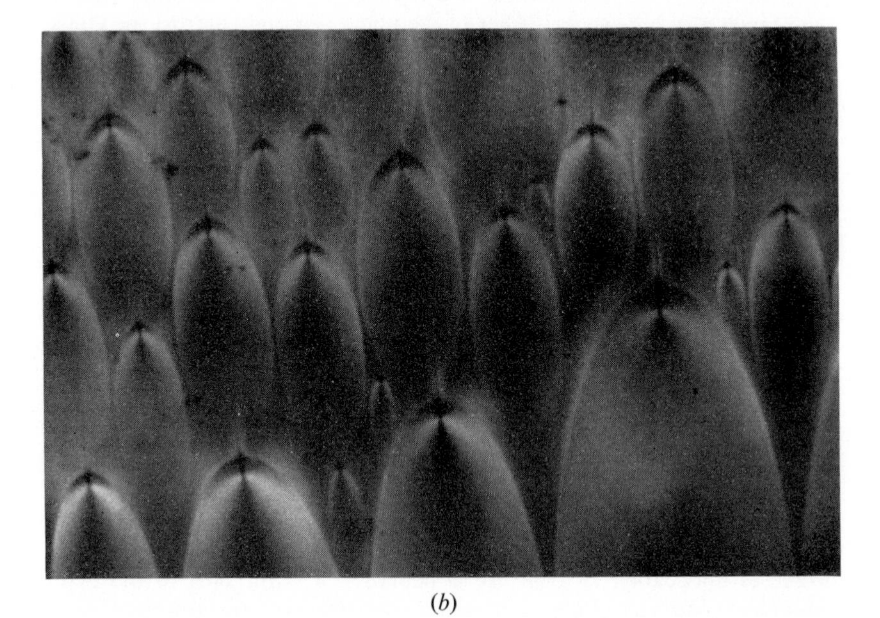

(*b*)

FIG. 2.1.12. Smectic *A* focal-conic texture of 4′-n-butyloxybenzylidene-4-aminopropiophenone; film between glass slides at 134°C; crossed polarizers, magnification ×320; (*a*) focused on lower interface, (*b*) focused on upper interface.

FIG. 2.1.13. Smectic *A* focal-conic texture of 4′-n-octyloxybenzylidene-4-chloroaniline; uncovered film on a glass slide at 96°C; focused on upper surface; crossed polarizers; magnification ×320.

FIG. 2.1.14. Same compound and conditions as for Fig. 2.1.12; crossed polarizers; magnification ×60.

Small smectic drops on a clean glass surface show sometimes a peculiar feature. Their thickness does not change steadily but in steps. The steps can be seen in Fig. 2.1.16 as lines. Similar stepped drops may be observed with other smectics. Larger drops usually show a focal-conic texture where the conic sections are circles and straight lines (Fig. 2.1.13). When the smectic is obtained directly from the normal amorphous isotropic liquid, elongated sharp-pointed smectic particles, the *bâtonnets*, are formed first on cooling (Fig. 2.1.17). They float in the normal liquid. On cooling they merge and may result in a fan texture.

FIG. 2.1.15. Smectic *A* fan texture of 4′-n-octyloxybenzylidene-4-chloroaniline; film between glass slides at 96°C; crossed polarizers; magnification ×96.

Biaxial Smectic Liquid Crystals with Unstructured Layers

(1) GENERAL PROPERTIES AND STRUCTURES

Figure 2.1.18(*a*) shows examples of smectics with C_{2h} symmetry. The two-fold symmetry axis is perpendicular to the plane of drawing which is itself a plane of symmetry. The essential difference to the smectic structures described previously consists in a tilt of the molecular orientation. The axis of preferred orientation for the long molecular axes is no longer normal to the layer but forms a "tilt" angle with the layer normal.

As a consequence of the reduced symmetry, the orientational order is, in general, no longer strictly uniaxial unless the molecule itself has an at least three-fold symmetry axis. Elongated planar molecules, for instance, with an anisotropy in the plane perpendicular to their long axes will preferably orient with their planes either perpendicular or parallel to the two-fold symmetry axis of the liquid crystal. In addition, the fluctuations of the long molecular axes about their preferred orientation have no

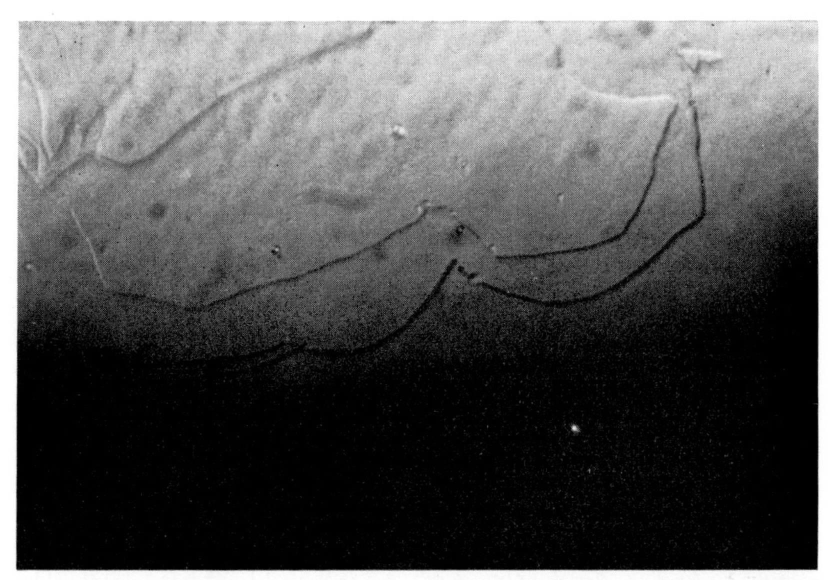

FIG. 2.1.16. Stepped drop formed on a glass slide by the smectic *A* phase of n-$C_4H_9{\cdot}OOC{\cdot}C_6H_4{\cdot}N{=}CH{\cdot}C_6H_4{\cdot}CH{=}NC_6H_4{\cdot}COO{\cdot}C_4H_9$-n at 180°C; crossed polarizers; magnification ×200.

FIG. 2.1.17. *Bâtonnets* at the phase transition normal liquid → smectic *A* for 4′-n-octyloxybenzylidene-4-chloroaniline; film between glass slides at 96·5°C; crossed polarizers; magnification ×94.

longer a rotational symmetry and a second parameter is required to describe their degree of order. No experimental data are available on the degree of these additional features of the orientational order.

Tilted smectic liquid crystals are in general optically biaxial because they lack a more than two-fold symmetry axis. A few studies have been made which directly confirm the biaxial properties for smectic *C* phases [21]; their biaxial character is not very pronounced.

Smectic *C* is the most common tilted smectic. Its structure consists of single molecular layers with a non-polar order (Fig. 2.1.18(*a*)). The phase adjacent to smectic *C* at higher temperatures is either a normal amorphous isotropic liquid, a nematic, a smectic *A* or the recently discovered smectic that is optically isotropic. The transition enthalpies (Tables 2.1.1 and

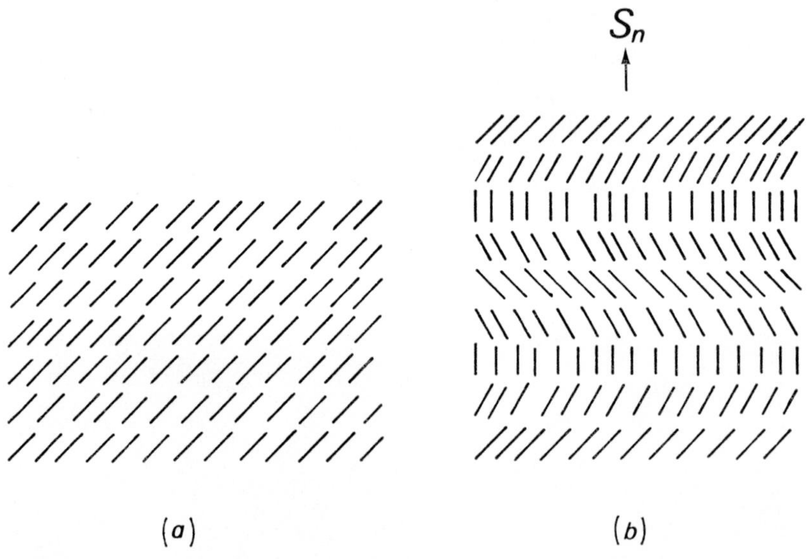

FIG. 2.1.18. Schematic structures of biaxial smectic liquid crystals with tilted unstructured layers. The lines represent molecules and indicate the preferred orientations.

(*a*) smectic *C*; single unstructured layers without polarity;

(*b*) twisted smectic *C*; the twist axis is perpendicular to the molecular layers.

2.1.2) in the first two cases have similar values to the corresponding transitions from smectic *A* phases. The transition from smectic *A* to smectic *C* can have a very low transition enthalpy and may perhaps be second order in some cases.

Some experimental studies of tilt angles have been made [21]–[23]. In the cases where smectic *C* is adjacent to a nematic, a large tilt angle (40–45°) was obtained which showed no observable temperature dependence [21]. For a smectic *C* adjacent to smectic *A*, in contrast, a strong temperature dependence of the tilt angle was found [22]. The tilt angle increased from zero or near zero at the transition point to about 25° at lower temperatures. When an optically active compound is added to a smectic *C*, a uniform alignment of the tilt direction in general no longer corresponds to an equilibrium structure because one symmetry element is lost, the mirror plane perpendicular to the two-fold symmetry axis. The equilibrium structure shows now a twist of the tilt directions (Fig. 2.1.18(*b*)) analogous to the twist of the preferred orientations in cholesterics. A tilted smectic with double layers can of course also assume a twisted structure. When the considered tilted smectics are formed by optically active compounds, they will accordingly spontaneously assume a twisted structure in the same way as cholesterics are formed by optically active compounds instead of nematics.

Only three active compounds have so far been reported that form a twisted smectic in addition to a cholesteric:

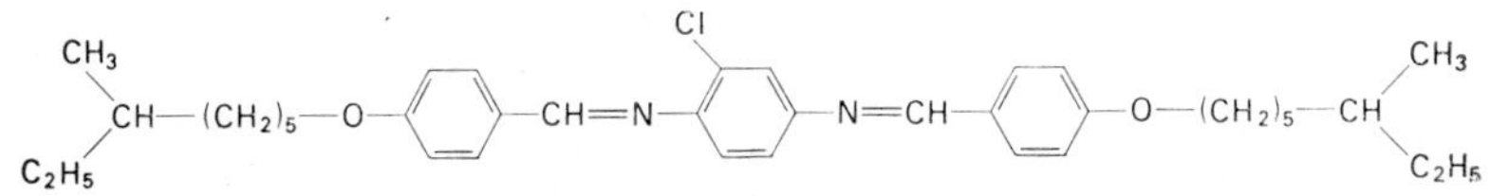

$$(CH_3)(C_2H_5)CH—(CH_2)_5—O—C_6H_4—CH{=}N—C_6H_3(Cl)—N{=}CH—C_6H_4—O—(CH_2)_5—CH(CH_3)(C_2H_5)$$

Transition temperatures: Solid-smectic 29°C; smectic-cholesteric 94·4°C; cholesteric-normal liquid 146·5°C [24].

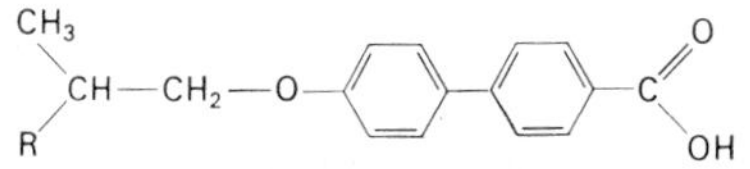

$$(CH_3)(R)CH—CH_2—O—C_6H_4—C_6H_4—C(=O)OH$$

Transition temperatures: R = C_4H_9; solid-smectic 165°C, smectic-cholesteric 215°C, cholesteric-normal liquid 229°C

R = C_6H_5-CH_2; solid-smectic 205°C, smectic-cholesteric 213°C; cholesteric-normal liquid 241°C [25].

In thin layers, a planar texture can be obtained corresponding to that of cholesterics; the molecular layers are parallel to the plane of the film and the twist axis normal to it. Their optical properties for perpendicular incidence are similar to those of planar cholesterics. The same relations are valid for optical activity and selective light reflection. Contrary to cholesterics, no Grandjean steps are formed in wedge shaped samples by a twisted smectic [24].

(2) Textures

Typical textures for smectic *C* phases are in Fig. 2.1.19–21. Figures 2.1.19 and 2.1.20 show the "broken" fan texture and the "broken" focal-conic texture. Compared to the corresponding textures in smectic *A*, they are less regular and disturbed by additional disclinations. Figure 2.1.21 shows the *Schlieren* texture of a smectic *C*; this can be very similar to a nematic *Schlieren* texture, but only singularities with four dark brushes between crossed polarizers can be present [18] [26].

FIG. 2.1.19. Smectic *C* "broken" fan texture of 4,4′-di-n-heptyloxyazoxybenzene, film between glass slides at 90°C; crossed polarizers; magnification ×200.

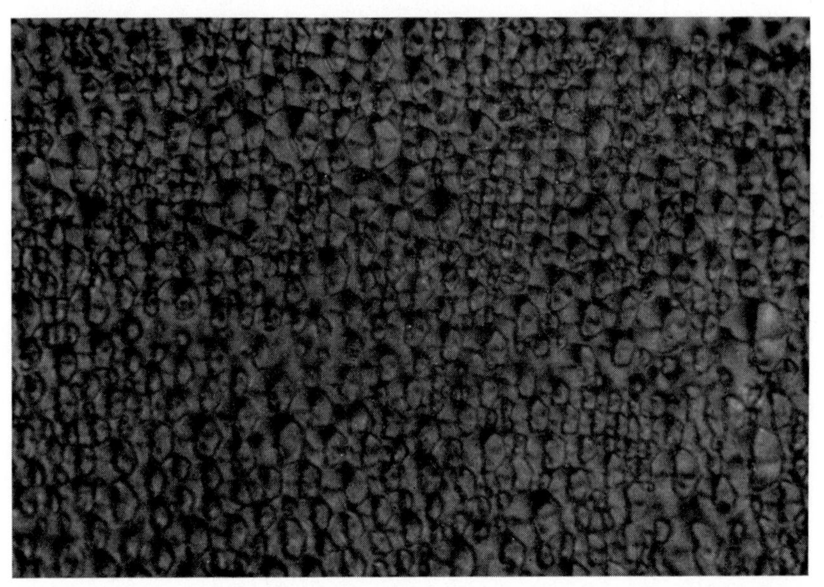

FIG. 2.1.20. Smectic *C* "broken" focal-conic texture of 4,4′-di-n-heptyloxyazoxybenzene; uncovered film on a glass slide at 93°C; crossed polarizers; magnification ×200.

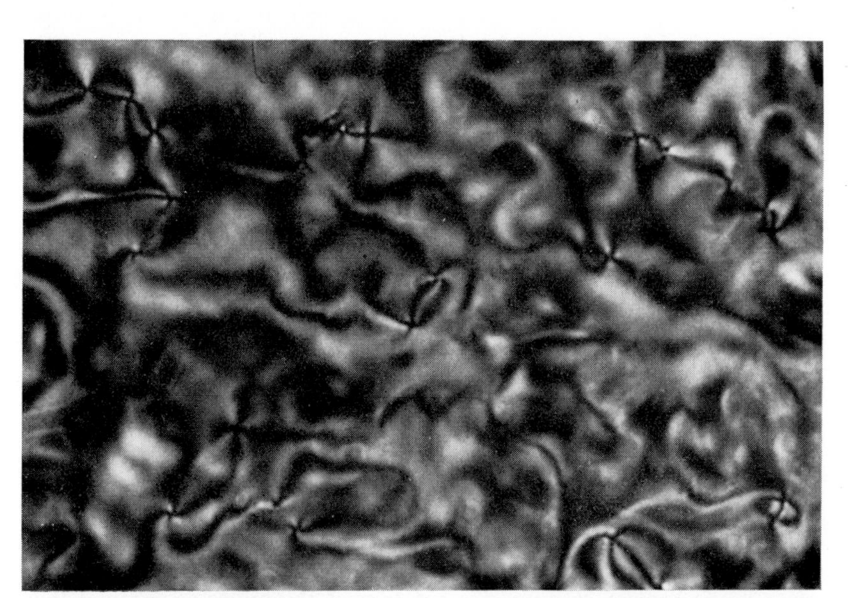

FIG. 2.1.21. Smectic *C Schlieren* texture of n-$C_4H_9 \cdot OOC \cdot C_6H_4 \cdot N = CH \cdot C_6H_4 \cdot CH = N \cdot C_6H_4 \cdot COO \cdot C_4H_9$-n with point-like alignment singularities; uncovered film on a glass slide at 130°C; crossed polarizers; magnification ×320.

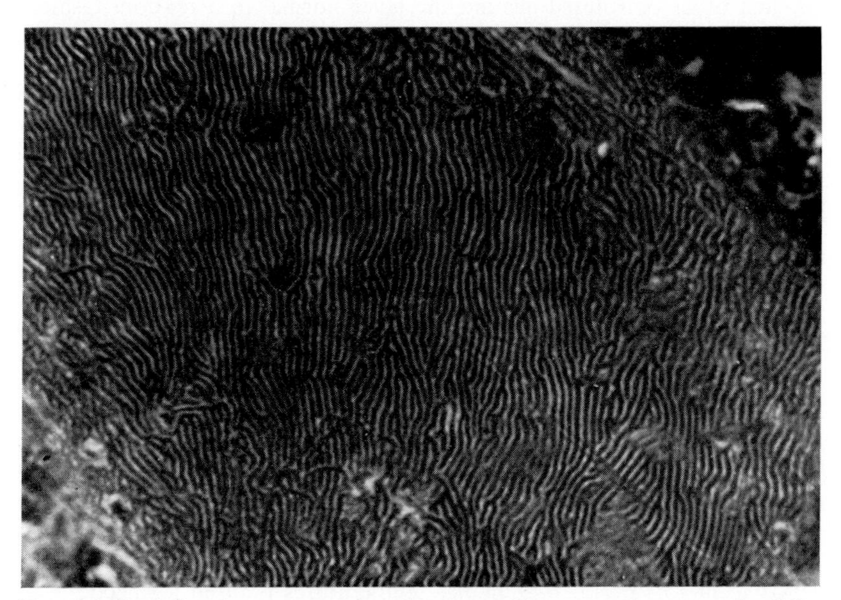

FIG. 2.1.22. Striated pattern at the phase transition nematic → smectic *C* for 4,4′-di-n-heptyloxyazoxybenzene; film between glass slides at 96°C; crossed polarizers; magnification ×96.

The *Schlieren* texture of smectic C corresponds to the pseudo-isotropic texture of smectic A; the molecular layers are parallel to the plane of the film. The *Schlieren* pattern is due to variations in the tilt direction. It is possible to obtain a uniform alignment of the tilt direction by rubbing the surfaces. The smectic C *Schlieren* texture shows light scattering similar to that shown by nematic *Schlieren* textures due to a directly observable Brownian motion. Shifting the cover glass causes strong optical effects due to flow alignment of the tilt direction.

At the transition from a normal isotropic liquid to smectic C, *bâtonnets* appear first similar to those of smectic A phases. At a nematic to smectic C transition and when the transition leads from a nematic to a smectic *Schlieren* texture, a striated texture may appear first (Fig. 2.1.22) that is replaced by the *Schlieren* texture only on further cooling.

Smectic Liquid Crystals with Structured Layers

(1) General Properties

Smectics with structured layers represent the liquid crystals of highest order. The molecules in the layers are arranged in a regular two-dimensional lattice. It is, however, generally assumed that the orientational order is incomplete because of a rotation of the molecules around their long axis. There is a negligible entropy difference between a regular arrangement of the structured layers so that a three-dimensional lattice results and an irregular arrangement where the layers are shifted parallel to each other or rotated around the layer normal in a random fashion. One could expect, therefore, that smectic crystals with structured layers build in fact a three-dimensional lattice.

The few X-ray studies that have been made on uniformly oriented samples or "single crystals" confirm in fact that the layers are all equally oriented and that there can be a more or less perfect three-dimensional lattice [27], [28]. The layers slip, however, easily past each other and the liquid crystals still flow and may even have a true viscous flow when it involves only parallel shifting of the layers.

There exist seventeen two-dimensional space groups [29], but so far only four different structured layer smectics have been established.

The most common smectic with structured layers is smectic B. It was studied first by Herrmann and Krummacher [29] who concluded from X-ray results that the lattice in the layers has hexagonal symmetry. When the long molecular axes are perpendicular to the layers the symmetry class of such a smectic B is accordingly D_{6h}. It has the highest symmetry class possible for structured smectics and is optically uniaxial. Tilted smectics with structured layers have also been established by X-ray studies [27], [28] and some of them are also classed as smectic B [28]. They can have only C_{2h} symmetry and are optically biaxial.

Adjacent to smectic B at higher temperatures is often a smectic A or a smectic C. The transition enthalpies listed in Table 2.1.1 indicate that the change in order at this transition is considerably larger than that from smectic A or smectic C to a nematic.

(2) TEXTURES

The characteristic texture for liquid crystals with structured layers is the "mosaic" texture (see Fig. 2.1.23). It has been observed with smectic *B* and, slightly modified, also with other smectics with structured layers. The mosaic texture is composed of uniform areas in which the orientation of the index ellipsoid is constant. At the boundaries the orientation changes abruptly forming what we may call a disclination surface. This is in contrast to smectics with unstructured layers and to nematics and cholesterics where only disclination lines form spontaneously. The existence of disclination surfaces indicates that structured smectics are

FIG. 2.1.23. Smectic *B* mosaic texture of terephthalylidene-bis-(4-n-butylaniline) between glass slides at about 130°C; crossed polarizers; magnification ×100 (according to T. R. Taylor).

more rigid and have deformation energies which are considerably higher. Nevertheless, a modification of fan textures and of *Schlieren* textures can also be observed with smectic *B* films that are obtained on cooling from a smectic *A* or a smectic *C*. These textures owe their appearance in smectic *B*, however, to the existence of the corresponding texture in the preceding state and are not stable. It has been observed in a number of cases that, under suitable conditions, they change gradually to the mosaic texture [2] [13]. In the continuously deformed focal-conic and *Schlieren* textures with a number of line singularities, the deformation energy exceeds accordingly the energy of the disclination surfaces in the mosaic texture.

A pseudo-isotropic texture can also be observed with smectic *B*. Here the liquid crystal is uniformly aligned with the layers parallel to the plane of the film. The existence of the pseudo-isotropic texture shows directly that the corresponding smectic *B* is uniaxial and has upright layers.

A Smectic Phase with an Isotropic Texture

A very unusual smectic has been observed with two compounds of the homologous series of the 4′-n-alkoxy-3′-nitrobiphenyl-4-carboxylic acids [30].

The 4′-n-hexadecyloxy compound gives on cooling of the normal amorphous liquid a smectic *A* at 202°C, which at 197°C transforms to another smectic, smectic *D*, and at 171°C this transforms to smectic *C*. The 4′-n-octadecyloxy compound forms on cooling first a smectic *D* at 195°C, and this then transforms to smectic *C* at 159°C.

Smectic *D* has a strong tendency to assume an optically isotropic texture [30]. This property is not reconcilable with a layered structure, in particular not when it is combined with a parallel orientation of the long molecular axes. X-ray studies show no indication of a layered structure, but indicate a cubic lattice with a spacing that corresponds to the length of the dimeric molecule.

Optically isotropic liquid crystal phases ("cubic" or "viscous isotropic") were previously known only for amphiphilic compounds, except for possibly the optically isotropic texture of the cholesteric mesophase formed by some cholesteryl esters. In some of these cases spherical micelles are arranged on a cubic lattice. It seems possible that in smectic *D* a similar "micelle" formation takes place and that these then arrange in cubic symmetry. The presence of the long paraffin tails may perhaps stabilize the micelle structure in the absence of a solvent. It could be possible then that other forms of mesophase formed by amphiphiles (with or without solvent) can also be realized by suitable non-amphiphilic compounds without the presence of a solvent.

REFERENCES

[1] Alexander, E. and Herrmann, K. *Z. Kristallogr. Kristallgeom. Abt.* **A69,** 285 (1928).

[2] Sackmann, H. and Demus, D. *Fortschr. chem. Forsch.* **12,** 349 (1969); *Molec. Crystals* **2,** 81 (1966).

[3] Arnold, H. Dissertation, Halle 1959, East Germany.

[4] Arnold, H. and Sackmann, H. *Z. Elektrochem.* **63,** 1171 (1959).

[5] de Vries, A. *Molec. Crystals Liqu. Crystals* **10,** 31 (1970); **10,** 219 (1970).

[6] Weygand, C. and Gabler, R. *Z. phys. Chem.* **B44,** 69 (1940).

[7] ARNOLD, H. *Z. phys. Chem.* **226,** 146 (1964).
[8] ARNOLD, H., JACOBS, J. and SONNTAG, O. *Z. phys. Chem.* **240,** 177 (1969).
[9] ARNOLD, H., EL-JAZAIRI, E. B. and KOENIG, H. *Z. phys. Chem.* **234,** 401 (1967).
[10] ARNOLD, H. and ROEDIGER, P. *Z. phys. Chem.* **231,** 407 (1966).
[11] MEIER, W. and SAUPE, A. *Z. Naturf.* **15a,** 287 (1960).
[12] SAUPE, A. *Angew. Chem. Int. Edn.* **7,** 97 (1968).
[13] DEMUS, D. and SACKMANN, H. *Z. phys. Chem.* **238**, 215 (1968).
[14] DREYER, J. F. Lecture on "Epitaxy of Nematic Liquid Crystals", 3rd Int. Liquid Crystal Conference, Berlin (1970).
[15] ARNOLD, H. and ROEDIGER, P. *Z. phys. Chem.* **239,** 283 (1968).
[16] KASSUBEK, P. and MAIER, G. *Molec. Crystals Liqu. Crystals* **8,** 305 (1969).
[17] KEATING, P. N. *Molec. Crystals Liqu. Crystals* **8,** 315 (1969).
[18] SAUPE, A. *Molec. Crystals Liqu. Crystals* **7,** 59 (1969).
[19] HERRMANN, K. *Trans. Faraday Soc.* **29,** 972 (1933).
[20] ARORA, S. L., TAYLOR, T. R. and FERGASON, J. L. *Liquid Crystals and Ordered Fluids* (edited by R. S. Porter and J. F. Johnson), Plenum Press, New York (1970), p. 321.
[21] TAYLOR, T. R., FERGASON, J. L. and ARORA, S. L. *Phys. Rev. Lett.* **24,** 359 (1970).
[22] TAYLOR, T. R., ARORA, S. L. and FERGASON, J. L. *Phys. Rev. Lett.* **25,** 722 (1970).
[23] DE VRIES, A. *Acta Crystallogr.* **A25,** 135 (1969).
[24] HELFRICH, W. and CHAN, S. OH. *Molec. Crystals Liqu. Crystals* **14,** 289 (1971).
[25] LECLERCQ, M., BILLARD, J. and JACQUES, J. *Molec. Crystals Liqu. Crystals* **8,** 367 (1969).
[26] NEHRING, J. and SAUPE, A. *J. chem. Soc. Faraday II* **68,** 1 (1972).
[27] LEVELUT, A.-M. and LAMBERT, M. *C. r. hebd. Séanc. Acad. Sci., Paris* **272B,** 1018 (1971).
[28] DE VRIES, A. *Molec. Crystals Liqu. Crystals*, to be published (1972).
[29] HERRMANN, K. and KRUMMACHER, A. H. *Z. Kristallogr. Kristallgeom. Abt.* **A81,** 317 (1932).
[30] DEMUS, D., KUNICKE, G., NEELSEN, J. and SACKMANN, H. *Z. Naturf.* **23a,** 84 (1968).
[31] BROOKS, J. D. and TAYLOR, G. H., *Carbon* **3,** 185 (1965).

2.2 NON-AMPHIPHILIC CUBIC MESOPHASES—"PLASTIC CRYSTALS"

P. A. Winsor

Non-amphiphilic cubic mesophases—"Plastic crystals"

For reasons which have been noted in the Preface and Introduction, the non-amphiphilic cubic mesophases, or "plastic crystals", have usually been treated in the literature of the solid state [1–7]. Yet, as will be seen from the following account, they are among the most molecularly disordered of all mesophases and, from this point of view, the most "liquid" in character.

On the other hand, plastic crystals separate in the crystal forms of the cubic system (rarely hexagonal) and to this extent resemble ordinary solid crystals. However, they show unusually low yield points. The most plastic (e.g. perfluorocyclohexane [3]) will flow under their own weight and although the majority are less soft, they may readily be cut with a knife or extruded through a small hole. The pressure required to produce flow is very considerably less (2–14 times) than that required to extrude the ordered solid crystal (usually anisotropic) produced by transition from the plastic crystal at a lower temperature [2, 3, 5]. Tertiary butyl alcohol, trimethylacetic acid (pivalic acid) and *d*-camphor provide familiar laboratory examples of plastic crystals.

Plastic crystals are thus neither true liquids wholly devoid of long-range molecular order nor true crystalline solids with molecules in regular long-range orientational and positional order, but constitute a further mesomorphic state of matter equivalent in status to Friedel's nematic and smectic mesophases.

The nematic and smectic mesophases are typically formed by relatively elongated and rather rigid lath-like molecules (Chap. 2.1) which, when undergoing rotatory displacements about their long axes, acquire effectively circular cylindrical symmetry. The cubic mesophases are typically constituted by rather compact globular molecules which are not markedly anisodimensional and which, when undergoing rotatory displacements about certain axes, acquire, effectively, close-to-spherical symmetry. These globular molecules in the stationary state may, however, depart quite appreciably from spherical symmetry. Thus, tetramethylmethane and the carbon tetrahalides (near-to-spherical), hexamethylethane (prolate spheroid), cyclohexane (oblate spheroid) and 2,2-dimethylbutane (pear-shaped) are among the very numerous compounds which form cubic mesophases or, as these are more commonly termed, plastic crystals.

Plastic crystals were first recognized by Timmermanns [2, 8, 9] as characteristically containing "globular" molecules and showing a low entropy of liquefaction, σ, usually below 5 entropy units (cal deg^{-1} mol^{-1}), although occasionally, as with sulphur hexafluoride ($\sigma = 5{\cdot}4$) and hexachloroethane ($\sigma = 5{\cdot}5$), slightly higher values may occur. In other cases values as low as 0·80, for cyclopentane, are found.

It was early recognized that the formation of plastic crystals is due to the capacity of the constituent molecules over a particular range of temperatures to arrange themselves in a cubic array while at the same time undergoing thermal rotatory displacements so that there is no long-range orientational order between the molecules. At the upper limit of this temperature range liquefaction occurs with breakdown of the cubic arrangement but with only a small increase in entropy and little increase in volume [3]. At the lower temperature limit a transition occurs (Crystal I $\rightarrow$ Crystal II), typically to an ordered anisotropic solid crystal. In this case there is a large decrease in entropy. Occasionally, however, there is a transition with only a small entropy decrease to a second (and sometimes even to a third) plastic crystalline form which undergoes a transition at a still lower temperature with a larger decrease in entropy to an ordinary solid crystal. The successive plastic crystalline phases presumably differ in the details of the permissible thermal displacements of the molecules and thus provide an analogy in the plastic mesophases to the polymorphism encountered in the smectic series. Examples of the behaviour observed are provided by Table 2.2.1 [3].

Although plastic crystals almost always contain globular molecules, not all molecules that are globular give rise to plastic crystals (*cf.* [3]). Thus, while methane, silane, germane and carbon tetrachloride all form plastic crystals, silicon tetrachloride does not. This may be because the longer Si–Cl bonds produce greater molecular interlocking and thus hinder rotatory displacements. Again, while adamantane (symmetrical tricyclodecane)

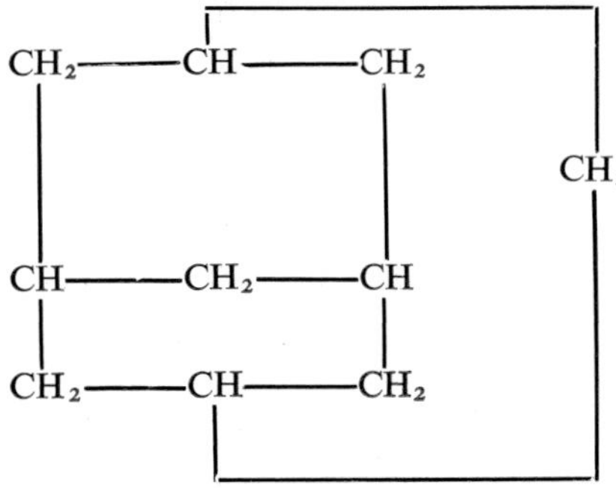

```
          ┌──────────────────────┐
          │                      │
CH2──────CH──────CH2             │
 │                │             CH2
 │                │              │
CH───────CH2─────CH              │
 │                │              │
CH2──────CH──────CH2             │
          │                      │
          └──────────────────────┘
```

forms a face-centred cubic plastic crystal phase, hexamethylenetetramine, which may be regarded as being derived from adamantane by replacement of the CH≡ groups by N≡ groups, forms an ordinary non-rotatory body-centred cubic crystal. This may be because molecular reorientation is made more difficult by the local polarity due to the nitrogen atoms [3].

The type of transition between a true crystal and the plastic mesophase

TABLE 2.2.1. *Temperatures of transition* (T_t) *and liquefaction* (T_l) *for some substances forming plastic crystals** [3]

Substance	T_t, °K	T_l, °K	ΔS_t	ΔS_l
Argon	—	83·85	—	3·35
HCl	98·4	158·9	2·89	2·99
H_2S	103·5	187·6	3·55	3·03
	126·0		0·96	
N_2	35·6	63·1	1·54	2·73
CH_4	20·5	90·7	0·76	2·48
SiH_4	63·45	88·5	2·61	1·80
GeH_4	73·2	107·3	1·78	1·86
	76·55		1·69	
CF_4	76·2	89·5	4·64	1·87
CCl_4	225·5	250·3	4·86	2·40
CBr_4	320·0	363·3	4·98	2·60
$C(CH_3)_4$	140·0	256·6	4·4	3·03
$C(CH_3)_3Cl$	219·0	248·0	8·7	2·0
$C(CH_3)Cl_3$	224·0	241·0	7·97	4·5
$C(CH_3)_3.C_2H_5$	126·8	174·2	10·17	0·80
	140·9		0·48	
$C(CH_3)_3SH$	151·6	274·4	6·41	2·16
	157·0		0·99	
	199·4		1·16	
C_2Cl_6	344·6	458·0	5·7	5·5
$Si_2(CH_3)_6$	221·9	287·6	10·5	2·51
$(CH_3)_2CH.CH(CH_3)_2$	136·1	145·2	11·41	1·32
Cyclohexanol	263·0	297·0	7·45	1·4
$C(CH_2OH)_4$	457·0	539·0	22·8	3·16
Cyclobutane	145·7	182·4	9·36	1·43
Cyclopentane	122·4	179·7	9·52	0·80
	138·1		0·60	
Cyclohexane	186·0	279·8	8·59	2·22
SF_6	94·3	222·5	4·07	5·40
PtF_6	276·0	334·4	7·7	3·2
Camphor	250·0	453·0	7·6	2·8
Ammonium nitrate†	398·0	442·0		2·94

* For gradual transitions, T_t is the temperature at which the heat capacity reaches a maximum. ΔS_t and ΔS_l are the entropies of transition and liquefaction respectively, in cal deg^{-1} mol^{-1}.

† Additional information from [21]

may vary from case to case [3, 4, 5]. It may be isothermal (first-order) or gradual (second-order), but it is not always easy to determine which, since impurities can make what is actually an isothermal transition appear anisothermal.

THE NATURE OF THE MOLECULAR MOTION IN THE PLASTIC CRYSTAL MESOPHASES

It is now known that the molecular motion in plastic crystals is rarely, if ever, free molecular rotation in the literal sense. Such motion is not generally found even in amorphous liquids. Rather, the barriers between positions of minimum potential energy are small enough to allow the molecules to tumble rapidly from one orientation to another, the orientations at distances of more than a few molecular diameters being randomly distributed throughout the phase.

The experimental evidence concerning the degree of molecular motion in plastic crystals is based mainly on investigations of thermal properties, X-ray diffraction, dielectric properties, nuclear magnetic resonance and diffusion.

Thermal properties

It was their peculiar thermal behaviour which led Timmermanns [2, 8, 9] to the recognition of plastic crystals as a unique intermediate state of matter. In addition to a very low entropy of liquefaction, plastic crystals usually also show a relatively high temperature of liquefaction as illustrated by Timmermanns [2] for the comparative behaviour of neopentane and *n*-pentane in Fig. 1.3. With the high liquefaction temperature goes a high vapour pressure which, at the triple point (vapour + liquid + plastic crystals), may be higher than atmospheric so that under ordinary pressure the plastic crystals may sublime directly without melting. This is found, for example, with sulphur hexafluoride, hexachloroethane and perfluorocyclohexane. Thus, with globular molecules which form plastic crystals the normal amorphous liquid phase as formed by the majority of non-globular and non-lath-like molecules is replaced, over a particular range of temperatures, by the cubic plastic mesophase much as, with lath-like molecules, it is replaced by the anisotropic smectic and/or nematic mesophases. This was illustrated by Timmermanns as in Fig. 1.4. The transition, crystalline solid → mesophase, is accompanied by a high entropy increase for both types of mesophase, while the entropy increase for the transition mesophase → amorphous liquid is, in both cases, relatively small. This points to a degree of molecular disorder in the mesophases closer to that in the amorphous liquid than to that in the solid crystal. A value for the number of permissible molecular orientations that can be adopted at random in the plastic crystals may be estimated in some instances from measurements of heat capacities and entropies of phase change [3, 4].

X-ray diffraction patterns

That there is considerable mobility of the molecules in plastic crystals, but that this does not amount to literally free rotation, is indicated by their X-ray diffraction patterns. The relatively high degree of disorder in plastic crystals is shown by the fact that they characteristically give only a small number of reflections with considerable background scattering. Thus, with carbon tetrachloride, the cubic plastic crystal phase "diffracts very poorly; exposures of three hours and more were needed to obtain usable diffraction patterns. All diagrams were marked by extremely heavy non-radial diffuse scattering" [10]. With cyclohexane there is evidence of a decrease in the number of lines with increasing temperature [11]. With cyclobutane, only the (110) and (200) reflections are evident, indicating a body-centred cubic unit cell [12]. In consequence of the above effects unit cell dimensions can often be estimated only approximately. Some crystallographic details of a number of plastic crystals, as listed by Dunning, are given in Table 2.2.2 [13]. The different forms of cubic crystal are presumably to be related to different degrees of restriction in the rotatory reorientation processes of the molecules within them. With completely unrestricted rotation one might expect a face-centred cubic lattice.

It is significant that the symmetry elements of many plastic crystals are incompatible with those of their constituent molecules if these are regarded as stationary [13]. Thus, the plastic mesophase of *t*-butyl chloride has a face-centred cubic lattice with one molecule per lattice point. This requires that effectively each molecule has a two-fold axis whereas the stationary molecule has only a three-fold axis. Such apparent discrepancies between the lattice symmetry and the molecular symmetry provide one of the reasons for thinking that the molecules, even if not literally rotating, are disordered in orientation.

From the known van de Waal's radii of the atoms of a molecule, the diameter of the sphere necessary to enclose the freely rotating molecule may be calculated. The diameter of this sphere is always found to be greater (by about 15–20%) than the distance between the lattice points in the plastic crystal [13]. Hence, the molecules cannot have room to rotate freely and change of orientation must require correlated movements by neighbouring molecules. Further, since the volume change on liquefaction is usually small, free rotation will not be possible even in the liquid. These relationships recall the conclusion reached on similar grounds that at least in some smectic and nematic mesophases, as well as in the corresponding amorphous liquid, literally free rotation of the lath-like molecules about their longitudinal axes is not geometrically possible [14].

Dielectric properties

The detection of molecular motion by studies of dielectric properties is possible in the case of molecules possessing a dipole moment [6]. In the case of compounds with such molecules there is a sharp drop in dielectric constant at the transition from the liquid to the solid crystalline state (Fig. 2.2.1). This is because orientation polarizability in an electric field drops to zero when the molecular dipoles become fixed in the crystal lattice [6, 15]. In the case of the plastic crystal mesophase, on the other hand, on account of rotational reorientation, the dielectric constant does not fall but retains a value of the same order as that for the liquid. At the transition from the plastic mesophase to the solid crystal with its orientationally immobile molecules a sharp drop in dielectric constant occurs. This behaviour is found, for example, with cyclopentanol, cyclohexanol,

TABLE 2.2.2. *Lattice characteristics for some plastic mesophases* [13]

Substance	Lattice structure	No. of molecules in unit cell	Lattice constant, Å
Tetrahedral molecules			
CCl_4	f.c.c.	4	8·34
CBr_4	$T_h^6(Pa3)$	8	11·34
	or $T_d^1(P\bar{4}3m)$	1	5·67
CI_4	O_h, T_d or O	1	9·14
$C(Me)_4$	$O_h^7(Fd3m)$	8	11·25
$C(SMe)_4$	b.c.c.	2	8·15
$(Me)_3CCl$	f.c.c.	4	8·40
$(Me)_3CBr$	f.c.c.	4	8·70
$(Me)_3CSH$	f.c.c.	4	8·82
$C(NO_2)_4$	$T_d^3(I\bar{4}3m)$	2	7·08
SiF_4	$T_d^3(I\bar{4}3m)$	2	5·41
SiI_4	$T_h^6(Pa3)$	8	11·99
Octahedral molecules			
$Cl_3C.CCl_3$	$O_h^9(Im3m)$	2	7·43
$Me_3C.CMe_3$	b.c.c.	2	7·69
$Me_3C.CMeCl_2$	b.c.c.	2	7·38
$Me_3C.CMe_2Cl$	b.c.c.	2	7·62
$Me_2ClC.CClMe_2$	b.c.c.	2	7·58
$Me_2ClC.CCl_3$	b.c.c.	2	7·4
$Br_3C.CBr_3$	b.c.c.	2	—
$Me_3Si.SiMe_3$	b.c.c.	2	8·47
$Me_3C.COOH$	$O_h^5(Fm3m)$	4	8·82
Cyclic molecules			
Cyclobutane	b.c.c.	2	6·06
Cyclopentane	hexagonal	2	5·83
			9·35
Cyclohexane	f.c.c.	4	8·76
Thiacyclohexane	f.c.c.	4	8·69
Cyclohexanol	$O_h^5(Fm3m)$	4	8·83
Cyclohexanone	f.c.c.	4	8·61
Chlorocyclohexane	b.c.c.		9·05
Cycloheptatriene	cubic	8	10·6
Cage molecules			
Chinuclidin	f.c.c.	4	8·977
sym-Tricyclodecane	$T_d^2(F\bar{4}3m)$	4	9·426
DL-Camphor	f.c.c.	4	10·1
DL-Camphene	b.c.c.	2	8·00
Borneol	f.c.c.	4	10·25
Bornyl chloride	f.c.c.	4	10·39

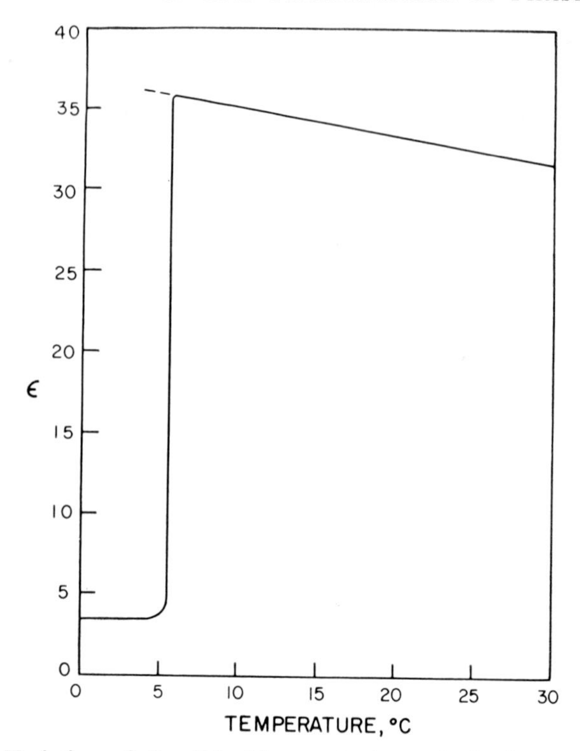

FIG. 2.2.1. Variation of the dielectric constant, ε, of nitrobenzene at 0·5 to 60 kc/s with temperature [6].

cyclohexanone, cyclohexyl chloride and with camphor and a number of its derivatives (*cf.* [6]).

A very interesting study of the rotational characteristics of a number of tetrasubstituted methanes has been made by Smyth [15] on the basis of their dielectric behaviour. The dipolar molecules, $CH_3.CCl_3$, $(CH_3)_2.CCl_2$ and $(CH_3)_3.CCl$ have been shown by dielectric constant measurements to rotate in the plastic crystal phase. $(CH_3)_3CBr$ and $(CH_3)_3CI$ show similar behaviour, although for the iodide the relaxation time for the rotation process is much longer.

The rotational behaviours of ten different halogenated tetra-substituted methanes were compared [16] in relation to the lengths of the carbon–halogen units, C–X, calculated as the sum of the two bond radii plus the van de Waal's radius of the halogen, that is as the distance from the carbon nucleus to the surface of the halogen atom. These lengths (Å), together with the corresponding length for $C–CH_3$ are as follows: C–F, 2·76; C–Cl, 3·56; C–Br, 3·86; C–I, 4·25 and $C–CH_3$, 3·54.

The closeness of the values for C–Cl and $C–CH_3$ indicates why the tetrasubstituted methylchloromethanes rotate so readily in the plastic crystal. $CClBr_3$, CCl_2Br_2 and CCl_3Br, where the ratio of the C–X values is 1·09, also form plastic crystals with molecular rotation, but CF_2Cl_2 and

$CFCl_3$, where the ratio is 1·29, and CF_2Br_2 and CF_3Br, where the ratio is 1·40, and CF_2BrCl show no rotation and thus do not form plastic crystals. Although CF_4, CCl_4, CBr_4 and CI_4, being non-polar, cannot be characterized by dielectric constant measurements, they are all known on other grounds to form plastic mesophases showing molecular rotatory motion.

From measurements of dielectric constants and losses at a wavelength of 3·2 cm and from the calculated relaxation times it has been shown [17] that for a number of compounds which form plastic crystals the internal viscosity or resistance to molecular rotation is less in the plastic crystal phase than in the amorphous liquid when both are close to the temperature of liquefaction. This was shown for $(CH_3)_3CCl$, $(CH_3)_2CCl_2$, CH_3CCl_3 and $(CH_3)_3CBr$. Decreasing the temperature of the plastic crystal raised the resistance to rotation so that at 20–25°C below the liquefaction temperature, the internal viscosity in the plastic crystal is about equal to that in the liquid just above the liquefaction temperature. These results recall the conclusions reached by Friedel concerning the nematic to amorphous liquid transition considered on p. 9.

Nuclear magnetic resonance studies and self-diffusion

Both molecular rotation and self-diffusion in plastic crystals have been investigated by nuclear magnetic resonance (NMR) studies of line widths and spin lattice relaxation times [3, 5, 7] but the method does not distinguish between literally free rotation and the capacity of the molecules to tumble rapidly from one orientation to another. At the transition from the solid crystal to the plastic crystal there is usually a decrease in line width and an increase in the spin-lattice relaxation time, T_1. This is found, for example, with methylchloroform [18] and triethylenediamine [19]. With neopentane [18], however, there is sufficient reorientation, even below the transition temperature, to narrow the line width to the extent that further change is not observed at the transition. There is, however, a sudden rise in T_1. It is frequently found that when the line width falls at the solid crystal to plastic crystal transition point it decreases further as the liquefaction temperature is approached [20, 21, 22]. Thus, for cyclohexane [20] the second moment falls at the transition at 186°K from about 6 G^2 to 1·4 G^2. It remains at this value until about 220°K after which it decreases, falling almost to zero at the liquefaction point. From this decrease it may be inferred that the molecular motion is becoming isotropic and, further, that self-diffusion is taking place. Andrew [23] has given a list of organic plastic crystals in which self-diffusion has been inferred from NMR studies. To account for some of the observed line narrowings, diffusion rates of at least 10^4 displacements per second were assumed by Rushworth [24]. However, Blum and Sherwood [22] have found that the activation energies for self-diffusion, as calculated in the usual manner from NMR measurements are, in general, much lower (approximately one-half) than those obtained from radio-tracer and creep studies [25]. They found further that, although the rate of self-diffusion in the plastic crystal phase of camphene, as evaluated from radio-tracer and creep measurements, was enhanced by impurities, the value as calculated from NMR measurements remained unaffected. They concluded therefore that NMR measurements (whether of line widths or of relaxation times) do not provide a reliable basis for the calculation of self-diffusion rates.

Plastic crystalline phases containing more than one type of unit: Plastic crystalline solutions

An interesting fact which may be of particular relevance in connection with the formation of the cubic phases in amphiphilic micellar systems (*cf.* Chap. 5), is that compounds that form plastic crystals are frequently miscible in the plastic phase, in some cases forming almost perfect solutions [26]. Such miscibility is found, for example, with methane and krypton; cyclopentane and 2,2-dimethylbutane; 2,2-dimethylbutane and 2,3-dimethylbutane; carbon tetrachloride and tertiary butyl chloride, and with mixtures of *d*- and *l*-nitrocamphor. Such miscibility is in accord with the mesomorphic character of the plastic crystal phase and is paralleled by the miscibility commonly found between amorphous liquids, between nematic mesophases and between smectic mesophases of corresponding types (A, C or B) [27, 28]. Miscibility of plastic crystalline phases is, however, not an absolutely general rule. Thus, cyclohexane and carbon tetrachloride, both of which form face-centred cubic plastic crystals with similar lattice constants, are reported not to form solid solutions at all in the plastic phase [26]. An interesting cubic plastic crystal phase, which may be regarded as a solution built up from two conformational isomers in equilibrium, is provided by succinonitrile (*cf.* [21]). Differences in infra-red spectra above and below the transition point were accounted for by assuming that below the transition point the succinonitrile molecule is present only in the *cis*-configuration, while above the transition point it is present as an equilibrium mixture of conformational isomers [29].

A number of high temperature cubic polymorphs of inorganic compounds are certainly mesomorphic in character [30] as was very early appreciated by Lehmann [31]. The diffusional mobility of the units within these cubic mesophases is shown in some cases by quite high electrical conductivities. An interesting example of this is provided by the cubic phase of ammonium nitrate [21] (*cf.* Table 2.2.1), which again can, in a sense, be regarded as a plastic crystalline phase containing two types of unit. From the measurements represented in Fig. 2.2.2 an activation enthalpy for diffusion in ammonium nitrate was calculated on the usual assumptions both from the NMR data and from the electrical conductivities. The values obtained were $11.3 \pm 0{\cdot}5$ kcal mol^{-1} and $16{\cdot}0 \pm 0{\cdot}3$ kcal mol^{-1} respectively, a ratio in interesting relationship to the results of Blum and Sherwood [22] already discussed.

Plastic crystals containing multi-molecular units

With some of the systems in which plastic crystal solutions are formed, it is thought that certain groups of molecules are able to rotate together as a whole. Thus, it is believed that two molecules of cyclopentane with

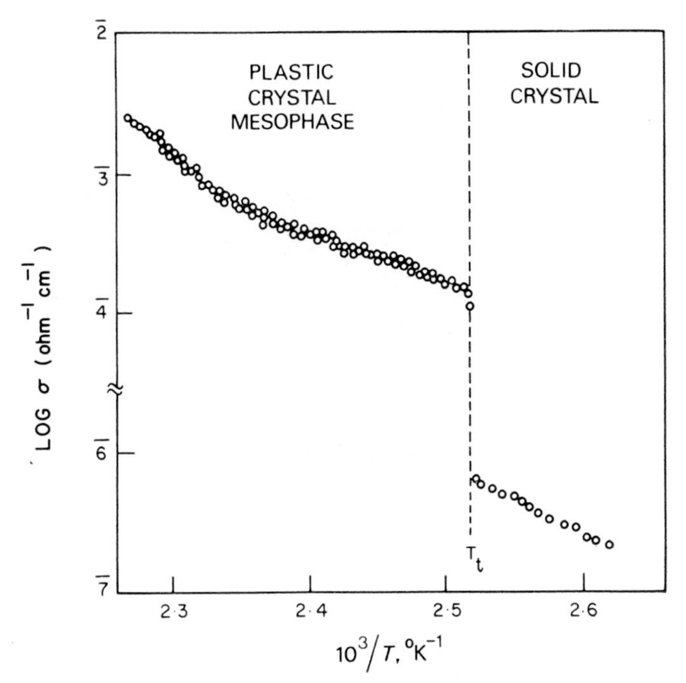

FIG. 2.2.2. Variation of the specific conductivity, σ, of ammonium nitrate with absolute temperature [21].

one molecule of 2,2-dimethylbutane can rotate as a unit because it is easier for the three to rotate together than separately. Similarly, dissociable units that rotate as a whole are formed from two molecules of 2,3-dimethylbutane and three molecules of 2,2-dimethylbutane [2, 5].

It seems possible that such rotation of multi-molecular units may account for the formation of cubic mesophases (Fig. 2.2.3) by 4′-n-hexadecyloxy- and 4′-n-octadecyloxy-3′-nitrobiphenyl-4-carboxylic acids recorded by Demus, Kunicke, Neelson and Sackmann [27, 28]. In the cubic phase the lattice spacing corresponds to about twice the molecular length. It therefore seems probable that the rotating units are here composed of globular assemblies ("cybotactic complexes") of parallel lath-like molecules, the optical isotropy of the phase arising from the statistically random orientation of these individually highly anisotropic units. The structure for this phase would then resemble the structures suggested for the amphiphilic cubic phases in Chap. 5. It is interesting that, like these mesophases, the smectic D phase from 4′-n-octadecyloxy-3′-nitrobiphenyl-4-carboxylic acid shows a very stiff consistency and marked flow birefringence (personal observation).

It seems possible that the high-temperature coloured "Modifications" formed by certain cholesterol derivatives [32–36] may also possess related constitutions. These phases are usually much more fluid than those just mentioned but may likewise show marked flow birefringence (p. 31) [36].

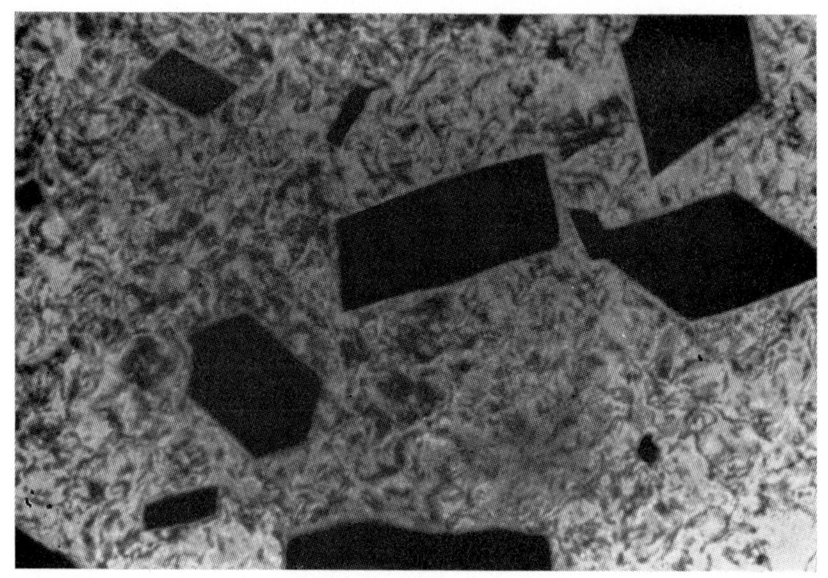

FIG. 2.2.3. The separation of the cubic mesophase from smectic C on rise of temperature with 4′-n-hexadecyloxy-3′-nitrobiphenyl-4-carboxylic acid. Crossed polars ×120 [27].

In these phases it would seem that the individual units must consist of small molecular assemblies each with a twist as in the cholesteric plane texture (p. 11). These units will be orientated at random but those suitably orientated in relation to the direction of vision should give a colour effect analogous to that of the plane texture itself. This would equally be possible whether the coloured phase represented an amorphous liquid or a plastic mesophase.

REFERENCES

[1] The nature of plastic crystals and of the associated molecular rotation were the subjects of a Symposium at Oxford University in 1960 the proceedings of which were published in *The Physics and Chemistry of Solids* **18,** (1961).

A further recent discussion has been given by Gray, G. W. and Winsor, P. A., *Molec. Crystals Liqu. Crystals* (1974), in press.

[2] TIMMERMANNS, J. *J. Phys. Chem. Solids*, **18,** 1 (1961).

[3] STAVELEY, L. A. K. *Ann. Rev. Physical Chem.* **13,** 351 (1962).

[4] WESTRUM, E. F., Jr. and MCCULLOUGH, J. F. *Physics and Chemistry of the Organic Solid State*, Interscience Publishers, London (1963), p. 89.

[5] ASTON, J. G. *ibid.*, p. 543.

[6] SMYTH, C. P. *ibid.*, p. 715.

[7] SMITH, G. W. *International Science and Technology* January 1967, p. 72.
[8] TIMMERMANNS, J. *Bull. Soc. Chem. Belg.* **44,** 17 (1935).
[9] TIMMERMANNS, J. *J. chim. Phys.* **35,** 331 (1938).
[10] POST, B. *Acta Crystallogr.*, **12,** 349 (1959).
[11] HASSEL, O. and SOMMERFELDT, A. M. *Z. phys. Chem.* **B40,** 391 (1938).
[12] CARTER, G. F. and TEMPLETON, D. H. *Acta Crystallogr.* **6,** 805 (1953).
[13] DUNNING, W. J. *J. Phys. Chem. Solids* **18,** 21 (1961).
[14] DE VRIES, A. *Molec. Crystals Liqu. Crystals* **11,** 361 (1970) (*cf.* Chap. 1 p. 8).
[15] SMYTH, C. P. *Dielectric Behaviour and Structure*, McGraw-Hill, New York (1955).
[16] MILLER, R. C. and SMYTH, C. P. *J. Am. chem. Soc.*, **79,** 20 (1957).
[17] POWLES, J. G., WILLIAMS, E. D. and SMYTH, C. P. *J. chem. Phys.* **21,** 136 (1953).
[18] POWLES, J. G. and GUTOWSKY, J. S. *J. chem. Phys.* **21,** 1695 (1953).
[19] SMITH, G. W. *J. chem. Phys.* **43**, 4325 (1965).
[20] ANDREW, E. R. and EADES, R. G. *Proc. R. Soc.* **A216,** 398 (1953).
[21] SUGA, H., SUGISAKI, M. and SEKI, S. *Molec. Crystals* **1,** 377 (1966).
[22] BLUM, H. and SHERWOOD, J. N. *Molec. Crystals Liqu. Crystals* **10,** 381 (1970).
[23] ANDREW, E. R. *J. Phys. Chem. Solids* **18,** 40 (1961).
[24] RUSHWORTH, F. A. *J. Phys. Chem. Solids* **18,** 77 (1961).
[25] HAWTHORNE, H. M. and SHERWOOD, J. N. *Trans. Faraday Soc.* **66,** 1783, 1792, 1799 (1970).
[26] ASTON, J. G. *J. Phys. Chem. Solids* **18,** 62 (1961).
[27] DEMUS, D., KUNICKE, G., NEELSEN, J. and SACKMANN, H. *Z. Naturf.* **23a,** 84 (1968).
[28] SACKMANN, H. and DEMUS, D., *Fortsch. chem. Forsch.* **12,** 349 (1969).
[29] FITZGERALD, W. E. and YANZ, G. J. *J. Molec. Spectroscopy* **1,** 49 (1957).
[30] *cf.* UBBELOHDE, A. R. *Melting and Crystal Structure*, Clarendon Press, Oxford (1965).
[31] LEHMANN, O., *Z. phys. Chem.* **4,** 462 (1889).
[32] REINITZER, F., *Mh. Chem.* **9,** 421 (1888).
[33] LEHMANN, O., *Z. phys. Chem.* **56,** 750 (1906).
[34] GRAY, G. W., *J. chem. Soc.* 3733 (1956).
[35] PRICE, F. P. and WENDORFF, J. H., *J. phys. Chem.* **75,** 2839 (1971).
[36] FERGASON, J. L. in *Liquid Crystals* (edited by G. H. Brown, G. J. Dienes and M. M. Labes), Gordon and Breach, New York (1967), p. 103.

2.3 MESOPHASES FORMED BY AMPHIPHILIC COMPOUNDS

P. A. Winsor

A number of types of mesophase, both optically anisotropic and optically isotropic, are formed by amphiphilic compounds either in the pure state, in admixture or in the presence of added solvents, e.g. water and/or organic compounds. The types of mesophase produced by a given amphiphilic compound depend both on the temperature and on the amount and character of any additional compounds present. In some instances an amphiphilic compound, which in the pure state does not form a mesophase, gives a mesophase in the presence of water. Examples of compounds showing this behaviour are n-octylamine and many ethers and esters of the polyoxyethylene glycols (Chap. 5). On account of the marked effect of solvents (in particular of water) on their formation, the amphiphilic mesophases have frequently been termed lyotropic mesophases" (*cf.* Preface p. x).

The influence of composition and temperature on the formation of amphiphilic mesophases is discussed in detail in Chapter 5, while the physical properties of the mesophases are considered in Vol. 2, Chapters 4 and 6. The present section gives a brief formal classification, on the basis of their micellar structures, of the various types of mesophase encountered. Literature references will not be cited, but are given in the more detailed Chapters quoted.

In the most familiar amphiphilic mesophases the rigidity of molecular arrangement characteristic of the crystalline state has been entirely lost through the effects of temperature or/and of added solvent. Such mesophases may conveniently be termed "fused mesophases" (Chap. 5).

On the other hand, as also considered in Chapter 5, in certain aqueous amphiphilic mesophases at not too high temperatures the hydrocarbon chains may collectively retain some of the order and rigidity found in the crystal, while the solvent water is in the fused state. Such mesophases may therefore be termed "semi-crystalline". A further group of apparently semi-crystalline mesophases arises on heating the crystalline alkali or alkaline earth soaps of the fatty acids. Here collective fusion of the hydrocarbon chains is thought to precede the fusion of the polar moieties giving a series of distinct semi-crystalline mesophases with rising temperature. Ultimately the fused "neat soap", or lamellar mesophase is produced and is followed at a still higher temperature by the amorphous melt. To the writer it appears that the detailed structural characters of both series of semi-crystalline mesophases are not unequivocally established (Chap. 5). Their further classification will therefore not be pursued here. Some structural models proposed by Luzzati and co-workers are discussed in Chapter 5.

The structural character of the various types of fused amphiphilic mesophase is better understood. These mesophases are formed by the ordered mutual arrangement, through the operation of intermicellar forces, of various types of "micelle" or "aggregates" of amphiphile molecules. These aggregates, or micelles, may arise with pure amphiphiles at suitable temperatures, with mixtures of amphiphiles or with pure or mixed amphiphiles in the presence of solvents (water and/or organic liquids) at suitable compositions and temperatures. A molecular theory of the influence of composition and temperature on the variety of shapes and sizes adopted by these aggregates is discussed in Chapter 5.

Briefly, according to the conditions, the aggregates, as indicated diagrammatically in Fig. 1.1, may be

(1) Spherical
 (*a*) with polar exterior, or
 (*b*) with hydrocarbon exterior.

(2) Cylindrical
 (*a*) with polar exterior, or
 (*b*) with hydrocarbon exterior

These cylindrical aggregates, under conditions of temperature and composition which favour their optimum development, are indefinitely long, but under less favourable conditions they may be more nearly isometric; these near-isometric or "globular" forms (Fig. 1.1) may be regarded as mobile and in thermo-kinetic transition with spherical and/or near-isometric sandwich micelles, as well as with the more extended forms.

(3) Sandwich-like
 (*a*) with polar exterior, or
 (*b*) with hydrocarbon exterior.

The sandwich-like aggregates, under conditions for their optimum development, form indefinitely extended "bi-molecular" lamellae, but under less favourable conditions they may be less extended, approaching near-isometric globular forms in thermo-kinetic transition with spherical micelles and/or near-isometric globular cylindrical micelles, as well as with more extended forms.

Under suitable conditions of temperature and proximity (the latter depending on concentration, i.e. on the proportion of intermicellar liquid) the above types of micelles (like the individual molecules in the non-amphiphilic series) arrange themselves into mesophases through the operation of inter-particle forces, e.g., van der Waal's forces, repulsive forces opposing the inter-penetration of diffuse electrical double layers, repulsive forces between oriented dipoles.

A considerable variety of mesophases arises in this way. Some of these are frequently encountered and stable over fairly broad ranges of composition and temperature. Others occur less frequently, are less well characterized and are stable only over comparatively narrow ranges of temperature and composition. The classification of the more frequently encountered mesophases will first be considered. The geometrical disposition of the micellar units in these mesophases has been determined directly by X-ray diffraction (see Vol. 2, Chap. 4), but the details of the structures of the micellar units themselves are often rather less certain (Chap. 5).

Classification of the Commoner Amphiphilic mesophases

The more commonly encountered amphiphilic mesophases may be classified as follows. The nomenclature used is that developed by the writer, but alternative equivalent nomenclatures adopted by other workers are listed in Table 5.17 (Chap. 5).

(1) Mesophases composed of parallel indefinitely extended fibrous micellar units in two dimensional hexagonal array. These optically anisotropic mesophases are termed the middle (M_1) and inverse middle (M_2) mesophases. In the M_1 mesophase the polar groups of the amphiphile lie on the outside of the cylindrical micellar aggregates. In the M_2 mesophase the polar groups comprise the cores of the micelles. Up to the present, the M_1 mesophase has been encountered only in the presence of water. This is incorporated as the intermicellar solution. The M_2 mesophase is formed by certain pure amphiphiles over the appropriate temperature range, but in such cases the mesophase can further incorporate considerable amounts of water and/or organic liquid (either into the micellar structure itself or as intermicellar solution) before conversion to an amorphous liquid or to a different mesophase. The M_2 mesophase may also arise in ternary aqueous systems containing mixtures of amphiphiles which do not individually give an M_2 phase (Chap. 5).

(2) Mesophases containing a succession of parallel indefinitely extended micellar units composed of bimolecular lamellae of amphiphilic molecules. These optically anisotropic mesophases are termed the lamellar, smectic or neat phases (G). More than one smectic mesophase may occasionally arise in a given system. These distinct smectic polymorphs apparently differ in the arrangement of the polar groups in the lamellae. As can be clearly seen from Fig. 1.1, the type of difference corresponding to that between the M_1 and M_2 phases cannot arise with these G phases with indefinitely extended lamellae since the two forms in this case are equivalent.

(3) Mesophases containing near-isometric globular micellar units. As indicated in Fig. 1.1, three types of near-isometric or "globular" micellar units apparently arise: spherical, cylindrical and sandwich. The approximately isometric, cylindrical and sandwich micelles form intermediate stages in the transition (due either to changes in composition or temperature) between the indefinitely extended cylindrical micelles of the M_1 or M_2 mesophases and the indefinitely extended lamellar micelles of the G mesophase. All three types of near-isometric unit may exist either with the polar groups directed outwards or directed inwards. As discussed in more detail in Chapter 5, these near-isometric micellar units can give rise to a number of optically isotropic mesophases in which the units are arranged in various forms of cubic lattice. Of these phases those designated S_{1c}, V_1 and V_2 have been well characterized. It should be noted that, except with truly spherical micellar units, the units themselves will not be strictly

optically isotropic, so that the arrangement of the units must involve mutually compensatory orientations which confer over-all optical anisotropy on the mesophase. It has been suggested by Luzzati and co-workers that the units in these mesophases are linked in various types of network, (*cf.* Vol. 2, Chapter 4). However a characteristic of all the cubic mesophases is that, in contrast to the M_1, G and M_2 mesophases, they show high resolution nuclear magnetic resonance spectra of the amphiphile molecules present (Vol. 2, Chap. 8). This suggests that the individual globular micellar units are undergoing thermal rotatory motion at the lattice points analogous to the thermal rotatory motion of the globular molecules at the lattice points in the optically isotropic, non-amphiphilic, cubic plastic crystals (Chapter 2.2).

An important point which may be noted here is that when, as a result of a progressive change in composition or temperature in a given system (for example, pure strontium myristate, sodium laurate + water or sodium caprylate + water + decanol-1), two or more of the above amphiphilic mesophases are formed in succession the sequence is always in the order S_{1c}, M_1, V_1, G, V_2, M_2, followed either from left to right or from right to left. However, the complete succession has apparently never been encountered in any single system. Reversible successions such as S_{1c}, M_1, V_1, G, or G, V_2, M_2 have been observed in several different systems, but out-of-order successions such as V_1, M_1, G or G, M_2, V_1 or their reverse have never been found.

Other Fused Amphiphilic Mesophases

In addition to the series of commoner amphiphilic mesophases S_{1c}, M_1, V_1, G, V_2 and M_2, certain other optically anisotropic mesophases have been observed at points intermediate in the mesophase succession between M_1 and V_1 (*cf.* Table 5.1, Chap. 5). These "intermediate" mesophases, in those systems where they occur, again always follow one another, interposed within the main series, in a regular order, although, as with the main mesophases, not all of them are represented in any particular system. As with the V_1 phase, the micellar units of these mesophases must in some way be intermediate in character between the indefinitely extended lamellae of the G phase and the indefinitely extended fibrous micelles of the M_1 phase. The detailed structure of these intermediate units is however uncertain and provides a fascinating subject for future investigation. Such studies will however be made difficult by the relatively narrow limits of stability of the intermediate mesophases (*cf.* Chapter 5).

3

Microstructures of Liquid Crystals

3.1 INTRODUCTION: SWARM THEORY AND THE THEORY OF THE ELASTIC CONTINUUM OF LIQUID CRYSTALS*

H. Zocher

The observations of Reinitzer [1] are usually considered to mark the discovery of liquid crystals or mesophases, athough myelin structures were known thirty-three years before this and were later included in the same class of system. Compared with the unique properties of the mesophases first observed by Reinitzer, the properties of myelin structures were more similar to those of microcrystalline aggregates and did not give rise to the outburst of research work and theoretical discussion which stemmed from Reinitzer's observations. Lehmann, an excellent experimentalist, contributed especially in exciting interest in these systems through his high-spirited publications. His introduction of the contradictory term "Liquid Crystals", later substituted by "mesomorphous states of aggregation" (Friedel [2]) or "mesophases" (Zocher [3]), caused lively discussions and led to attempts to create a more satisfactory theoretical basis to describe these states.

The Swarm Theory proposed by Bose [4] found approval and acceptance by many authors, and until a few years ago was used as a basis for the interpretation of experimental results. The theory was elaborated by Ornstein and his collaborators [5]–[9], although they ignored completely the great wealth of factual information being accumulated by Friedel [2]

* Just before his death in October 1969, Professor Zocher wrote the following pages which represent the first part of a section originally intended to trace the historical development of the Swarm Theory and the Theory of the Elastic Continuum. Although incomplete, this article is included as a small tribute to Professor Zocher's contributions to knowledge relating to Liquid Crystals. Had he lived, Professor Zocher would clearly have followed his critical analysis of the Swarm Theory by an account of the development of the Continuum Theory which he first proposed in 1927 and which provides the basis for the remainder of this chapter on Microstructures of Liquid Crystals (Editors).

and others who were investigating mesophases by means of the polarizing microscope.

Bose's concept started from the observation of turbidity in typical liquid crystals, and in this he saw an analogy to the light-scattering properties of liquids near the critical point. He recognized that the turbidity could not be caused by differences in optical density, but must be a consequence of varying orientations of birefringent volume elements. He assumed these elements were swarms comprised of molecules which were in more or less parallel orientation, and that these anisotropic swarms were in vigorous thermal motion so that the mass as a whole would be isotropic. In consequence, the swarms had to be of colloidal dimensions, with a diameter less than the wavelength of light. Evidently, this concept can be applied only to systems of high mobility, and all smectic mesophases and also nematic mesophases of high viscosity are excluded from consideration. Nevertheless, many examples are not excluded, and for these the theory offers an easy way to interpret the influence of external fields on their properties.

Although the thermal motion of a swarm must be less than that of a molecule, Bose's concept carries the surprising implication that the forces between molecules in the small swarms are able to produce a rather high degree of order and, consequently, a high birefringence, whereas the forces between swarms are so weak that the mass as a whole is isotropic and completely disordered. Since the formation of a mesophase requires that the molecules are elongated, the effective forces must be of the apolar van der Waal's type, acting only between neighbouring molecules.

In connection with this problem, there arose doubts as to whether systems comprised of swarms could (*a*) be considered as phases according to the thermodynamic definition, (*b*) be capable of exhibiting a sharply defined surface in contact with the amorphous isotropic phase, and (*c*) be characterized by sharp transition temperatures. Nernst, for example, proposed a special explanation for the well-defined transition temperature marking the change from the liquid crystalline to the amorphous state. Ornstein and Kast [9] concluded, on the other hand, that there was no contradiction with thermodynamic principles. According to definition, a phase in the sense of Gibbs, is a homogeneous system in thermodynamic equilibrium. The concept of homogeneity depends of course upon the size of the elements from which the system is built up; common phases, for example, are composed of atoms or molecules smaller than colloidal particles. However, it is quite possible for a phase to be composed of particles of colloidal dimensions, and the present author has himself found a series of phases denominated "phases of higher order" or "superphases". However, the particles in these phases differ from Bose's swarms in that they are separated by water layers, giving rise to electrolytic dissociation and surface charges. In this way, a thermodynamic equilibrium is established in phases which may be amorphous, nematic, smectic or crystalline, whereas the elementary particles may have the structure of any state of aggregation and may even be viruses. These phases give sharply defined boundaries of demarcation with the amorphous phase, and these are clearly visible under the microscope.

Phases corresponding in properties to Bose's model could never be observed microscopically; thermotropic as well as hydrotropic nematic phases are always anisotropic in dimensions much greater than Bose's swarms. Indeed, van Wyk [10] came to the conclusion that swarms must orient one another too, i.e., that they are united into a continuum and lose their separate existence.

Authors defending the Swarm Theory tried to explain the far reaching anisotropy in terms of the influence of surrounding surfaces and walls exerted through forces acting over distances of a few hundredths of a millimetre only. The nature of these forces could never be explained, and a range of some tens of microns is certainly much too high for molecular forces. According to observations under the microscope, the amorphous isotropic phase does not produce any change on the anisotropy of included droplets of mesophase which are not in contact with the microscope slide and cover slip. Moreover, as Friedel showed, the same continuous anisotropy is observed even in layers of mesophase stretched over a hole in the supporting surface.

Fürth and Sitte [11] tried to prove Swarm Theory and to calculate the swarm diameter using observations made by Tropper [12] on liquid crystal films of *p*-acetoxybenzalazine mounted between a slide and a cover glass. They believed that nematic films of this substance represented the isotropic condition of the mesophase claimed by Bose. However, an anisotropy which sometimes appeared in the films could not be explained. Further, they overlooked the anisotropic character of a nematic phase when this is oriented with the long molecular axes perpendicular to the slide surface, in the position of "pseudo-isotropy" as Lehmann [13] called it. That the system is in fact anisotropic is readily confirmed when it is observed, in convergent light, in directions oblique to the glass surface. Moreover, this anisotropy may become visible in the direction of the microscope axis in regions where the optic axis is bent into rather stable, closed deformations [14]. In this way, Tropper's unexplained observations can be understood.

Summarizing the situation, one must state that the Swarm Theory does not reconcile the discrepancy between the submicroscopical anisotropy and the alleged microscopical isotropy of the nematic state. Neither does it explain innumerable detailed microscopic observations concerning the anisotropy which extends uniformly over large regions of mesophase, observations which in fact gave rise to the discovery of mesophases and from which most of our knowledge of mesophases has come. Furthermore, there exists a considerable number of observations proving that the alleged isotropy of the mesophase does not exist.

3.2 PRINCIPLES OF THE CONTINUUM THEORY

P. G. de Gennes

Weak Distortions in Nematics

A nematic phase is characterized by a certain type of *long range order*: in an ideal single crystal of a uniaxial nematic, the molecules are (on the average) aligned along one common direction, labelled by a unit vector $\mathbf{n}$ (the director). The amount of alignment along $\mathbf{n}$ is characterized by an order parameter S.

However, in most practical circumstances, this ideal conformation is not compatible with the constraints imposed by the walls of the container, and by external fields (electric, magnetic, . . .): then there will be some modification both of $\mathbf{n}$ and of S at each point.

(*i*) the modifications of $\mathbf{n}$ are interesting, because they take place over macroscopic distances (typically a few microns). Thus, they are easily observed optically. Distortions of this sort can be induced by rather weak external perturbations (voltages of the order 10 volts, for instance): thus, they provide ideal means to convert various types of signals into optically visible signals. From a more theoretical point of view, these macroscopic distortions can be described in terms of a *continuum theory*, which is the analogue for liquid crystals of classical elasticity for solids.

(*ii*) the modifications of S are less interesting, because they do not persist over long lengths in space. To understand this statement more precisely, consider for instance the effects of a *small floating object* in a nematic single crystal (Fig. 3.2.1). It imposes some distortions to the alignment

$$\mathbf{n}(\mathbf{r}) = \mathbf{n}_0 + \delta\mathbf{n}(\mathbf{r})$$

where $\mathbf{n}_0$ is the unperturbed director and $\mathbf{r}$ is the distance to the centre of the object. If the object is acted upon by external torques, e.g., if it is a magnetic needle, $\delta\mathbf{n}(\mathbf{r})$ *decreases only slowly with* r (in fact like $1/r$: see reference 15).

Similarly S is modified near the object

$$S(r) = S_0 + \delta S(r)$$

but $\delta S(r)$ decreases *rapidly* with r: in a mean field approximation for instance, the asymptotic behaviour would be $\delta S \approx \frac{1}{r} e^{-r/a}$ where a is a molecular length. Thus, if we are interested only in the long range part of the effects due to the floating object, we may disregard δS entirely, and keep only $\delta\mathbf{n}$.

More generally, in a weakly distorted nematic, i.e., when space variations are slow on the scale of a, we may still picture the material as being

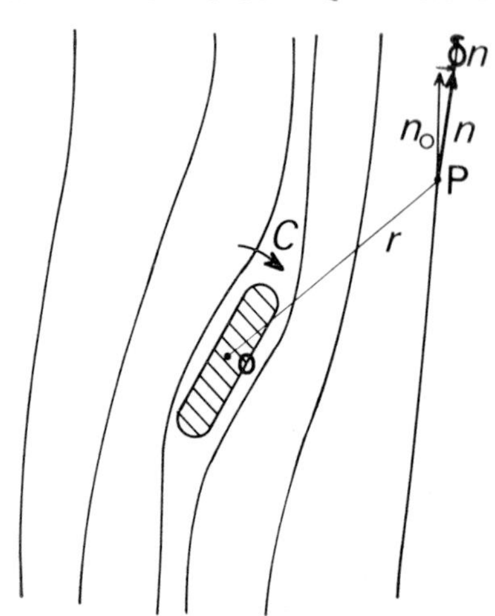

FIG. 3.2.1. Long range distortions around a floating object, in a nematic single crystal. When a non-zero external torque C is applied to the object, the angular deflection $\delta\mathbf{n}$ decreases like $\frac{1}{r}$. The figure has been drawn for tangential boundary conditions at the surface of the object, but the result is more general.

locally uniaxial, with a fixed order parameter S, but with variable optic axis $\mathbf{n}(\mathbf{r})$: this is the basis of the so called continuum theory of nematics, as it was set by Zocher [16], Oseen [17] and Frank [18].

CONTRIBUTIONS TO THE FREE ENERGY

There are three types of distortions in nematics: splay, twist and bend (Fig. 3.2.2). With each of these types is associated one elastic constant

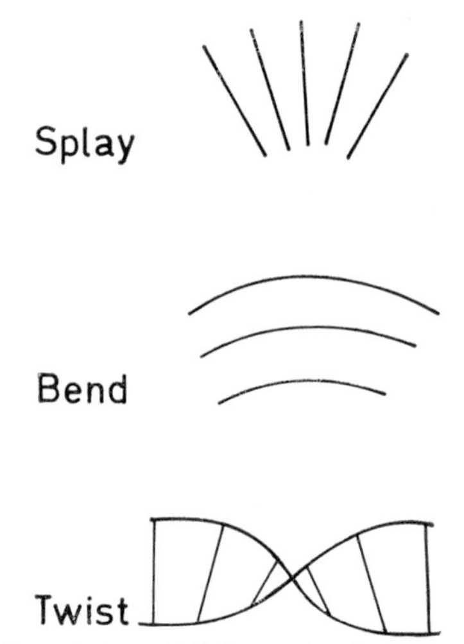

FIG. 3.2.2. The three types of deformation in a nematic liquid crystal.

Here **h** is what we call the *molecular field* (a notation inspired from magnetism) and is given explicitly by

$$h_\beta = -\frac{\partial F}{\partial n_\beta} + \frac{\partial}{\partial x_\alpha}\Pi_{\alpha\beta} \tag{3.2.5}$$

$$\Pi_{\alpha\beta} = \frac{\partial F}{\partial g_{\alpha\beta}}$$

The second integral in Equation 3.2.4 is taken on the sample surface, and dS_α is the corresponding (vector) surface element. At equilibrium $\delta\mathscr{F}$ must vanish for all acceptable variations $\delta\mathbf{n}(\mathbf{r})$, i.e., for variations such that (*a*) **n** remains a unit vector, (*b*) the boundary conditions are satisfied.

In the bulk of the sample, this requires *that* **h**(**r**) *be collinear to* **n**(**r**) at all points. At the surface, if the boundary conditions are non-degenerate, the surface integral drops out (since **n** is fixed). If the boundary conditions are degenerate, if, for instance, we have a cone of easy directions around the normal **u** to the interface, we must put

$$\delta\mathbf{n} = \delta\phi\mathbf{u} \times \mathbf{n}$$

in the second term of (Equation 3.2.5), where $\delta\phi$ is an (arbitrary) small rotation angle. Imposing $\delta\mathscr{F} = O$ then gives the following condition, on the limiting surface:

$$u_\alpha\,\Pi_{\alpha\beta}(\mathbf{u} \wedge \mathbf{n})_\beta = O \tag{3.2.6}$$

A mathematical discussion of the equilibrium equations in the bulk, and the types of possible solutions, has been given by Ericksen [26].

Measurements of Elastic Constants in Nematics

There seem to be at least three approaches which lead to measurements of the constants K_1, K_2, K_3.

Mechanical Measurements

For instance, if a nematic single crystal is held between two glass plates, polished by the method of Chatelain [31], one may rotate one plate in its plane and twist the nematic: the restoring torque is proportionnal to K_2. This approach is, however, rather difficult experimentally.

Competition Between the Alignments Imposed by a Wall and a Magnetic Field

This is the most classical method. Consider the case in Fig. 3.2.3 with a polished wall having an easy direction along z, while **H** is applied along x. Far from the wall, the field dominates, and the optic axis is along x. Near the wall, there is a transition layer of thickness

$$\xi_2(H) = \sqrt{\frac{K_2}{\chi_a}}\,\frac{1}{H} \tag{3.2.7}$$

Typically $K_2 = 10^{-6}$, $\chi_a = 10^{-7}$, and $\xi_2 = 3\,\mu$ for $H = 10^4$ oersteds. We call ξ_2 the (twist) *magnetic coherence length*. If χ_a is measured inde-

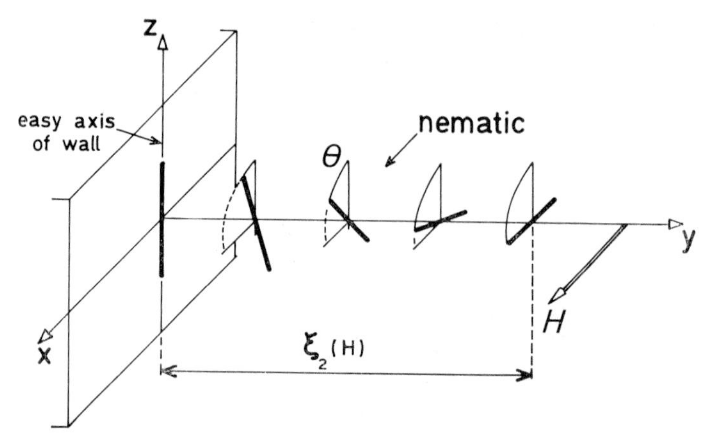

FIG. 3.2.3. Competition between wall alignment and field alignment.

pendently, an optical measurement of the thickness of this transition layer gives K_2. In practice, geometries involving two parallel walls are more convenient and have been used for some time by the Russian school [22, 23]. Theoretical calculations for various cases are given [27, 28], and complications which may occur if the anchoring at the walls is weak are discussed [29]. Speculations on the effects of magnetic fields at the free surface are given in reference [30].

STUDY OF SPONTANEOUS FLUCTUATIONS BY LIGHT SCATTERING

A nematic single crystal gives rise to a very strong scattering of light; the emergent light is concentrated at small angles from the incident beam, and strongly depolarised [31]. These effects have been interpreted in terms of small fluctuations of the optic axis [32]. The theoretical form of the scattered intensity I, for a scattering wave vector $\mathbf{q}$, an incident polarisation $\mathbf{i}$, an emergent polarisation $\mathbf{f}$, is

$$I = \text{const.} \sum_{\alpha=1,2} \frac{T}{K_3 q_{\parallel}^2 + K_\alpha q_{\perp}^2 + \chi_a H^2} (i_\alpha f_{\parallel} + i_{\parallel} f_\alpha)^2 \qquad (3.2.8)$$

The index α serves to define two unit vectors: $\mathbf{e}_1$ and $\mathbf{e}_2$. $\mathbf{e}_2$ is normal to the optic axis $\mathbf{n}_0$ and to $\mathbf{q}$; $\mathbf{e}_1$ is normal to $\mathbf{n}_0$ and to $\mathbf{e}_2$; i_α is the component of $\mathbf{i}$ along $\mathbf{e}_\alpha$, etc.; $\mathbf{q}_{\parallel}$ and $q_{\perp}$ are the components of $\mathbf{q}$ along $\mathbf{n}_0$ and $\mathbf{e}_1$. Equation 3.2.8 shows that relative measurements of intensities, performed at various $\mathbf{q}$ values, under a finite field H, are enough to determine the three elastic constants (assuming again that χ_a is known). Experiments of this type are currently under way [33].

A REMARK ON "SWARMS"

As explained in the introduction by H. Zocher, several attempts have been made in the early literature to explain the properites of nematics, and in particular the light scattering, in terms of swarms, i.e., small ordered regions (with a diameter D in the micron range), different swarms having uncorrelated orientations. A more rigorous analysis of fluctuations, based on the continuum theory, allows us to see when this concept may, or may not, work. The results are as follows:

(*i*) in the nematic phase, and in zero magnetic field, the fluctuations $\delta\mathbf{n}$ of the director $\mathbf{n}$, at two points 1 and 2, are correlated according to a law of the form [30]:

$$\langle \delta\mathbf{n}(1).\delta\mathbf{n}(2)\rangle \simeq \frac{k_B T}{2\pi K R_{12}} \qquad (R_{12} \gg a)$$

With such a slow decrease ($\approx 1/R_{12}$) it is not possible to define a distance D above which correlations are negligible: swarms are meaningless. For instance, if we tried to define a swarm diameter D by the angular distribution of the scattered light, the value which we would obtain for D would depend on the wavelength used!

(*ii*) on the other hand, above the nematic to isotropic transition point T_c, if we look at the fluctuations $\delta\varepsilon$ of the dielectric constant we find (omitting all tensor indices) [34]

$$\langle \delta\varepsilon(1).\delta\varepsilon(2)\rangle \simeq \text{const.}\ \frac{1}{R_{12}} \exp\left(-R_{12}/\xi(T)\right)$$

where the "thermal coherence length" $\xi(T)$ is finite (of the order ten molecular lengths just above T_c, and smaller at higher temperatures).

Thus, in the isotropic phase, it is not entirely impossible to use the notion of swarms (with diameter $D \approx \xi(T)$) to describe short range order effects. However, even in this region, the notion is not very precise, and its consequences must always be checked by comparison with the complete fluctuation theory.

Other Mesomorphic Phases

Cholesterics

In a cholesteric, of pitch $\frac{2\pi}{q_0}$, the distortion energy F_d becomes [18]

$$F_d = \tfrac{1}{2} K_1(\operatorname{div}\mathbf{n})^2 + \tfrac{1}{2}K_2\{\mathbf{n}.\operatorname{curl}\mathbf{n} + q_0\}^2 + \tfrac{1}{2}K_3(\mathbf{n}\times\operatorname{curl}\mathbf{n})^2 \qquad (3.2.9)$$

(More general forms for F_d have been discussed by Jenkins [35], but they are not different from Equation 3.2.9 in the (usual) situation where the pitch is much larger than the molecular length, a).

The magnetic and electric terms (F_m and F_e) retain their previous structure. Provided that χ_a (or ε_a) is positive, a sufficiently large field H (or E) will untwist the helical structure [36, 37]. Experiments bearing on this point are listed under [38]. They sometimes allow us to measure K_2 in a nematic, using the following procedure: one adds to the nematic a small concentration c of optically active solutes, obtaining a cholesteric of large pitch (experimentally the inverse pitch changes linearly with concentration). In the most usual situation, where $K_2 < \frac{\pi^2}{4} K_3$, the critical field is given by [36, 37]

$$H_c = \frac{\pi}{2} \sqrt{\frac{K_2}{\chi_a}}\, q_0 \tag{3.2.10}$$

K_2 and χ_a differ from their values for the pure nematic only by small terms of the order c; thus from measurements of q_0 and H_c one can extract K_2 for the nematic matrix.

The above picture of field effects applies only to *bulk samples* (with positive χ_a) where the initial helix is free to orient with respect to **H**. A different process occurs if the helix is prepared as a *planar texture* between two polished glass plates. Then (for $\chi_a > 0$) a field H, larger than a certain threshold value H_d, and normal to the plates, induces a remarkable periodic distortion: looking through the plates one sees a square lattice. This effect (visualized first by Helfrich in a slightly different context) has been demonstrated by Rondelez *et al.* [79].

The continuum theory is also useful in providing a microscopic picture of such dilute solutions of chiral molecules in nematic solvents [15]. The method is to consider first one molecule of solute, treat it as a "floating object", and derive the long range distortions which it induces in a nematic matrix. In an approximation where the three elastic constants are taken to be equal, the result is particularly simple; at each point **r** (far from the floating object) the director **n** is rotated by:*

$$\boldsymbol{\omega}(\mathbf{r}) = \beta \operatorname{grad} \left(\frac{1}{r}\right) \tag{3.2.11}$$

β is a constant (with the dimensions of an area) and differs from zero only if the solute is optically active. Having established Equation 3.2.11, one can then proceed to the case of a finite concentration c of solute molecules, and add up their distortions ω; the result is a helical structure with wave vector [15]:

$$q_0 = 4\pi\beta c \tag{3.2.12}$$

Equation 3.2.12 gives a linear relation between q_0 and c, as is indeed found experimentally.

To derive the elastic constants in a cholesteric material, one could also study the diffuse scattering of light (in the vicinity of the Bragg peaks); this is discussed in reference [39].

Smectics A

These correspond to layer structures, with one optic axis normal to the layers, and complete rotational symmetry in the plane of the layers. In

* Note the difference with the case discussed under Weak Distortions in Nematics (p. 68); there an external torque was applied to the object, and the distortion decreased like $1/r$. In the present case, no torque is applied, and the distortion decreases like $1/r^2$.

such a system, it does not cost much energy to curve the layers, but it requires very much energy to change the interlayer spacing. Thus, if one is interested primarily in the orientation of the layers, one may still describe the state of the system, at each point **r**, simply by a unit vector, or director **n**(**r**) (normal to the layers) giving the direction of the optic axis.

Since in this approximation the layer thickness d is constant, the curvilinear integral

$$\frac{1}{d}\int_{A}^{B} \mathbf{n}.\mathbf{dr}$$

gives the number of layers between point A and point B, and must be independent of the path chosen between A and B: this imposes the condition

$$\text{curl}\ \mathbf{n} \equiv 0$$

The distortion free energy then reduces to [17]:

$$F_d = \tfrac{1}{2}K_1(\text{div}\ \mathbf{n})^2 \qquad (3.2.13)$$

If, on the other hand, one wishes to make a more complete description of a smectic A, allowing for changes of d, one must (at least) include at the same time possible changes of the overall density ρ. The state of affairs at each point is then specified by d, ρ, and the unit vector **n**. An elastic theory based on these ingredients has been constructed [40]. This predicts *two* branches of acoustical waves: "first sound" (a density oscillation) and "second sound" (a modulation of d). Second sound is slower, and more strongly damped than first sound, but it has been observed by Pershan *et al.* using Brillouin scattering [80].

Smectics C

These biaxial smectics have more degrees of freedom than the smectics A, and their elasticity is more complex; the most interesting distortions have been discussed by Saupe [41], and a more general description has been given in [42].

3.3 DEFECTS

M. Kléman

Introduction

Mesophases frequently show various singularities in the distribution of the molecules, which are characteristics of the structural features of the perfectly ordered phase. In fact, it was a study of these which led Friedel [43] to his well-known classification of mesophases, and his article remains the best introduction to a description of defects. The existence of these defects is related to topological constraints imposed, for example, by boundary conditions (as illustrated in Fig. 3.3.1 here), and appearing

FIG. 3.3.1. Disclination line in a cylinder. The molecules are tangential to the boundary.

necessarily in any growth process of a single nematic liquid crystal, or when an external factor, for example an electric field, violently disturbs the perfect arrangement. These defects are mainly linear defects, and their generic name (disclination lines) has been derived by analogy with the dislocation lines of the theory of solid crystals [44], to which they are related. Disclination points exist as well. The existence of disclination walls is less probable. These singularities may gather, in simple experimental situations, in more or less regular sets, forming liquid crystal textures. The present text is devoted more to the elements of these textures than to the textures themselves. Smectic phases and cholesteric phases also give the well-known focal–conic domains, which do not enter into the disclination scheme. The last section will emphasize the relationship between dislocation theory concepts and the concept of disclination.

Defects in a Nematic Phase

DISCLINATION LINE IN A NEMATIC

A rotation through an angle which is a multiple of π, about any axis χ perpendicular to the average molecular direction $\mathbf{n}_0$, is a symmetry operation of the perfect nematic phase. Consider then a surface Σ drawn in the nematic medium, bounded by a line L, and rotate each molecule on Σ through an angle which is a multiple of π, supposing that this movement is transmitted to the matter on one side of Σ, leaving the other side in the vicinity of Σ unaffected. Σ is the cut surface. After this operation is completed, the intermolecular bonds are re-established on Σ and the entire medium allowed to relax. A singularity of the conformation subsists on L, the so-called disclination line; the final distribution of the molecules and the energy of the line depends on the values of the physical constants K_1, K_2, K_3, but the topological constraint is defined unambiguously by the above process.

Screw Disclinations

The simplest case is provided by a line L coincident with the axis χ, and has been treated by Frank [18]. The singularities have translational symmetry along the line. The nature of the topological singularity is easily established by assuming the three elastic constants to be equal. If θ is a polar angle in a plane perpendicular to χ and ϕ the angle of $\mathbf{n}(\mathbf{r})$ with a given direction of that plane, the following equation

$$\phi = S\theta + \theta_0 \tag{3.3.1}$$

minimizes the free energy

$$F = \tfrac{1}{2}K\int \text{grad } \theta)^2 \, \mathrm{d}V \tag{3.3.2}$$

(where $\mathrm{d}V$ is a volume element) and corresponds to a rotation $2S\pi$ of the molecules around any circuit enclosing the singularity line. Therefore S must be an integer or an half-integer. Fig. 3.3.2, taken from Frank's

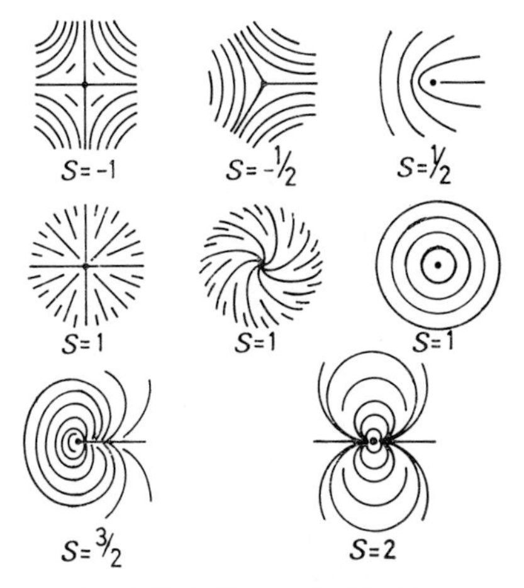

FIG. 3.3.2. Screw disclinations.

article, shows some typical configurations. The configurations $S = \pm\frac{1}{2}$ and $S = \pm 1$ are frequently observed in optical microscopy studies and correspond, when perpendicular to the preparation, to the "nuclei" of

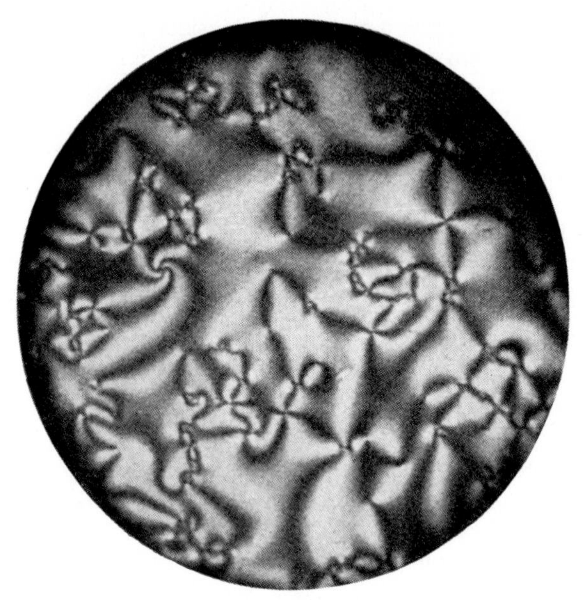

PLATE 3.1. Observation of nuclei between crossed polars ($S = \pm 1$, four black branches; $S = \pm 1/2$, two black branches). Nematic phase of *p*-azoxyphenetole (after G. Friedel [43]).

Friedel (*plages à noyaux*) (cf., Plate 3.1). The reader will easily check that the process described under Disclination Line in a Nematic leads to the topological features of Fig. 3.3.2. S is called the strength of the disclination.

The energy of a screw disclination is given by Equation 3.3.2. Inserting from Equation 3.3.1 gives for the energy per unit length of line:

$$W = \pi K S^2 \ln \frac{\rho_{\max}}{a} + \text{core energy} \tag{3.3.3}$$

where $\rho_{\max}$ is a typical width of the specimen, or an average distance between lines, and a the radius of the core, typically a few molecular distances. The energy of the core cannot greatly exceed the energy of the completely disordered nematic, so that the total energy per unit length (the so-called line tension) is of the order of K. Equation 3.3.3 is reminiscent of the line tension of a screw dislocation of Burgers' vector $b = S\sqrt{(\pi K/2)}$ [45]. It shows that $S = \pm\frac{1}{2}$ disclinations are strongly favoured. It is probable that in most cases [46] the core of $S = \pm 1$ disclinations is "dissociated", reducing the total energy of the line.

Parallel screw disclinations of the same sign of S repel themselves; they attract themselves when of opposite sign. The interaction between a screw disclination parallel to the surface boundary, at a distance d from it, and the surface, depends on the boundary conditions for the molecules. If the wall imposes tangential or normal boundary conditions, the interaction is similar to that which would result from the presence of a screw disclination of the same sign on the other side of the boundary and at an equal distance from it (image disclination). The disclination is then repelled by the wall, by a force f per unit length

$$f = \frac{4\pi K S^2}{d} \tag{3.3.4}$$

A discussion of the line tension of a disclination and of configurations of disclination groups may be found in Ericksen's article [47].

General Disclination

Figure 3.3.3 represents some possible configurations for an edge disclination, i.e., a straight disclination perpendicular to the axis of rotation. Figure 3.3.3(a) corresponds to a planar configuration $\phi = S\theta$ likely to occur when the twist elastic coefficient K_2 is very small; Fig. 3.3.3(b) is a more complex case in which there is no bending energy.

The general case of a closed loop has been treated by Friedel and

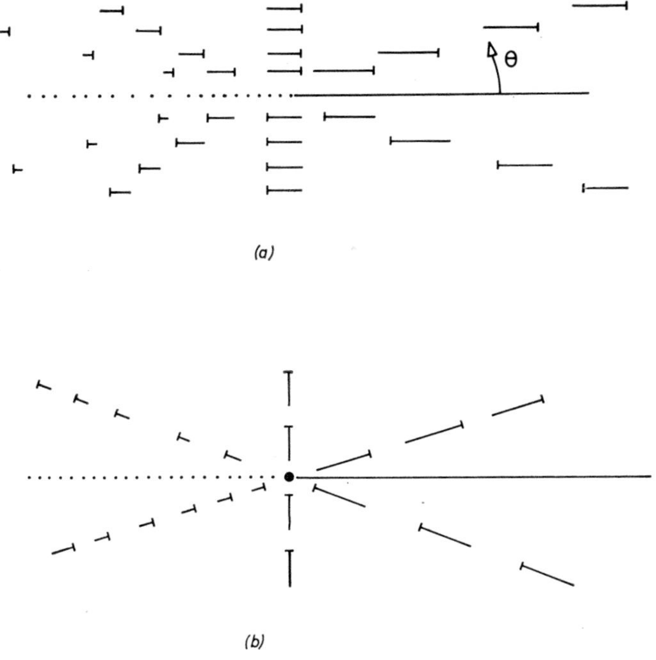

Fig. 3.3.3. Edge disclinations seen end-on. Dots represent molecules viewed end-on, and the continuous line represents molecules in the plane of the diagram. Molecules making angles other than 0° or 90° with the plane of the diagram are represented by "nails". The head of the "nail" is behind the plane of the drawing, and the nail points towards the observer. The length is proportional to the angle with the normal to the drawing

(a) K_2 small; planar configuration, $\phi = S\theta$ ($S = \frac{1}{2}$)

(b) $\psi = S\theta$; $\mathbf{n} \times \text{rot}\, \mathbf{n} = 0$ (no bending energy).

de Gennes [48]. Assuming the three elastic coefficients equal and a planar configuration, the azimuthal angle ϕ is given by

$$\phi(\mathbf{r}) = S\frac{\Omega_s(\mathbf{r})}{2} + \ldots$$

where Ω_s is the solid angle under which the line is seen from $\mathbf{r}$.

The general disclinations give rise to the well-known threaded texture (*plages à filaments*).

DISCLINATION POINTS

Disclination points have been recently observed at interfaces between the nematic and the isotropic phases.

Near the clearing point the isotropic phase produces nearly spherical nematic droplets which often show two singular points at both ends

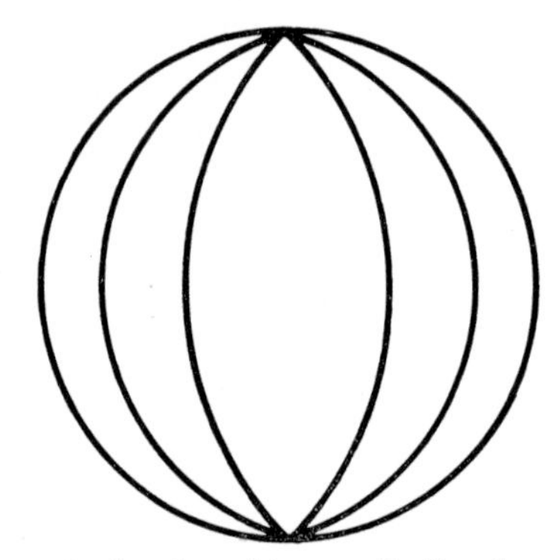

FIG. 3.3.4. Bipolar nematic droplet with two disclination points ($S = 1$).

of the same diameter [20]. The energy of such a bipolar conformation (Fig. 3.3.4), assuming tangential boundary conditions, is $\approx 5\pi KR$ (R is the radius of the droplet), which is smaller than the energy of a droplet containing a diametral disclination $\left(\approx 2\pi KR \ln \frac{R}{a}\right)$. R is bigger than $0{\cdot}1\ \mu$. This figure corresponds to experimental values [49].

Meyer has also observed [50] disclination points at a planar interface

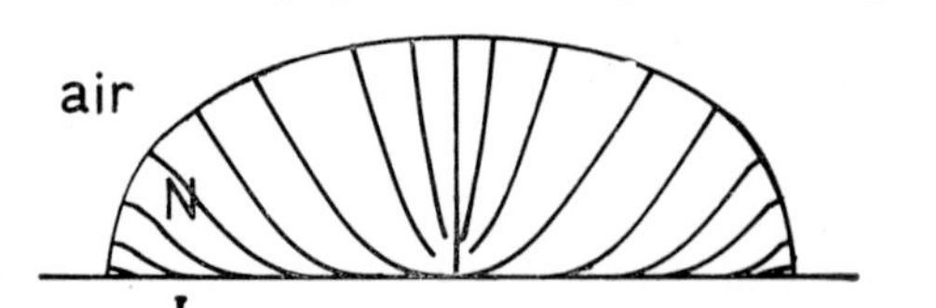

FIG. 3.3.5. Disclination point at a planar interface of a nematic (N) droplet with the isotropic liquid (I).

between the nematic and the isotropic phase, which he explains by tangential boundary conditions at the interface and homeotropic conditions at the free surface (Fig. 3.3.5).

Defects in a Cholesteric Phase

There are two types of disclinations in cholesteric mesophases: those corresponding to rotations which are multiples of π about the cholesteric axis (χ disclinations), and those corresponding to rotations which are multiples of π about the direction of the molecule (λ disclinations) or about the normal to that direction in the cholesteric plane (τ disclinations).

χ DISCLINATIONS

These disclinations are similar to the nuclei or the filaments in nematics and may be built by the same process. When screw-like in character, the configuration of the molecules is similar to that obtained—see Fig. 3.3.2, this configuration rotating with the pitch of the cholesteric along the axis χ. Notice that the configuration $S = +1$ is the only one which shows various configurations (see Fig. 3.3.2) along the axis. It is possible to pass from one to another by a local homogeneous rotation of the molecules about the local χ axis.

Fig. 3.3.6 shows the conformation for an edge χ ($S = \pm\frac{1}{2}$). It is apparent that there is an extra half-pitch inserted in the perfect cholesteric, so that

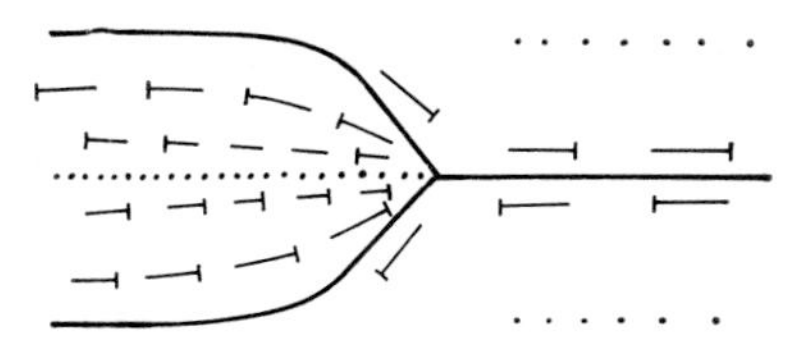

FIG. 3.3.6. Edge χ disclination in a right-handed cholesteric. See caption to Fig. 3.3.3 for an explanation of the symbols used.

the χ edge disclination may be considered as an edge translation dislocation of Burgers' vector $|Sp|$. Moreover this is true of a general χ loop and is related to the possibility of creating disclination loops with a displacement α along the axis χ followed by a rotation $+\alpha/q_0$.* Such dislocation loops, as soon as created, tend to disappear by viscous relaxation, because α may decrease continuously to zero. Add therefore a disclination $\left(-\alpha, -\frac{\alpha}{q_0}\right)$, $\left(\alpha = \frac{p}{2}\right)$ to a χ. The rotation part of this disclination cancels the rotation part of the χ, which subsists as a translation dislocation $\left(-\alpha = -\frac{p}{2}, 0\right)$.

According to the model of Pryce and Frank, the spherulites of PBLG observed by Robinson [52] possess a radial screw disclination of a χ type, with a strength $S = 2$.

The energy of an edge χ has been calculated by de Gennes [53] for an isotropic approximation and later refined [54] by taking into account the difference between the elastic constants.

* The plus sign occurs for a right-handed cholesteric ($\theta = q_0 z$ in a right-hand system of reference).

λ AND τ DISCLINATIONS

The creation of such disclinations may be visualized as follows. Consider a line L (Fig. 3.3.7(*a*)) perpendicular to the molecules and lying in a cholesteric plane, and a cut surface Σ limited by L. Rotate aside the two lips Σ_1 and Σ_2 of Σ by an angle π about an axis $\mathbf{v}$ lying on L (Fig. 3.3.7(*b*).

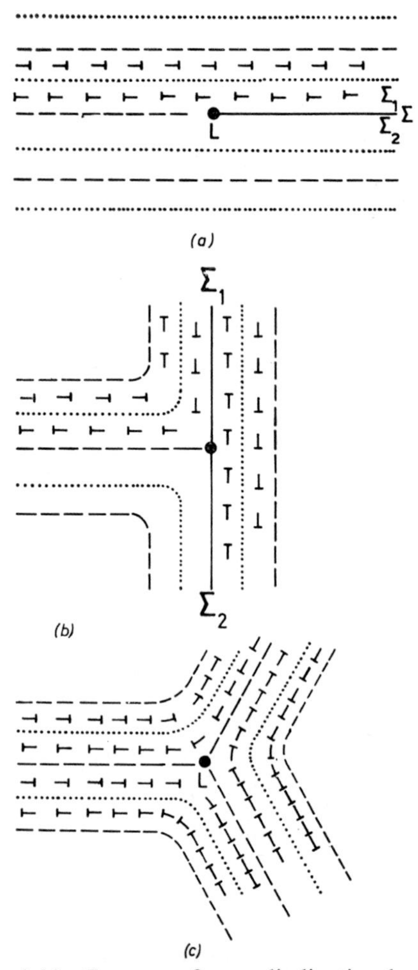

FIG. 3.3.7(*a*), (*b*) and (*c*). Creation of a τ^- disclination by a Volterra process. The dotted or full lines represent molecules perpendicular to or in the plane of the diagram. Other symbols are as explained in the caption to Fig. 3.3.3 and in the text.

The empty space may be filled with perfect cholesteric, which matches perfectly the distorted part along Σ_1 Σ_2. After relaxation, the whole system leads to the configuration in Fig. 3.3.7(c), which we shall call a τ^- (τ for transverse, minus because of the similarity with a configuration $S = -\frac{1}{2}$. In fact, similar configurations may be obtained for all the values of S).

Fig. 3.3.8(*a*), (*b*) and (*c*) show the resulting configurations for a λ^- (λ for longitudinal), a τ^+ and a λ^+ disclination, respectively; these different configurations are illustrated in Plate 3.2. It is worth noticing that the core energies are very different for λ's and τ's; the core is singular only in the last case, and most of the existing disclinations are probably λ's.

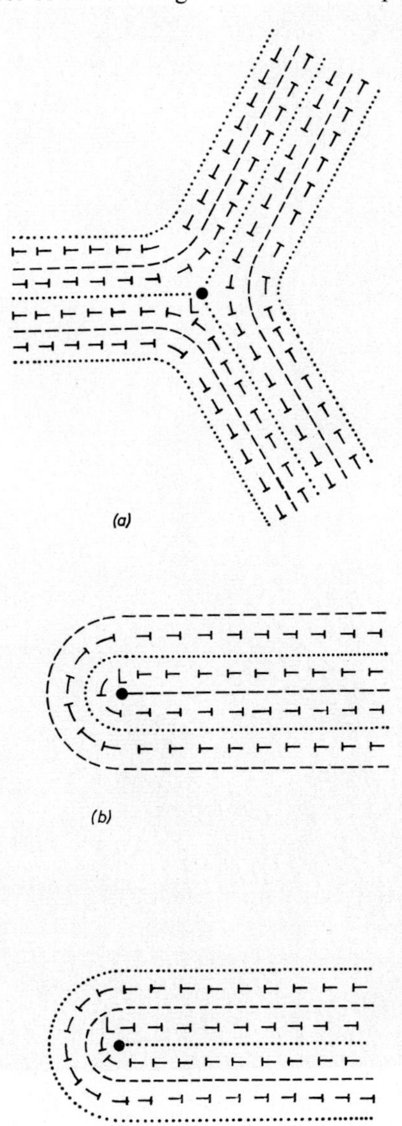

FIG. 3.3.8(*a*), (*b*) and (*c*). (*a*) : λ^- disclination,
(*b*) : τ^+ disclination,
(*c*) : λ^+ disclination.

PLATE 3.2. High pitch cholesteric phase. $P = 16\ \mu$. The cholesteric axis is roughly "in the plane" (by courtesy of J. Rault).

Curved λ and τ disclinations cannot be easily obtained by the process just described, which would lead to macroscopic deformations (and therefore prohibitive energies) on those regions of the core which are not along the axis of rotation, in contradistinction with the χ disclinations.

Pairs of Disclinations in Cholesterics

It may be shown [51] that the pairing of λ's and τ's of opposite signs enables them to take any shape; this pairing must be such that the opposite disclinations stay at a constant distance d, a multiple of $p/4$, measured along the perpendicular to the common rotation axis $\boldsymbol{\nu}$.

Figure 3.3.9 shows the simplest case, when the disclinations are straight. The total configuration is clearly equivalent at a distance to a χ of Burgers'

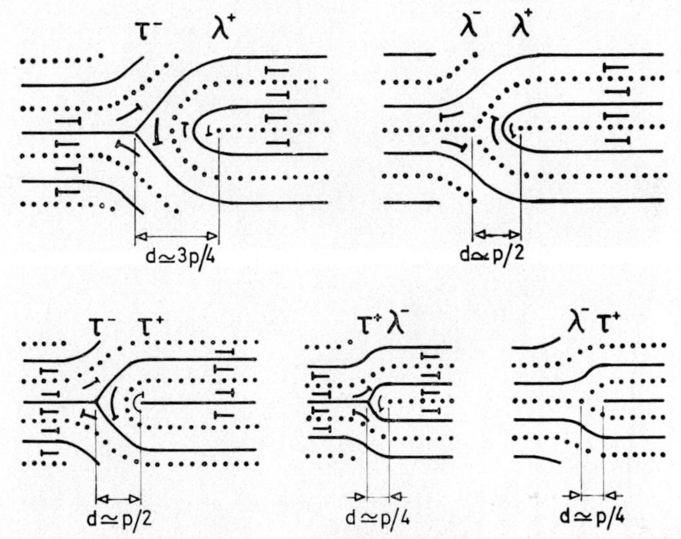

Fig. 3.3.9. Pairings of λ's and τ's are equivalent to χ's at a distance.

vector $b = 2d$, and can be referred to as a dissociated χ. The energy of the core is certainly smaller in that case than for the equivalent χ. The Grandjean–Cano walls are most probably pairs of disclinations of these types, and the configurations in Fig. 3.3.9 explain quite satisfactorily their zig-zag behaviour under an applied magnetic field [55] (cf., Plate 3.3.).

A less simple pairing is indicated in Fig. 3.3.10(a). The shaded area represents a cholesteric pattern containing a disclination pair, which is rotated helically about an axis with a pitch P, a multiple of $p/2$; notice that the pitch of the cholesteric obtained is different inside and outside the helical pair. It is possible in that way to part a nematic or isotropic phase from a cholesteric medium (Fig. 3.3.10(b)). The operation of rotation may also leave a singularity (of the χ type) along the helical axis (Fig. 3.3.10(c)). The cholesteric droplets observed in the isotropic phase are probably related to these configurations. A detailed theory of these helical pairs may be found in reference [56].

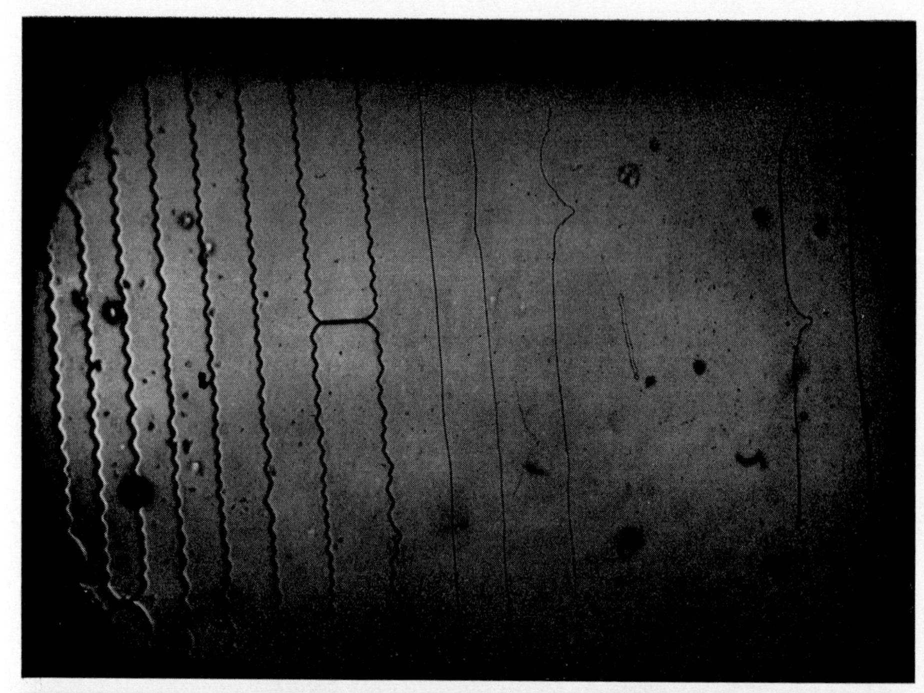

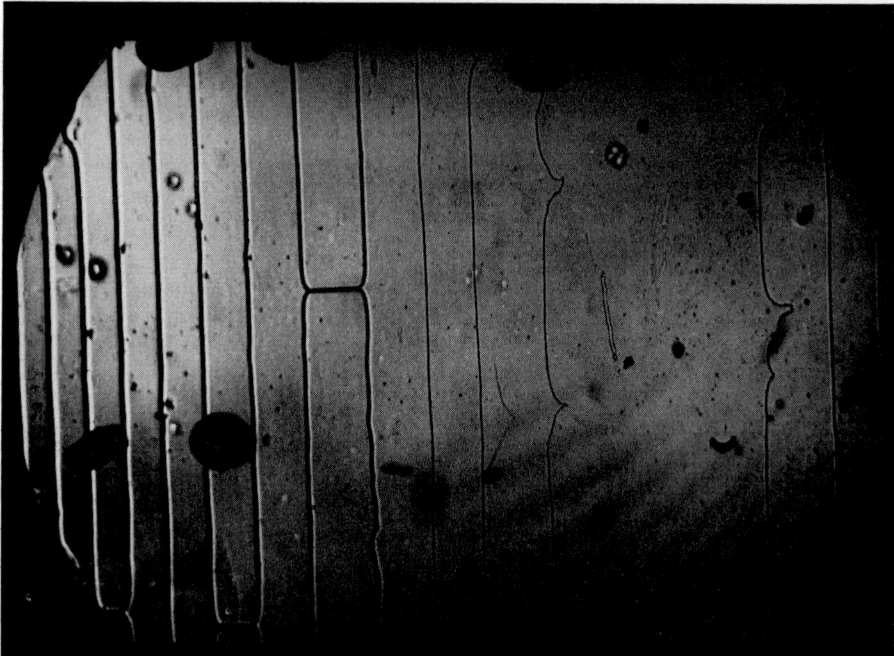

PLATE 3.3. High pitch cholesteric phase. Observation of Grandjean walls. These walls show (upper part of Plate) a characteristic zig-zag instability under an applied field of the order of $H_c/2$ when the core width of the disclination pair is of the order of $P/2$ (by courtesy of the Orsay Liquid Crystal Group).

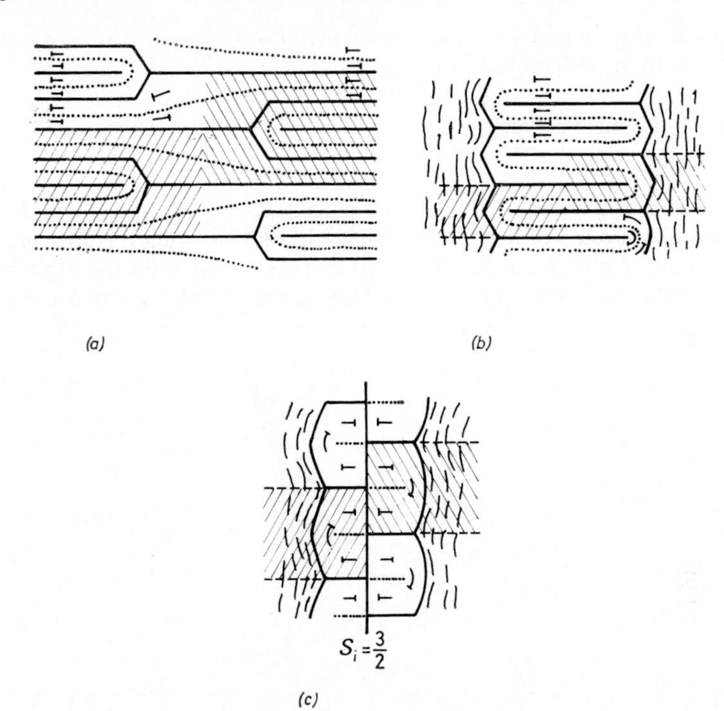

FIG. 3.3.10. Examples ((*a*) and (*b*)) of how to create a helical pair: the shaded pattern is rotated along a right-handed helix (right-handed cholesteric); no singularity appears on the axis. In (*c*) the pattern is rotated along a right-handed helix (left-handed cholesteric); a singularity $S_i = \frac{3}{2}$ appears on the axis.

Glide and Climb of Disclinations

Any movement of a χ disclination or a pair may be considered as a combination of *glide* along a direction parallel to the Burgers' vector and *climb* along a direction of the plane perpendicular to the Burgers' vector. Glide is difficult for a straight pair, because the intermediate positions of the lines between two equilibrium positions at a distance of $p/4$ are not allowed for the process described above for creation of a λ or a τ disclination. There is therefore a large deformation of the core. Glide of a χ disclination has the same anisotropic character, due to the anisotropy of the Frank elastic coefficients (this anisotropy is reminiscent of the Peierls–Nabarro [45] forces of dislocation theory). Climb is more favoured, for χ's as well as for pairs, and is easily observed in specimens, in which it is induced by any small movement communicated to the glass plate. More generally, χ's and pairs may be deformed to any shape as long as they do not intersect.

Other Textural Features of Cholesterics

Near the clearing point, most cholesteric phases show very metastable textures, closely resembling smectic textures (Fig. 3.3.11). The article of

G. Friedel [43] is still the best source of information on these textures, which are little understood in the main.

Defects in a Smectic Phase

Focal-conic Domains [43, 51] in Smectics A

A focal–conic domain (Fig. 3.3.11(*a*)) consists of two focal–conics, an ellipse and a branch of hyperbola. The molecules lie along the lines drawn between any point A on the ellipse (E) and any point B on the hyperbola (H). All the lines starting from A and describing (H) form a cone of

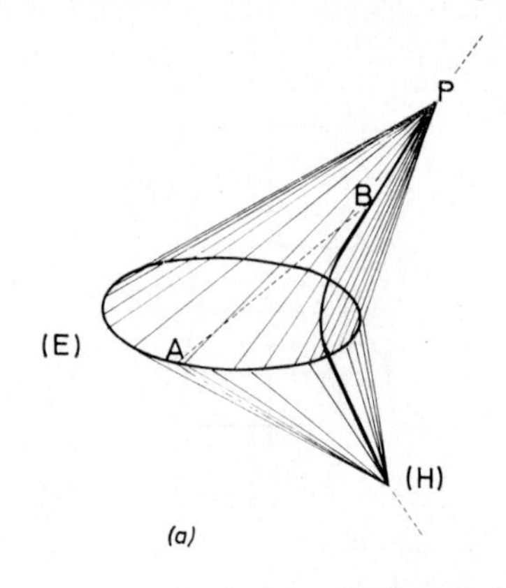

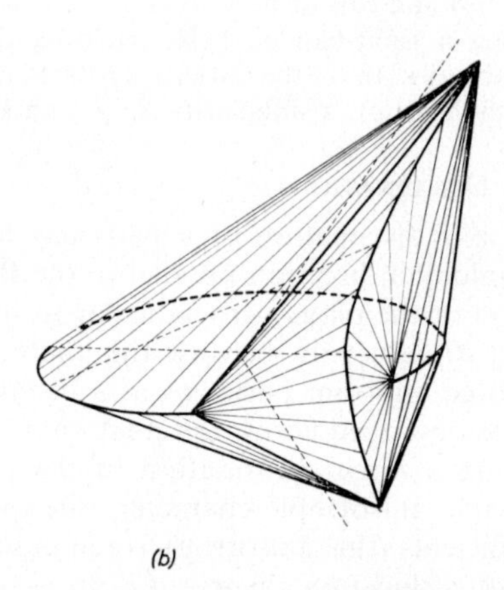

Fig. 3.3.11(*a*) and (*b*). Scheme of focal–conic domains. Straight lines such as AB joining points on the ellipse to points on the hyperbola mark the direction of **n** in the case of smectics, and the direction of the helical axis in the case of cholesterics.

revolution and similarly for the lines starting from any point on (H) and describing (E). The smectic layers are perpendicular to these lines and form a family of Dupin cyclides. The focal-conic domains are limited by the cones tapering to the limiting points on (E) and (H), and are tangential to the adjacent focal–conic domains along these cones.

The occurrence of focal–conic domains may be easily understood on the basis of a layered structure. The layers glide easily upon each other, but deform less easily along the smectic axis. Any distorted structure should then, to first order, preserve layers of equal thicknesses, which means that the layered structures have common normals and the same centres of curvature along the same normal. These centres of curvature describe the so-called focal surfaces (Fig. 3.3.11(*b*)), which are singularities of the

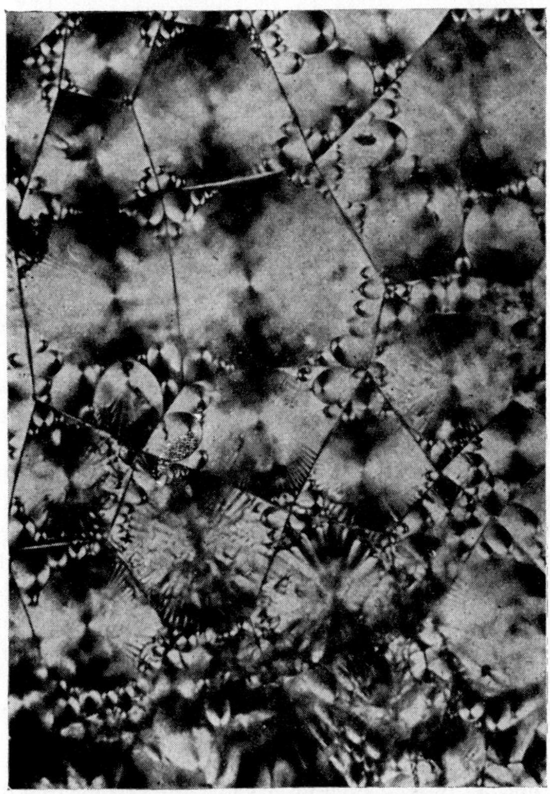

PLATE 3.4. Focal–conic domains, forming a characteristic polygonal texture (polarizer only) (after G. Friedel [43]).

structure on which the smectic phase must stop (cf., a similar discussion in Nye [57]). But, the smectic phase being a liquid with no empty space, the surface singularities are reduced to line singularities. This means that the radii of curvature of the smectic layers are constant along curvature lines, which are therefore circles. Hence the smectic layers are Dupin cyclides, and the line singularities focal–conics.

The focal–conic domains lead to beautiful arrangements described by G. Friedel [43] (Plate 3.4) as feathers, fan-shaped textures, *etc.* . . . A free droplet of smectic phase very often crystallizes in terraced layers (perfect smectic) bordered by tiny focal–conic domains on the free surface.

SMECTICS B AND C

On the basis of the occurrence of different textures, not clearly explained today, Sackmann and Demus [58] have been able to recognize the existence of smectics B and C. The latter has been recently investigated by Fergason *et al.* [59]. The nature of the defects in smectics C is still largely unknown.

Disclination Lines and Dislocation Theory

We have so far used two different processes of creation of a dislocation line. The process used for nematics and χ lines in cholesterics has been proposed by de Gennes [53]; that used for λ and τ lines is due to Volterra (cf., [55]). We show now that this last process is of more general use.

Consider a line L and a surface Σ bounded by L. The general Volterra process consists in displacing the two lips Σ_1 and Σ_2 of the cut surface Σ, one relatively to the other, by an amount

$$\mathbf{d}(\mathbf{r}) = \mathbf{b} + 2 \sin \frac{\Omega}{2} \boldsymbol{\nu} \times \mathbf{r} \tag{3.3.5}$$

b is a translation symmetry element of the lattice (Burgers' vector), $(\Omega, \boldsymbol{\nu})$ is a rotation symmetry element and **r** is the distance to a point on $\boldsymbol{\nu}$. The displaced part **d** is either removed or filled in with extra matter of the perfect crystal, the intermolecular bonds are re-established and the medium is allowed to relax.

In solid crystals, dislocations are *translation* dislocations. Disclinations (rotation dislocations) would lead to abnormal situations (Moebius crystals) and, furthermore, their energy would be prohibitive; but they are common in liquid crystals. The preceding definition, if it does not forbid *curved* disclination lines from a topological point of view, favours, from an energy point of view, only straight lines, when the line L itself is coincident with the axis $\boldsymbol{\nu}$ of the rotation. In that case the displacement $\mathbf{d}(\mathbf{r})$ of the molecules in the core of the disclination is zero. But curved lines may be stabilized by infinitesimal translation dislocations, in the following way.

Consider, on a curved line L, generated by a rotation of axis $\boldsymbol{\nu}$, angle Ω, two points **A** and $\mathbf{A}' = \mathbf{A} + \mathrm{d}\mathbf{l}$ (Fig. 3.3.12), and assume that the displacement of the cut surface may be described with the axis of rotation either in A or in A′. The difference $\boldsymbol{\delta}$ between the displacements **d** and **d**′ of a given point **M** of Σ is independent of **M** and given by

$$\boldsymbol{\delta} = 2 \sin \frac{\Omega}{2} . \boldsymbol{\nu} \times \mathrm{d}\mathbf{l} \tag{3.3.6}$$

The difference between the two disclinations L(A) and L(A′) is therefore a translation dislocation, starting from A and going to infinity or to another defect; if such translation dislocations of Burgers' vector $\boldsymbol{\delta}$ are allowed by the medium, we may imagine that the line L is provided with a continuum of *infinitesimal* translation dislocations. This continuum is two-dimensional but it may expand in a three-dimensional continuum (with a one-dimen-

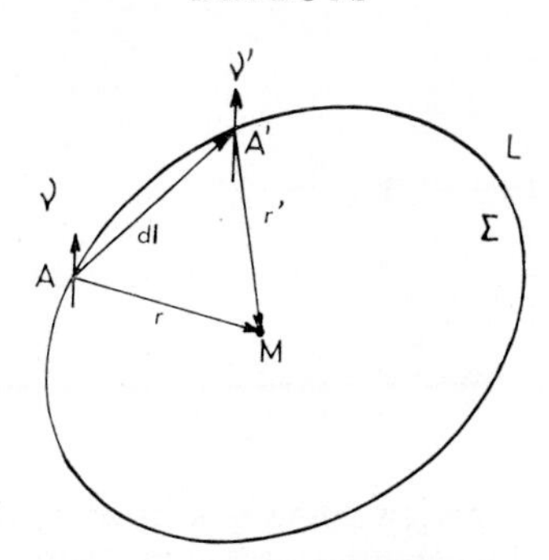

FIG. 3.3.12. Introduction of infinitesimal translation dislocations on a curved disclination (see text).

sional singularity L) by letting each dislocation **δ** dissociate. Let us remember that any translation is a symmetry element in a nematic, and that it is the same for translations along the cholesteric plane in cholesterics.

The de Gennes' process provides a line with a three-dimensional set of infinitesimal translation dislocations. To each segment **AA**′ of the *cut surface* is now attached a dislocation **δ**. But it is essential to notice that such a set of dislocations relaxes viscously very easily: the energy of each dislocation decreases continuously to zero by a continuous decrease to zero of the Burgers' vector, so that such a three-dimensional set should condense on the disclination line in a two-dimensionnal set, if it is not stabilized by other defects and favourable elastic constants.

3.4 DYNAMICAL EFFECTS

P. G. de Gennes

In this section we shall review some aspects of motion, at *low frequencies* and *long wavelengths*, in nematics only—the data on other mesophases are still too scanty. The fundamental work on the theory of these effects is due to Ericksen [25], Leslie [60], and Martin, Parodi and Pershan [80].

The Friction Coefficients

To describe the system of forces in a fluid, we must specify first the nature of the *stresses*. In a nematic liquid crystal, under homogeneous external fields, the stress tensor has the form:

$$\sigma_{\beta\alpha} = -p\,\delta_{\beta\alpha} + \sigma_{\beta\alpha} + \sigma'_{\beta\alpha} \tag{3.4.1}$$

where p is the scalar pressure. σ^0 is the stress related to the distortion free energy, and is given by [25]

$$\sigma^0_{\beta\alpha} = -\Pi_{\beta\gamma}\frac{\partial}{\partial x_\alpha}n_\gamma \tag{3.4.2}$$

$$\Pi_{\beta\gamma} = \frac{\partial F}{\partial\left(\dfrac{\partial}{\partial x_\beta}n_\gamma\right)}$$

σ' describes viscous effects; it is a linear function of the velocity gradients

$$A_{\alpha\beta} = \frac{1}{2}\left(\frac{\partial}{\partial x_\alpha}v_\beta + \frac{\partial}{\partial x_\beta}v_\alpha\right) \tag{3.4.3}$$

($\mathbf{v}$ being the flow velocity). σ' also contains terms which are linear in the rate of change of $\mathbf{n}$ with respect to the surrounding fluid. This rate of change is defined by the vector

$$\mathbf{N} = \frac{\mathrm{d}\mathbf{n}}{\mathrm{d}t} - \omega \times \mathbf{n} \tag{3.4.4}$$

where $\boldsymbol{\omega} = \frac{1}{2}\,\mathrm{curl}\,\mathbf{v}$ is the local rotation velocity of the fluid.

We shall write down the tensor σ' for the comparatively simple case of an incompressible nematic.* Then the most general form for σ' is [60]:

$$\begin{aligned}\sigma'_{\beta\alpha} = {} & \alpha_1 n_\mu n_\rho A_{\mu\rho} n_\beta n_\alpha + \alpha_2 n_\beta N_\alpha + \alpha_3 n_\alpha N_\beta \\ & + \alpha_4 A_{\beta\alpha} + \alpha_5 n_\beta n_\mu A_{\mu\alpha} + \alpha_6 n_\alpha n_\mu A_{\mu\beta}\end{aligned} \tag{3.4.5}$$

* More general formulae are given in reference [60], and have been applied to the problem of attenuation of longitudinal sound waves.

The coefficients $\alpha_1 \ldots \alpha_6$ have the dimension of a viscosity, and are called friction coefficients. The number of independent coefficients is reduced to five by an Onsager relation [61]:

$$\alpha_2 + \alpha_3 = \alpha_6 - \alpha_5 \tag{3.4.6}$$

It is apparent from Equation 3.4.5 that the tensor $\sigma'_{\beta\alpha}$ is not symmetric:

$$\sigma_{yx} - \sigma_{xy} \equiv \Gamma_z \neq 0 \tag{3.4.7}$$

According to general theorems of mechanics [62], the vector $\mathbf{\Gamma}$ which is thus defined represents the torque (per unit volume) transferred from other degrees of freedom (here : internal rotation) to the overall hydrodynamic motion. Inserting (3.4.5) into (3.4.7) one may put $\mathbf{\Gamma}$ in the form

$$\left.\begin{aligned} \mathbf{\Gamma} &= \mathbf{n} \times \{\gamma_1 \mathbf{N} + \gamma_2 \mathbf{A}:\mathbf{n}\} \\ \gamma_1 &\equiv \alpha_3 - \alpha_2 \\ \gamma_2 &\equiv \alpha_6 - \alpha_5 = \alpha_2 + \alpha_3 \end{aligned}\right\} \tag{3.4.8}$$

The torque (per cm^3) on the molecules, due to external fields and to surrounding distortions, is simply $\mathbf{n} \times \mathbf{h}$, where the molecular field $\mathbf{h}$ has been defined in Equation 3.2.5. Since we are concerned here with motions which are *slow* on the molecular scale, the rotational inertia of the molecules is negligible : if the molecules receive the torque $\mathbf{n} \times \mathbf{h}$ and generate the torque $\mathbf{\Gamma}$, these two torques must be equal:

$$\mathbf{\Gamma} = \mathbf{n} \times \mathbf{h} \tag{3.4.9}$$

Having now specified the stresses and torques, we may write down the acceleration equation for the fluid*

$$\rho \frac{dv_\alpha}{dt} = \frac{\partial}{\partial x_\beta} \sigma_{\beta\alpha} \tag{3.4.10}$$

Together with the incompressibility condition:

$$\operatorname{div} \mathbf{v} = 0 \tag{3.4.11}$$

Equations 3.4.9, 3.4.10 and 3.4.11 may then be solved (more or less painfully!) for the unknowns p, $\mathbf{v}$, and $\mathbf{n}$. We shall now discuss this for a few typical cases which are experimentally important.

Simple Types of Laminar Flow

Shear Flow Under a Magnetic Field

One of the most clear experimental situations is obtained when the molecules are firmly aligned by an external field H. The word "firmly" implies two requirements:

(*i*) the lateral walls limiting the flow must not disrupt the alignment induced by H. Sometimes, e.g., in a circular capillary under a transverse field H, it is in fact impossible to prevent a certain competition between wall alignment and field alignment. Then we require that the field dominates; the diameter D of the capillary must be much larger than the magnetic coherence length $\xi(H)$ defined in Equation 3.2.7.

* The order of the indices in Equation 3.4.10 follows the convention adopted by Leslie (reference [60]).

(*ii*) the flow itself must not disrupt the alignment. This implies the inequality

$$\omega \ll \frac{\chi_a H^2}{\bar{\eta}}$$

where $\boldsymbol{\omega}$ is always defined as $\frac{1}{2}$ curl $\mathbf{v}$, and $\bar{\eta}$ is an average viscosity (a certain average of the friction coefficients) typically of order 0·1 poise.

Let us now assume that conditions (*i*) and (*ii*) are satisfied. Depending on the relative orientations of the field, the flow, and the flow gradient,

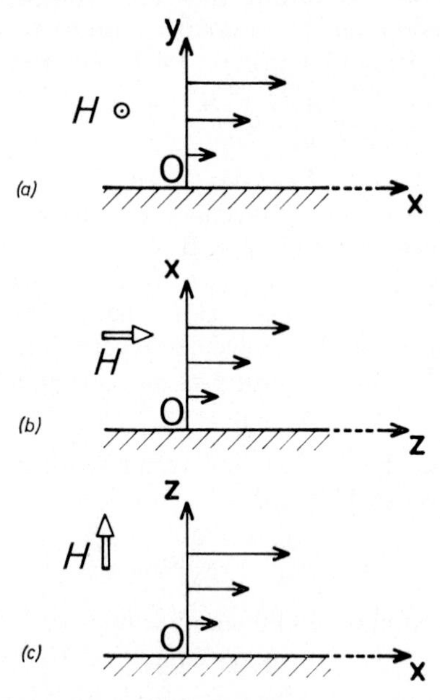

FIG. 3.4.1(*a*), (*b*) and (*c*). The three basic types of shear flow in a nematic fluid under a field **H** fixing the optic axis (parallel to Oz).

we find three typical geometries for simple shear, as shown in Fig. 3.4.1(*a*), (*b*),(*c*). Using Equation 3.4.5 one finds that, for each of these cases, there is a well defined effective viscosity, given by:

$$\left.\begin{aligned} \eta_a &= \tfrac{1}{2}\,\alpha_4 \\ \eta_b &= \tfrac{1}{2}\,(\alpha_3 + \alpha_4 + \alpha_6) \\ \eta_c &= \tfrac{1}{2}\,(-\,\alpha_2 + \alpha_4 + \alpha_5) \end{aligned}\right\} \qquad (3.4.12)$$

Measurements of these effective viscosities were carried out long ago by Miesowicz [63] on *p*-azoxyanisole.

FLOW IN THE ABSENCE OF EXTERNAL FIELDS

This class of experiments is much more delicate:

(*i*) The boundary conditions at the walls are very important, and must be defined by careful surface treatments.

(*ii*) Disclination lines coming from up stream may complicate the flow pattern. To be sure that these effects are properly taken care of, optical observations on the fluid vein are necessary: this in turn requires thin samples, and the whole set up becomes complex [64]. However, an excellent experimental study has been made by Fisher and Wahl [81]. The problem has been discussed theoretically both for flow between parallel plates [60, 65], and for Couette flow, i.e., flow between concentric cylinders [66, 67].

The expected behaviour depends critically on the ratio $\left|\frac{\gamma_2}{\gamma_1}\right|$.

(*i*) If $\left|\frac{\gamma_2}{\gamma_1}\right| \geqslant 1$ there is a certain angle θ, between the optic axis and the direction of flow (defined by $\cos 2\theta = -\gamma_1/\gamma_2$) for which the torque Γ (as defined by Equation 3.4.8) vanishes. Then, far from the walls, the molecules tend to lie precisely at this angle. In the vicinity of the walls, since the molecules must adjust to a certain boundary condition, there is a transition layer, of thickness

$$\delta \approx \sqrt{\frac{K}{\bar{\eta}\omega}}$$

(K = elastic constant; $\bar{\eta}$ = viscosity).

(*ii*) If $\left|\frac{\gamma_2}{\gamma_1}\right| < 1$, the torque Γ cannot vanish: this generally implies that the nematic structure is strongly twisted. For shear flow between two plates separated by a gap D, the director, as seen by an observer moving from one plate to the other, rotates by many turns: the pitch of the resultant "cycloidal structure" would be of order δ^2/D. In practice, these cycloidal structures have not been seen. It appears that $\frac{-\gamma_2}{\gamma_1}$ is very close to unity in all cases, i.e., that the long nematic molecules tend to align along the direction of flow ($\theta = 0$). Arguments suggesting that $\frac{-\gamma_2}{\gamma_1}$ must be slightly larger than one have been presented, but the correct answer may well depend on the detailed structure of the molecules. It might be possible to realize the cycloidal texture of case (*ii*), in a pure shear flow, even if $\left|\frac{\gamma_2}{\gamma_1}\right| > 1$, by imposing a magnetic field **H**, at 45° from the flow lines [68].

Ultrasonic Attenuation of Shear Waves

The principle of the experiment is shown in Fig. 3.4.2(*a*),(*b*) and (*c*): a transverse ultrasonic wave of frequency Ω propagates in a quartz crystal, then penetrates slightly into a nematic sample. The penetration thickness is $l \approx \left(\frac{\bar{\eta}}{\rho\Omega}\right)^{1/2}$ as it is in ordinary fluids, and is of the order of one micron for frequencies in the megacycle range. Finally the wave is reflected and the attenuation is measured. At first sight this situation appears similar to the uniform flows discussed under shear flow under a magnetic field; however there is a difference. In cases (*b*) and (*c*) of Fig. 3.4.2, the director **n** is locally deflected (in the thickness l) by the flow gradient.* The

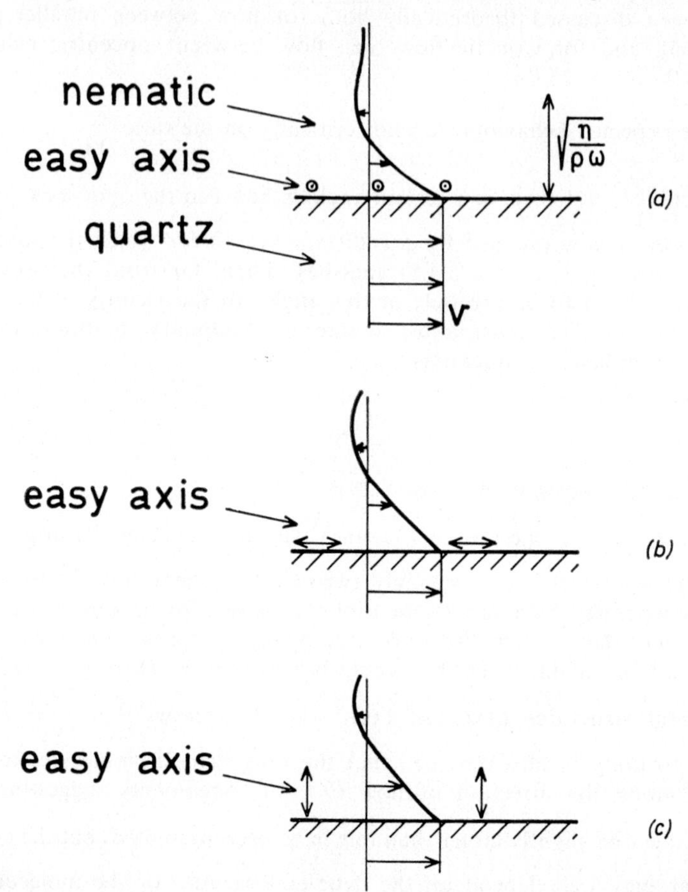

FIG. 3.4.2(*a*), (*b*) and (*c*). Three typical geometries for the study of shear wave attenuation in nematics. The ultrasonic wave propagates upwards in the quartz and is reflected at the interface.

analysis of the surface acoustic impedance depends critically on a parameter

$$\mu = \frac{K\rho}{\eta^2} \tag{3.4.13}$$

Typically $K = 10^{-6}$ dynes, $\rho = 1$ g/cm^3, $\eta = 10^{-1}$ poise, and $\mu \sim 10^{-4}$. When $\mu \ll 1$, a number of simplifications occur. The final result is that

* This occurs in spite of the fact that **n** is held rigidly by a boundary condition at the quartz surface: one may show that the effects of the boundary condition extend only to a thickness $\left(\frac{K}{\eta\Omega}\right)^{1/2}$ which is much smaller than l.

the nematic behaves acoustically like an ordinary fluid, but with the following effective viscosities [69]

$$\tilde{\eta}_a = \eta_a$$

$$\tilde{\eta}_b = \eta_b - \frac{\gamma_1 + \gamma_2}{2\gamma_1}\alpha_3 = \tilde{\eta}_c \tag{3.4.14}$$

for cases (*a*), (*b*) and (*c*) respectively. (η_a η_b η_c have been defined in Equation 3.4.12).

Measurements of the imaginary part of the acoustic impedance have been carried out by Candau and Martinoty [70], for the cases (*a*) and (*b*) above. It is very much to be hoped that case (*c*) will also be feasible.

One final remark on Equation 3.4.14. One might be tempted to clamp the optic axis (in the thickness l) by a magnetic field applied collinear to the easy direction at the quartz surface; then the effective viscosities would be η_a, η_b, η_c. However, this is not feasible in practice, since it would require fields of the order $\left(\frac{\eta\Omega}{\chi_a}\right)^{1/2}$ which are prohibitively large.

Dynamics of Fluctuations

Inelastic Scattering of Light

As explained on p. 72, the light scattered by a nematic single crystal probes the fluctuations of the director $\delta\mathbf{n}$. More precisely, if the scattering wave vector is $\mathbf{q}$, what is studied is one Fourier component $\delta\mathbf{n}_q$ of $\delta\mathbf{n}$. If the frequency distribution of the emergent light (for a monochromatic incident beam) is analysed, it gives us some information on the changes in time of $\delta\mathbf{n}_q$—i.e., on the relaxation of the fluctuations. A theoretical analysis of this effect, based on the Ericksen–Leslie hydrodynamic equations, gives the following results [71]:

(*i*) when the parameter μ (defined by Equation 3.4.13) is much smaller than one (as it is indeed in most cases), the relaxation of $\delta\mathbf{n}_q$ is expected to be purely viscous—no oscillations. This is confirmed by the experiments on *p*-azoxyanisole [72].

(*ii*) in zero magnetic field, the relaxation rate of $\delta\mathbf{n}_q$ is of the form:

$$\frac{1}{\tau_q} = \frac{K_{\text{eff}}q^2}{\eta_{\text{eff}}} \tag{3.4.15}$$

where K_{eff} and η_{eff} are averages of the elastic constants and of the friction coefficients. The detailed forms of K_{eff} and η_{eff} depend on the orientation of $\mathbf{q}$ with respect to the optic axis, and also on polarization indices [71].

Typical values of $1/\tau_q$ (for wave lengths $(2\pi/q)$ in the optical range) are 10^3–10^5 s^{-1}, and are thus very well suited for measurements using the photon beat methods. This has been carried out for *p*-azoxyanisole [72]; from the detailed angular dependences of η_{eff}, it has been possible to extract 4 out of the 5 friction coefficients: the method requires some care, but is powerful.

Nuclear Relaxation

The relaxation of a nuclear spin in the nematic is controlled by fluctuations of the local alignment: at first sight such a local property does not seem to be predictable from the continuum theory. However, what is involved in the relaxation is only one low frequency Fourier component (at the nuclear frequency ω_n) of the local alignment: this involves mainly the relaxation modes (Equation 3.4.15) for which

$$\frac{1}{\tau_q} \simeq \omega_n$$

Inserting Equation 3.4.15 for $1/\tau_q$, and taking $\omega_n = 10^7$, $K = 10^{-6}$, $\eta = 10^{-1}$ gives wave lengths $(2\pi/q) \sim 600$ Å, which are still large in comparison with the molecular length a. In this situation, a continuum description is not unreasonable; it has been carried out by Pincus [73], and improved by Doane and Johnson [74] and Lubensky [75]. The main result is that the spin lattice relaxation rate $1/T_1$ is strongly dependent on frequency

$$1/T_1 = A + B\,\omega_n^{1/2}$$

the coefficient B being nearly independent of temperature. Frequency dependences of $1/T_1$ close to a one-half power law have indeed been observed, although some complicating effects are often present [76].

3.5 CONCLUDING REMARKS

1. The static and dynamic continuum theories give a reasonable description of liquid crystal phases at long wavelengths and low frequencies. However, even for the simplest case of nematics, measurements have been too rare, and we have very few checks on the internal consistency of the theory (showing, for instance, that one elastic constant measured by different methods always has the same value).

On the other hand, there are some recent, spectacular successes of the continuum theory, which were purposely omitted from our review, since they will be covered in other chapters of this book. For instance, the Helfrich theory of instabilities, under electric fields [77, 78], describes these instabilities remarkably well (when complications due to carrier injection are suitably removed).

2. To make the discussion more complete, we should also have included the properties of nematics and cholesterics *above the clearing point* T_c, where remarkable short range order effects are observed. Since just above T_c, the "coherence length", $\xi(T)$, is still much larger than a, a continuum description is still useful. The language, however, is quite different [34], since both the magnitude and the direction of the optical anisotropy fluctuate widely.

3. Ultimately we would like to interpret the data on elastic constants, friction coefficients, etc., in terms of molecular properties of the liquid crystal, but this will represent a very long program, and only a few serious attempts in this direction have been made [75, 68].

Acknowledgments

The two Orsay contributors (P.G.G. and M.K.) have greatly benefited from discussions with other members of the Orsay Group, and also with R. B. Meyer, J. Fergason and S. Candau.

REFERENCES

[1] Reinitzer, F. *Mh. Chem.* **9,** 421 (1888).
[2] Friedel, G. *Annls Phys.* **18,** 273 (1922).
[3] Zocher, H. and Birstein, V. *Z. phys. Chem.* **A142,** 113 (1929).
[4] Bose, E. *Phys. Z.* **10,** 32 and 230 (1909).
[5] Ornstein, L. S. and Zernicke, F. *Proc. Acad. Sci. Amersterdam* **21,** 115 (1917).
[6] Ornstein, L. S. and Zernicke, F. *Phys. Z.* **19,** 134 (1918).
[7] Ornstein, L. S. *Z. Kristallogr. Kristallgeom.* **79,** 90 (1931).

[8] ORNSTEIN, L. S. *Kolloidzeitschrift* **69,** 137 (1934).
[9] ORNSTEIN, L. S. and KAST, W. *Trans. Faraday Soc.* **29,** 931 (1933).
[10] VAN WYK, A. *Annln Phys.* **3,** 879 (1929).
[11] FÜRTH, R. and SITTE, K. *Annln. Phys* **30,** 388 (1937).
[12] TROPPER, H. *Annln Phys.* **30,** 371 (1937).
[13] LEHMANN, O. *Annln Phys.* **12,** 311 (1903).
[14] ZOCHER, H. and UNGAR, G. *Z. phys. Chem.* **110,** 529 (1938).
[15] BROCHARD, F. and DE GENNES, P. G. *J. Phys., Paris*, to be published.
[16] ZOCHER, H. *Trans. Faraday Soc.* **29,** 915 (1933).
[17] OSEEN, C. W. *Trans. Faraday Soc.* **29,** 883 (1933).
[18] FRANK, F. C. *Discuss. Faraday Soc.* **25,** 19 (1958).
[19] ERICKSEN, J. L. *Archs ration. Mech. Analysis* **10,** 189 (1962).
[20] MEYER, R. B. *Phys. Rev. Lett.* **22,** 918 (1969).
[21] ORSAY GROUP ON LIQUID CRYSTALS in *Liquid Crystals and Ordered Fluids* (edited by J. F. Johnson and R. S. Porter) Plenum Press, New York (1970), p. 195.
[22] FREDERIKS, V. and ZWETKOFF, V. *Phys. J. Soviet Union* **6,** 490 (1934).
[23] ZWETKOFF, V. *Acta phys.-chim. URSS* **6,** 865 (1937).
[24] SAUPE, A. *Z. Naturf.* **15a,** 815 (1960).
[25] ERICKSEN, J. L. *Archs ration. Mech. Analysis* **4,** 231 (1960); *Archs ration. Mech. Analysis* **9,** 371 (1962).
[26] ERICKSEN, J. L. *Trans. Soc. Rheol.* **11,** 5 (1967).
[27] DAFERMOS, C. M. *J. Appl. Math. (SIAM)* **16,** 1305 (1968).
[28] LESLIE, F. M. *J. Phys. D* **3,** 889 (1970).
[29] PAPOULAR, M. and RAPINI, A. *J. Phys. Paris* **30,** C4, No. 11–12, 54 (1969).
[30] DE GENNES, P. G. *Solid St. Commun.* **8,** 213 (1970).
[31] CHATELAIN, P. *Acta Crystallogr.* **1,** 315 (1948).
[32] DE GENNES, P. G. *C. r. hebd. Séanc. Acad. Sci., Paris* **B266,** 15 (1968); *Molec. Crystals Liqu. Crystals* **7,** 325 (1969).
[33] MARTINAND, J. L. and DURAND, G. *Solid St. Commun.* **10,** 815 (1972).
[34] DE GENNES, P. G. *Molec. Crystals Liqu. Crystals* **12,** 193 (1971).
[35] JENKINS, J., Doctoral Thesis, Johns Hopkins University (1969).
[36] MEYER, R. B. *Appl. Phys. Lett.* **12,** 281 (1968); Thesis, Harvard University (1969).
[37] DE GENNES, P. G. *Solid St. Commun.* **6,** 163 (1968); *Molec. Crystals* **7,** 325 (1969).
[38] BAESSLER, H. and LABES, M. *Phys. Rev. Lett.* **21,** 1791 (1968).
ORSAY GROUP ON LIQUID CRYSTALS, *Phys. Rev. Lett.* **22,** 227 (1969).
BAESSLER, H. and LABES, M. *J. chem. Phys.* **51,** 1846, 3213, 5397 (1969).
LUCKHURST, G. R. and SMITH, H. J. *Molec. Crystals Liqu. Crystals* **20,** 319 (1973).
[39] PINCUS, P. *C. r. hebd. Séanc. Acad. Sci., Paris* **B267,** 1290 (1968).
[40] DE GENNES, P. G. *J. Phys., Paris* **30,** C4, No. 11–12, 65 (1969).
[41] SAUPE, A. *Molec. Crystals Liqu. Crystals* **7,** 59 (1969).
[42] ORSAY GROUP ON LIQUID CRYSTALS, *Solid St. Commun.* **9,** 653 (1971).
[43] FRIEDEL, G. *Annls Phys.* **18,** 273 (1922).

[44] See for example Nabarro, F. R. N. *Theory of Crystal Dislocations*, Oxford University Press (1967).
[45] Friedel, J. *Dislocations*, Pergamon Press, (1964).
[46] Meyer, R. B., private communication.
[47] Ericksen, J. L. in *Liquid Crystals and Ordered Fluids* (edited by J. F. Johnson and R. S. Porter) Plenum Press, New York (1970), p. 181.
[48] Friedel, J. and de Gennes, P. G. *C. r. hebd. Séanc. Acad. Sci., Paris* **B268,** 257 (1969).
[49] Dubois-Violette, E. and Parodi, O. *J. Phys., Paris* **30,** C4, No. 11–12, 57 (1969).
[50] Meyer, R. B., Proc. 3rd Int. Conf. on Liquid Crystals, Berlin (1970), Abstract 7–14.
[51] Kléman, M. and Friedel, J. *J. Phys., Paris* **30,** C4, No. 11–12, 43 (1969).
Friedel, J. and Kléman, M., Conf. on the Fundamental Aspects of the Theory of Dislocations, Gaithersburg (Maryland) (1969).
[52] Robinson, C., Ward, J. C. and Beevers, R. B. *Discuss. Faraday Soc.* **25,** 29 (1958).
Robinson, C. *Molec. Crystals* **1,** 467 (1966).
[53] de Gennes, P. G. *C. r. hebd. Séanc. Acad. Sci., Paris* **B266,** 571 (1968).
[54] Caroli, C. and Dubois-Violette, E. *Solid St. Commun.* **7,** 799 (1969).
[55] Orsay Group on Liquid Crystals, *J. Phys., Paris* **30,** C4, No. 11–12, 38 (1969).
[56] Bouligand, Y. and Kléman, M. *J. Phys., Paris* **31,** 1041 (1970).
[57] Nye, J. F. *Acta Metall.* **1,** 153 (1953).
[58] Sackmann, H. and Demus, D. *Molec. Crystals* **2,** 81 (1966).
[59] Taylor, T. R., Fergason, J. L. and Arora, S. L. *Phys. Rev. Lett.* **24,** 359 (1970).
[60] Leslie, F. M. *Archs ration. Mech. Analysis* **28,** 265 (1968).
[61] Parodi, O. *J. Phys., Paris* **31,** 581 (1970).
[62] See for instance Landau, L. D. and Lifshits, I. M., *Theory of Elasticity*, Pergamon Press, London (1959), Chap. I.
[63] Miesowicz, M. *Bull. int. Acad. pol. Sci. Lett.* **A,** 228 (1936); *Nature*, **158,** 27 (1946).
[64] A general discussion on the empirical aspects of these flows can be found in Porter, R. S., Barrall, E. M. and Johnson, J. F. *J. chem. Phys.* **45,** 1452 (1966).
[65] Ericksen, J. L. *Trans. Soc. Rheol.* **13,** 9 (1969).
[66] Atkin, R. J. and Leslie, F. M. *Q. Jl Mech. appl. Math.* **23,** 53 (1970).
[67] Currie, P. K. *Archs ration. Mech. Analysis*, **37,** 222 (1970).
[68] de Gennes, P. G. *Phys. Lett.* **41A**, 479 (1972).
[69] Rapini, A. and de Gennes, P. G., hitherto unpublished work.
[70] Candau, S. and Martinoty, P. *Molec. Crystals Liqu. Crystals* **14,** 243 (1971).
Martinoty, P., Thèse 3e cycle, Strasbourg (1970).

[71] ORSAY GROUP ON LIQUID CRYSTALS, *J. chem. Phys.* **51,** 816 (1969). See also *Liquid Crystals and Ordered Fluids* (edited by J. F. Johnson and R. S. Porter) Plenum Press, New York (1968), p. 195.
[72] ORSAY GROUP ON LIQUID CRYSTALS, *Phys. Rev. Lett.* **22,** 1361 (1969); *Molec. Crystals Liqu. Crystals* **13,** 187 (1971).
[73] PINCUS, P. *Solid St. Commun.* **7,** 415 (1969). See also Blinc, R. *et al. Phys. Rev. Lett.* **23,** 969 (1969).
[74] DOANE, J. W. and JOHNSON, D. L. *Chem. Phys. Lett.* **6,** 291 (1970).
[75] LUBENSKY, T. *Phys. Rev.* **A2,** 2497 (1970).
[76] WEGER, M. and CABANE, B. *J. Phys., Paris* **30,** C4 No. 11–12, 72, (1969).
DOANE, J. W. and VISINTAINER, J. J. *Phys. Rev. Lett.* **23,** 1421 (1969).
DOANE, J. W., FISHEL, D. and VISINTAINER, J. J., *Molec. Crystals Liqu. Crystals* **13,** 69 (1971).
[77] HELFRICH, W. *J. chem. Phys.* **51,** 4092 (1969).
[78] ORSAY GROUP ON LIQUID CRYSTALS, *Molec. Crystals Liqu. Crystals* **12,** 251 (1971).
DUBOIS-VIOLETTE, E., DE GENNES, P. G. and PARODI, O. *J. Phys. Paris*, **32,** 305 (1971).
[79] RONDELEZ, F. and HULIN, J. P. *Solid St. Commun.* **10,** 1009 (1972).
[80] LIAO, Y., CLARK, N. A. and PERSHAN, P. S. *Phys. Rev. Lett.* **30,** 639 (1973); MARTIN, P. C., PARODI, O. and PERSHAN, P. S. *Phys. Rev.* **A6,** 2401 (1972).
[81] FISHER, F. and WAHL, J. *Optics Commun.* **5,** 341 (1972).

4

Influence of Composition and Structure on the Liquid Crystals formed by Non-Amphiphilic Systems

4.1 INFLUENCE OF MOLECULAR STRUCTURE ON THE LIQUID CRYSTALS FORMED BY SINGLE COMPONENT NON-AMPHIPHILIC SYSTEMS

G. W. Gray

This section reviews the way in which the structures of certain lath-like molecules influence the properties of liquid crystals produced either by heating pure crystalline non-amphiphilic compounds, i.e., enantiotropic liquid crystals, or by supercooling the amorphous isotropic liquids or the enantiotropic liquid crystals formed by heating such crystalline compounds, i.e., monotropic liquid crystals. The liquid crystals formed under these circumstances have been classified broadly as smectic, nematic and cholesteric, and for all three classes evidence exists to show that they possess a degree of molecular organization intermediate between the three dimensional order of the crystal and the comparative disorder of the amorphous isotropic liquid (see Chap. 1 and 2.1).

Evidence suggesting that a cholesteric liquid crystal is a modified nematic liquid crystal was first obtained by Friedel [1], and this idea was later substantiated by Robinson's studies [2, 3] of polypeptide solutions exhibiting liquid crystalline behaviour. More recent studies [4–9] leave no doubt that this is the case and show that in a cholesteric, a helical twist is imposed on a molecular organization which is essentially nematic in type. This condition arises when (*a*) the molecules are disposed to arrange themselves in a nematic-type order, and (*b*) the compound is optically active. Therefore, whereas cholesterics and nematics are closely related, smectics are quite different. For the purposes of this section, we can therefore think broadly in terms of only two classes of liquid crystal—smectic and nematic/cholesteric.

On changing the molecular structure of a compound, many molecular parameters may be affected, e.g., the size and shape of the molecule, the dipolarity and polarizability of the molecule. The manner in which the molecules pack together in an ordered manner and the thermal stability of the ordered arrangement may therefore be changed appreciably by a structural alteration. Even though the change in molecular structure is quite small, the effects on liquid crystal behaviour may be far reaching and the reasons may be subtle to define.

Thus, the *racemic* system corresponding to structure (I) is purely nematogenic [8, 9], but if we replace the terminal $-NO_2$ group by $-NMe_2$, the racemic system (II) is now purely smectogenic [9]. Replacement of a single terminal substituent by another has therefore changed the liquid crystal behaviour of the system very markedly.

I $O_2N-C_6H_4-CH{=}N-C_6H_4-CH{=}CH\cdot CO\cdot O\cdot CH_2\cdot \overset{*}{C}HMe\cdot CH_2Me$

II $Me_2N-C_6H_4-CH{=}N-C_6H_4-CH{=}CH\cdot CO\cdot O\cdot CH_2\cdot \overset{*}{C}HMe\cdot CH_2Me$

The compounds (I) and (II) contain centres of asymmetry marked by an asterisk, and of course, the pure *optically active* isomers corresponding to (I) and (II) are cholesterogenic and smectogenic, respectively [8, 9].

The large effects of relatively small structural changes are also illustrated by reference to compounds (III) and (IV) [10]. Movement of the 3′-bromo-substituent in compound (III) to the 3-position giving compound (IV) changes the purely smectogenic compound to a purely nematogenic compound, with a marked change in the liquid crystal—amorphous isotropic liquid transition temperature.

III (3-Br)$C_6H_4-C_6H_4-N{=}CH-C_6H_4-OC_{10}H_{21}$-n

IV $C_6H_5-C_6H_3$(Br)$-N{=}CH-C_6H_4-OC_{10}H_{21}$-n

When compounds with widely different molecular structures are considered, even greater problems are encountered in interpreting the effects of the structural differences on the liquid crystal properties. The tables of liquid crystal systems compiled by Kast and published in Landolt–Börnstein [11] emphasize the extent of this problem, and the following examples* have been chosen to illustrate the diversity of structure which

* In this list, the first two examples represent amphiphiles. Although this section is concerned with non-amphiphilic compounds, the two structures of amphiphiles have been included to give the reader a general impression of the structural variations that may be encountered over a range of mesogenic compounds.

may be encountered amongst compounds reported in the literature to exhibit liquid crystal behaviour.

Thallium (I) decanoate [12] $Me(CH_2)_8 \cdot CO_2^- Tl^+$

N-n-Dodecylpyridinium chloride [13]

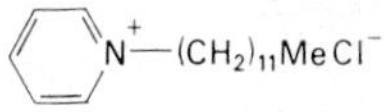

2-*p*-Methoxybenzylideneamino-phenanthrene [14]

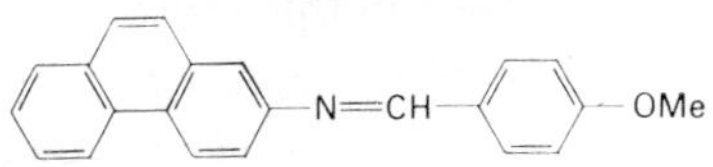

Cholesteryl chloride [15]

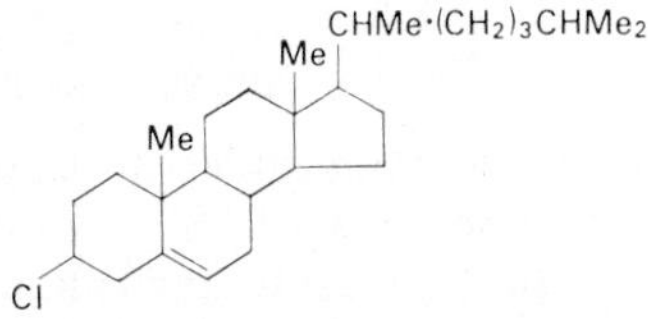

p-Pentaphenyl [16]

2,5-Di-*p*-methoxybenzylidenecyclo-pentanone [17, 18]

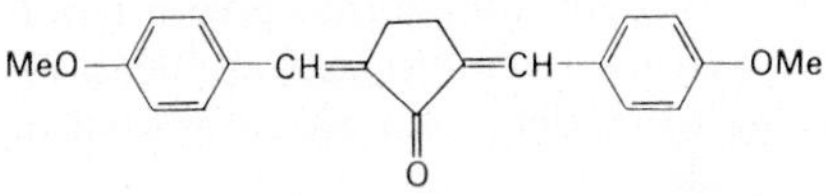

Bis-(4-*p*-methylbenzylideneamino-phenyl) mercury (II) [19]

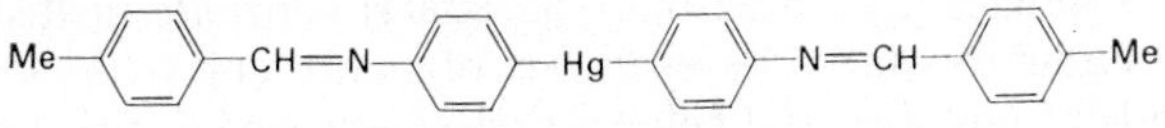

Ethyl *p*-azoxybenzoate [20]

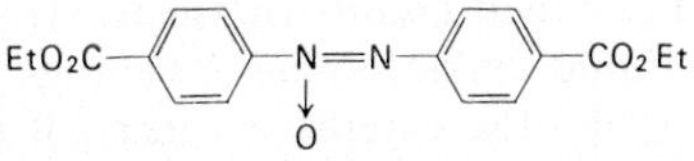

Systematic studies of a number of types of mesogenic compound have however established that certain systematic trends in liquid crystal behaviour accompany particular changes in molecular structure, and this suggests that it would be profitable to discuss three important topics.

(1) The features of molecular structure which are conducive to the formation of liquid crystals in general.

(2) The way in which changes in molecular structure affect the thermal stabilities of liquid crystals, i.e., the temperatures at which liquid crystals change to the amorphous isotropic liquid or at which smectic liquid crystals change to nematic/cholesteric liquid crystals.

(3) The features of molecular structure which favour the occurrence of smectic or nematic/cholesteric mesophases.

The subject of molecular structure and the properties of liquid crystals was reviewed in a monograph [21] published in 1962, and it is the writer's intention in this section to bring the information on this subject up to date and to present it in a more concise form. Throughout this section, it will be necessary to make reference to the transitions which occur between crystalline and liquid crystalline states, between different liquid crystalline states, and between liquid crystalline states and the amorphous isotropic liquid state. It is now convenient to define the following abbreviations, proposed [22] at the First International Conference on Liquid Crystals in 1965, that are used in this section:

C — crystal;
S — smectic liquid crystal;
N — nematic liquid crystal;
Ch — cholesteric liquid crystal;
I — amorphous isotropic liquid.

Transitions between different states are therefore indicated in the following manner, e.g., C–S: crystal to smectic liquid crystal, Ch–I: cholesteric liquid crystal to amorphous isotropic liquid.

It should be emphasized that it is not the author's intention to review factors which may influence the temperatures at which mesophases are formed from crystalline solids. This is not a subject about which much rationale yet exists, although it is of importance with regard to the availability of compounds that give mesophases in the room temperature range for various technological applications. A recent review [140] may be of value to readers who are interested in this subject.

Features of Molecular Structure Conducive to Formation of Liquid Crystals

In 1888, Reinitzer [23] noted that cholesteryl benzoate melted to give a cholesteric liquid crystal. The synthesis of many compounds exhibiting various kinds of liquid crystal behaviour was achieved during the next few years, and much credit for this early work that extended the number of known liquid crystal systems belongs to D. Vorländer and O. Lehmann. As shown by the extensive nature of the tables of compounds that form liquid crystals compiled by Kast [11], a large number of such systems was known by 1960, and since then the number and variety of compounds that form liquid crystals have grown steadily.

As already mentioned, mesogens embrace a diversity of structure, but with the non-amphiphilic compounds under discussion the molecules do share a common feature in being markedly geometrically anisotropic, i.e., the molecules are elongated and lath-like in shape. This is understandable, because the production of a liquid crystal requires that the crystal breaks down in stages on heating. Since the molecules are highly geometrically anisotropic, there may be a pronounced anisotropy of the forces maintaining order in the crystal, and, on heating, these forces may weaken more readily in certain planes or directions than in others. A stage-wise breakdown of the crystal would then occur giving states, i.e., liquid crystalline states, with degrees of order intermediate between the three dimensional order of the crystal and the disorder of the amorphous isotropic liquid.

Bernal and Crowfoot [24] examined several compounds which gave liquid crystals on heating, and by X-ray crystallography they established that the molecules adopted the expected packing in the crystals, with the long axes parallel. As discussed in earlier chapters, this parallel arrangement persists statistically in smectic and nematic liquid crystals.* The stage-wise melting process may therefore be represented simply as shown in Fig. 4.1.1, starting with one of two possible crystal lattices comprised of parallel, lath-like molecules, i.e., a layer crystal lattice or a non-layer crystal lattice.

Consider first the layer lattice. If the forces holding the planes of the layers together in the rigid crystal are weak relative to the attractions between molecules within a given layer, then at T_1, the layers may become free to slide (and rotate) over one another, giving a smectic liquid crystal. Although the layers are preserved, the side to side arrangement of the molecules within the layers may be affected to a greater or lesser extent, and the degree to which the order is changed is responsible, at least in some cases, for the existence of different polymorphic forms of the smectic state.

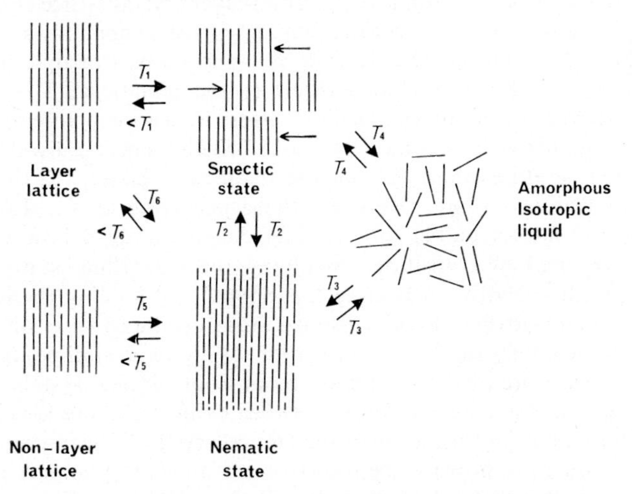

FIG. 4.1.1. Diagrammatic representation of the arrangements of lath-like molecules in various states of matter.

At least seven such polymorphic forms (A, B, C, D, E, F, and G) have been studied by Sackmann, Arnold and Demus [25–33, 131] and Fergason [132]. In the smectic A and B phases, the molecules lie with their long axes statistically perpendicular to the planes of the layers, but evidence now exists for one smectic B phase in which the molecules are tilted at an angle less than 90° to the planes of the layers [138]. This tilted smectic B phase gives rise to a smectic C on heating or is formed from a smectic C on cooling, and it has been shown that in smectic C phases the long axes of the molecules are again inclined to the layer planes at angles less than 90°

* In the cholesteric mesophase, the statistical parallelism of neighbouring molecules is somewhat deviated either to left or right according to the nature of the asymmetry ((+) or (−)) of the molecules.

[34–37]. More recently it has been observed that for some compounds, smectic C forms may exhibit nematic morphology, and it has been suggested that this may arise when there is an unusually large angle of tilt [119].

A fuller discussion of the molecular organization in smectic A–D phases is given in Chapter 2.1, but the other polymorphic forms are in general less well understood; quite recently de Vries [137] has proposed an eighth, possible smectic polymorphic mesophase—smectic H. Evidence that more than one type of nematic liquid crystal may exist has also been presented by de Vries [120].

On heating the smectic to a higher temperature, T_2, the intermolecular forces between the sides of the molecules may weaken sufficiently for the molecules to slide out of the layers. This gives the nematic in which the statistically parallel orientation is maintained, without alignment of the ends of the molecules. The nematic may then pass to the amorphous isotropic liquid at a higher temperature, T_3.

Alternatively, for a different system, the smectic may change directly to the amorphous isotropic liquid at T_4. A non-layer crystal lattice (Fig. 4.1.1) cannot give a smectic on heating, but may give a nematic at say T_5; further heating would eventually give the amorphous isotropic liquid. A layer lattice could however change direct to the nematic at T_6 through a suitable weakening of intermolecular forces. When two different liquid crystals occur for a compound, the more ordered smectic always occurs at lower temperatures than the nematic/cholesteric phase. All the above-mentioned transitions are reversible, those occurring at T_2, T_3 and T_4 doing so without detectable supercooling. Supercooling is however common for reversal of transitions involving the crystalline state. Liquid crystals produced both on heating the crystal and cooling the isotropic liquid are enantiotropic liquid crystals; those produced only on cooling the supercooled melts (amorphous isotropic or liquid crystalline) below the melting temperature of the crystal are monotropic liquid crystals.

Anisotropy of the cohesive forces between elongated molecules is therefore a requirement for formation of such liquid crystals. The intermolecular attractions must not however be too strong. If they are, the melting point of the crystals may be so high that thermal motion prevents the existence of an ordered liquid crystal after melting occurs. If the intermolecular forces are very weak, the same result may arise; although the crystals may melt at a low temperature, the cohesive forces will again be inadequate to maintain order in the fluid state. In both instances, normal melting will occur. For a liquid crystal to be formed, the cohesive forces operating between elongated molecules must be both anisotropic and of suitable magnitude.

It is necessary to recognize a further criterion for the formation of liquid crystals, and this relates to the rigidity of the elongated molecule. If the molecule lacks rigidity, flexing may occur along its length and this may in itself prevent liquid crystal formation. Thus, a parallel arrangement of elongated molecules may break down through one of two effects, or a combination of both. The system may become so thermally disturbed that the molecules move out of alignment but remain individually lath-like in shape, or the elongated molecules may bend and flex thereby destroying the possibility of parallel arrangement. If the molecules resist bending and flexing, i.e., if the molecules or appreciable parts of the molecules have lath- or rod-like properties, then breakdown of the order through this second effect will be less likely. Examples of this can now be quoted.

Long chain n-alkanes can adopt elongated conformations, but the flexible alkyl chains may coil and bend, and no liquid crystals are formed by these hydrocarbons. Neither are liquid crystals given by the long chain n-alkanoic acids (V); although a short, rod-like central unit now exists through hydrogen bonding between carboxyl groups, a high proportion of the dimeric molecule consists of potentially flexible alkyl chains.

$$Me(CH_2)_x—C\begin{matrix}=O\cdots H—O\\ —O—H\cdots O=\end{matrix}C—(CH_2)_xMe$$

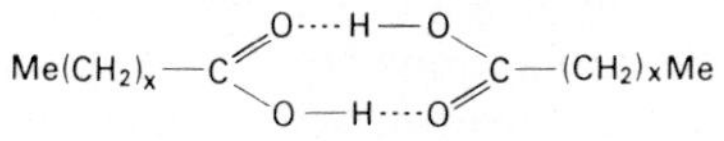

V

In both examples, the intermolecular forces will be quite weak, since the molecules are neither highly dipolar nor readily polarizable; the low melting points of the n-alkanes and n-alkanoic acids confirm that this is the case. Consequently, no strong constraints will be imposed on the potentially flexible alkyl chains to maintain these in an extended conformation, and the systems do not give liquid crystals.

However, if we decrease the flexibility of the alkanoic acid molecule by introducing a pair of conjugated double bonds, we may obtain the alkane-2,4-dienoic acids (VI). With suitable alkyl groups R, these acids form liquid crystals (nematic), the simplest being that in which R is ethyl [38]. Because of the sp^2 hybridized bonds of the conjugated parts of the dimer, rotation

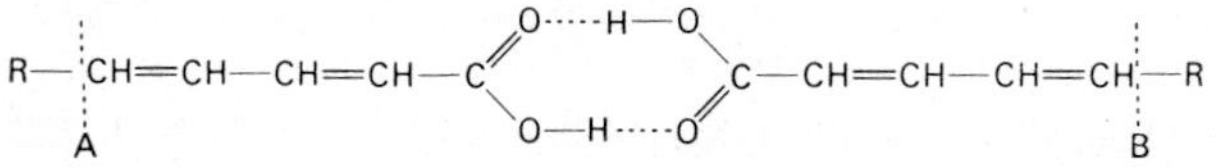

$$R—\underset{A}{\vdots}CH=CH—CH=CH—C\begin{matrix}=O\cdots H—O\\ —O—H\cdots O=\end{matrix}C—CH=CH—CH=CH\underset{B}{\vdots}—R$$

VI

and bending will be restricted, and between A and B there is a relatively rigid core, which for R = ethyl is an extensive part of the molecule. In addition, the conjugated molecule will be more polarizable than the saturated system, and stronger intermolecular forces will operate. A combination of these factors explains the formation of liquid crystals by these acids.

For such reasons, liquid crystals are not commonly given by purely aliphatic, non-amphiphilic compounds. One other example is provided by vinyl oleate [39–41] which again contains two double bonds in the molecule, although these are not in conjugation with one another; liquid crystalline behaviour occurs only at low temperatures. In the case of amphiphilic salts of n-alkanoic acids, e.g., thallium (I) decanoate [12] (p. 105), strong intermolecular attractions must stem from the polar groups, for the melting points are quite high. Many of these salts are mesogenic [11], and similar effects may operate for long chain quaternary salts of pyridine, e.g., *N*-n-dodecylpyridinium chloride [13].

The majority of substances that give liquid crystals on heating are however aromatic. Aromatic nuclei are polarizable, planar, and rigid, and if suitable substituents are correctly positioned in the aromatic rings, we obtain elongated molecules between which reasonably strong intermolecular forces should operate. For benzene nuclei, substituents must occupy *p*-positions and be of such a kind that they link up at least one other benzene ring which also carries a *p*-substituent. It is preferable if the central group which joins the rings is itself of a rigid nature, i.e., the two rings and the central group should constitute a lath- or rod-like core of the molecule. Central groups usually contain a multiple bond or a system of conjugated double bonds, or involve a ring formed by dimerization of carboxyl groups, as in the case of *p*-substituted benzoic acids and cinnamic acids. Structures VII, VIII and IX give typical examples—the significance of terminal substituents will be discussed later.

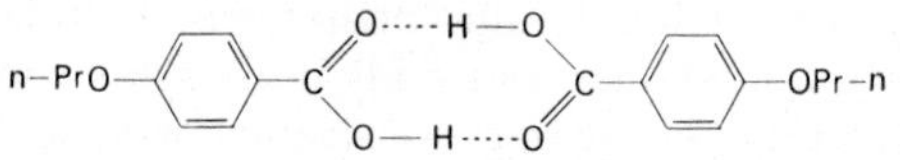

O·····H—O
n–PrO—C$_6$H$_4$—C　　C—C$_6$H$_4$—OPr–n　　VII
O—H·····O

p-n-propyloxybenzoic acid (C–N, 145°C; N–I, 154°C) [42]

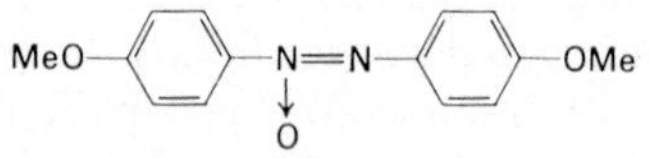

MeO—C$_6$H$_4$—N=N—C$_6$H$_4$—OMe　　VIII
↓
O

p-azoxyanisole (C–N, 118°C; N–I, 136°C) [11]

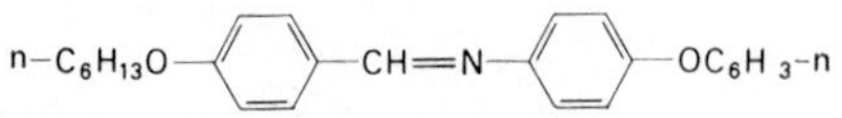

n–C_6H_{13}O—C$_6$H$_4$—CH=N—C$_6$H$_4$—OC$_6$H $_3$–n　　IX

p'-n-hexyloxybenzylidene-*p*-n-hexyloxyaniline (C–N, 97·5°C; N–I, 122·5°C) [11]

These and many other compounds which form liquid crystals fall into the general category represented below (X), where the rectangle represents the central group which should ideally extend the rigidity and linearity associated with the *p*-disubstituted benzene rings.

X—C$_6$H$_4$—▭—C$_6$H$_4$—Y

X

Only three of the following examples of central groups do not link the rings through at least one multiple bond or a hydrogen bonded ring system. In these cases (—CO·O—, —OCH_2·CH_2O— and —CH_2·CH_2—), the stereochemistry of the units does however preserve the linearity of the molecule, but in a simple system such as (X), with appropriate substituents

X and Y, these flexible links usually result in less thermally stable mesophases, the liquid crystals frequently being monotropic.

—CH=CH—; —(CH=CH)$_{2-3}$; —C≡C—; —CH=N—; —N=N—;

—N(→O)=N—; —CH=CH—C(=O)—O—; —CH=N(→O)—*;

—C(=O)—O— —(CH=CH)$_2$—C(=O)—(CH=CH)$_2$—

—CH=N—N=CH—; —CH$_2$—CH$_2$—; —O—CH$_2$—CH$_2$—O—;

(—C(=O)O—H···)$_2$; (—CH=CH—C(=O)O—H···)$_2$

When more than two benzene rings are linked (XI) through more than one central group, the liquid crystalline properties are enhanced most when all the rings are conjugated, i.e., the liquid crystal transition temperatures are highest when the entire system is linked through central groups involving multiple bonds. If at least one of the central groups is of this nature,

X—C$_6$H$_4$—(□—C$_6$H$_4$)$_n$—□—C$_6$H$_4$—Y

XI

e.g., —CH=CH— or —CH=N—, a wider range of units (see below) of the more flexible variety or of such a kind that conjugation is less extended in the system, may be found as additional central groups in mesogenic compounds. Such units again usually result in less thermally stable liquid crystals than those which occur if all the central groups are of a kind which permit conjugative interactions.

—O—CH$_2$—; —C—(CH$_2$)$_3$—O—; —(CH$_2$)$_4$—; —CH$_2$—NH—; —NH—;

—NH—(CH$_2$)$_{2-3}$NH—; —C(=O)—; —C(=O)—C(=O)—; —O—C(=O)—O—;

—C(=O)—NH—; —C(=O)—(CH$_2$)$_{1-8}$C(=O)—; —(CH$_2$)$_{1-4}$C(=O)—O—;

—C(=O)—O—(CH$_2$)$_2$—O—C(=O)—

For further exemplification of central groups, the reader is referred to articles by Brown and Shaw [43] and Brown [44], and to examples given by Kast [11].

* Aldonitrones have been shown to exhibit liquid crystalline properties [122].

The linkage of benzene rings through the *o*- or *m*-positions is not favourable to liquid crystal formation, and the *o*- or *m*-substituted analogues of (VII), (VIII) and (IX) do not form liquid crystals. However, additional *o*- or *m*-substituents may be introduced into already *p*-substituted systems. An example is afforded by the mesogenic acid (XII), but normally such lateral substitution diminishes the liquid crystal thermal stabilities.

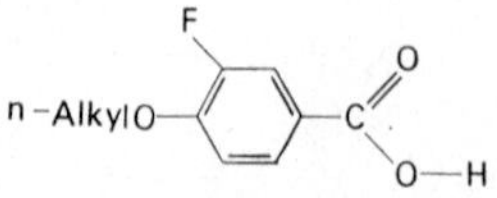

F
n-AlkylO—C6H3—C(=O)—O—H

(dimer) 3-fluoro-4-n-alkoxybenzoic acid [45]

XII

As discussed above, the more extended the conjugated, rigid, core part of the molecule, the more thermally stable the liquid crystals will be. The N–I transition temperatures for compounds (XIII) and (XIV) [11] illustrate this point.

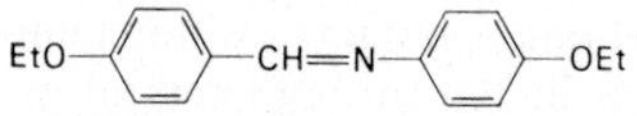

EtO—C6H4—CH=N—C6H4—OEt

C–I, 148°C; (I–N, 143°C)*

XIII

EtO—C6H4—CH=N—N=CH—C6H4—OEt

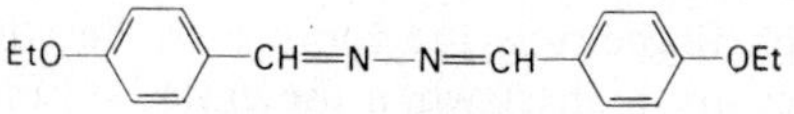

C–N, 172°C; N–I, 199°C

XIV

A way further to extend the linearity and rigidity of the system is to link up benzene rings directly. 4-Substituted and 4,4′-disubstituted biphenyls provide a rich source of liquid crystals which are more thermally stable than those of the benzene analogues, cf., the N–I temperatures for compounds (XV) and (XVI) [11].

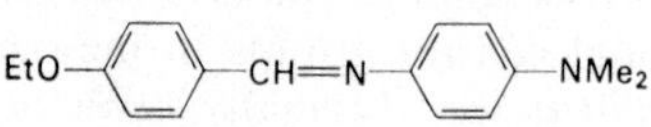

EtO—C6H4—CH=N—C6H4—NMe2

C–I, 140°C; (I–N, 102°C)*

XV

EtO—C6H4—CH=N—C6H4—C6H4—NMe2

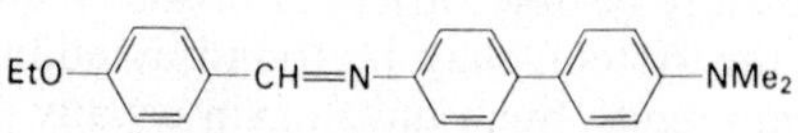

C–N, 208°C; N–I, 274°C

XVI

With biphenyl derivatives, the tendencies to form liquid crystals are so marked that derivatives with simpler 4,4′-substituents give liquid crystals, e.g., compound (XVII) [46], and also bulky lateral substituents may be introduced without destroying liquid crystal formation, e.g., compound (XVIII) [47]; see also [141, 142].

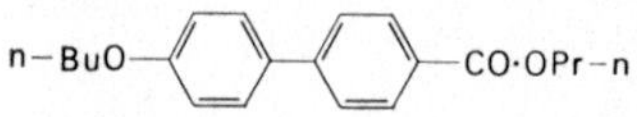

n-BuO—C6H4—C6H4—CO·OPr-n

C–S, 97°C; S–I, 113·5°C

XVII

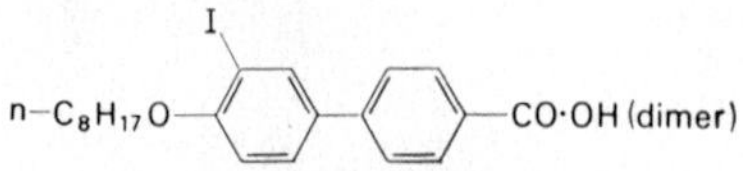

I
n-C_8H_{17}O—C6H3—C6H4—CO·OH (dimer)

C–S, 180·5°C; S–N, 192·5°C;
N–I, 214°C

XVIII

* The convention is adopted of writing monotropic transition temperatures in parenthesis.

With derivatives of *p*-terphenyl and *p*-quaterphenyl, liquid crystals of high thermal stability are found with simple end groups, e.g., compounds (XIX) and (XX) [48] and the unsubstituted acid (XXI) [49].

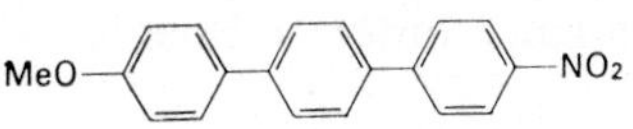

C–N, 210–212°C; N–I, 259°C

XIX

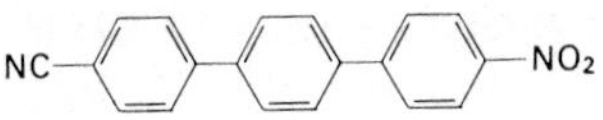

C–N, 232°C; N–I, 275°C

XX

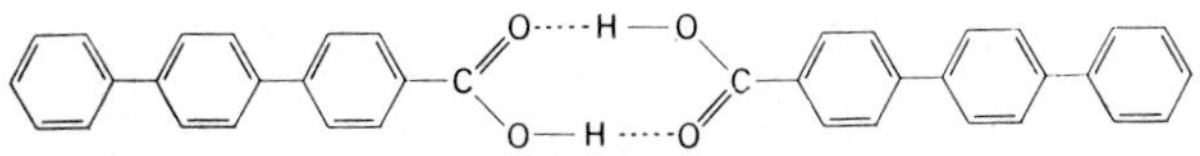

C–N, 306–309°C; N–I, 331°C

XXI

A problem is that such rigid molecules tend to pack efficiently and the compound may have such a high melting point that no liquid crystals are formed; melting points of compounds are perhaps the most unpredictable factor connected with assessing the potential of a new system to form liquid crystals. However, despite its very high melting point (401°C) and the absence of terminal substituents, the hydrocarbon *p*-pentaphenyl (p. 105) [16] gives a liquid crystal (N–I, 445°C).

Other ring systems may of course be employed, provided that molecular linearity is preserved. Examples are 1,4-, 1,5- and 2,6-disubstituted naphthalenes [50–52], 2-substituted and 2,7-disubstituted fluorenes and fluorenones [53, 54] and 2-substituted phenanthrenes [14]. The ring systems may be heterocyclic [143] as in compounds (XXII) [55] and (XXIII) [56], or alicyclic as in the derivative of cyclopentanone (p. 105) [17, 18] and in related cyclohexanones, and in the important derivatives of cholesterol, e.g., cholesteryl chloride (p. 105) [15].

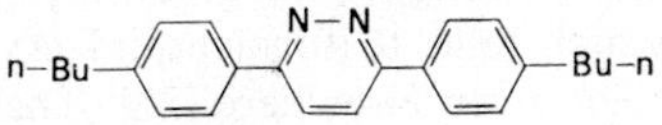

XXII

C–S, 207°C; S–N, 213°C; N–I, 220°C

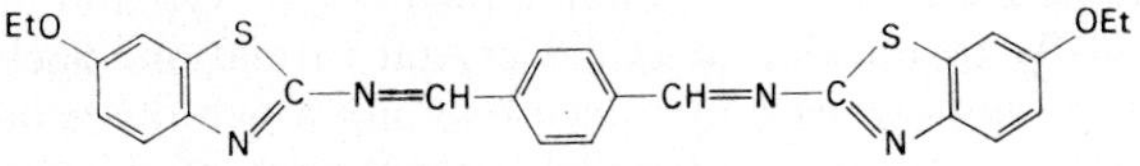

XXIII

C–N, 226°C; N–I, >325°C

However, Dewar and Goldberg [117] have shown that replacement of aromatic rings by saturated alicyclic rings gives reductions, often marked, in the nematic thermal stabilities of esters of the type (XXIV). When phenylene ring B is replaced by the *trans*-1,4-cyclohexylene ring (XXIV(*a*)), then N–I temperatures are lowered despite the collinearity of the ring–O links in the favoured diequatorial chair conformation. This emphasizes the importance of the rigid nature of the linear geometry of 1,4-substituted phenylene rings in stabilizing such liquid crystals. Replacement of one, two or all of the aromatic rings in (XXIV) by 1,4-bicyclo-[2,2,2]octylene rings (XXIV(*b*)) also gives much lower N–I temperatures,

AlkylO—(A)—CO·O—(B)—O·OC—(C)—OAlkyl

XXIV

XXIV(*a*)

XXIV(*b*)

particularly when rings A and/or C are involved. The bicyclo-octylene ring, though thicker, has the same width as a phenylene ring and is much more rigid than the cyclohexylene ring; again the ring–O bonds are collinear. Thus, phenylene rings play more than a purely geometrical rôle in stabilizing these nematic liquid crystals. It seems that the π-electron system of aromatic rings, being polarizable and permitting conjugative interactions in the elongated molecule, must give rise to suitably strong intermolecular attractions which are anisotropic and stabilize the ordered arrangement of molecules.

These results emphasize the particular suitability for liquid crystal formation of aromatic and presumably also of heteroaromatic rings compared with alicyclic rings, even when the latter give a linear bond distribution and possess considerable rigidity.

It is also known [83] that replacement of the terminal phenyl group of a phenyl ester by cyclohexyl markedly reduces nematic thermal stability, but results so far available suggest that a smectic mesophase may be stabilized by this particular change.

The above discussion is designed to give only general guidance as to types of compound which tend to form liquid crystals; a fuller analysis of the situation has been given elsewhere [21]. The exemplification is not therefore intended to be exhaustive or to take into account exceptional examples of liquid crystalline behaviour [57].

It should be pointed out now that *terminal* hydroxyl- and amino-groups are not usually favourable to liquid crystal formation. Such groups are capable of giving polymeric hydrogen bonding which raises melting points and may also favour a non-parallel arrangement of the molecules [21]. Thus, 6-hydroxy-2-naphthoic acid [51] does not form liquid crystals, whereas the methyl ether [51] does. Some examples of liquid crystalline compounds with terminal hydroxyl- or amino-groups are of course known, but in these cases polymeric hydrogen bonding may for certain reasons be disfavoured or restricted. For example, in compound (XXV) [48], the amino-group is partly involved in intramolecular hydrogen bonding. Relative to 4-amino-4″-nitro-*p*-terphenyl (m.p. 300–301°C), the melting point of compound (XXV) is quite low (218–219°C) and it gives a monotropic liquid crystal.

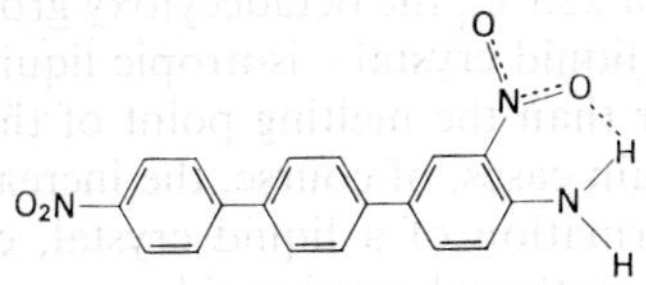

XXV

On the other hand, terminal —CO·OH and —CO·NH_2 [58] groups give linear dimers by intermolecular hydrogen bonding, and this favours liquid crystal formation. See also [144].

These examples emphasize the many factors which must be taken into account when attempting to assess the potential liquid crystal behaviour of a system.

The Effect of Terminal Substituents on Liquid Crystal Behaviour

Since it first became possible to compare the liquid crystal properties of series of related compounds, the influence of terminal substituents, X and Y, on systems such as (XXVI) has attracted interest. Early reviews state that molecules of mesogenic compounds should contain moderately dipolar terminal groups, and it is important to establish the function of

XXVI

such substituents more clearly. Evidence will be put forward to show that replacement of a terminal hydrogen in a molecule by another substituent enhances the potential of the system to form liquid crystals. If the unsubstituted compound forms liquid crystals, then the substituted compound will form liquid crystals which are, in the majority of cases, more thermally stable; only for smectic liquid crystals do certain terminal substituents reduce thermal stability. The role of the terminal substituent is not therefore simply to lower the melting point and reveal liquid crystal behaviour latent in the parent system. Indeed, terminal substituents usually raise melting points, but they increase liquid crystal thermal stability even more. Moreover, in many cases in which the substituent does not lower the melting point, the liquid crystal–amorphous isotropic liquid transition temperature is higher than the melting point of the unsubstituted compound, i.e., liquid crystal thermal stability is again enhanced. The following examples illustrate these points.

Biphenyl-4-carboxylic acid [49]	C–I, 223°C
4′-Methoxybiphenyl-4-carboxylic acid [46]	C–N, 258°C; N–I, 300°C
4′-n-Octadecylyloxybiphenyl-4-carboxylic acid [46]	C–S, 150°C; S–I, 238°C

The methoxy group raises the melting point by 35°C, but the liquid crystal – isotropic liquid transition temperature is 77°C higher than the original melting point of 223°C; the octadecyloxy group lowers the melting point by 73°C, but the liquid crystal – isotropic liquid transition temperature is still 15°C higher than the melting point of the unsubstituted compound (223°C). In certain cases, of course, the increase in melting point is too great to permit formation of a liquid crystal, e.g., 62°C on passing from benzoic acid to *p*-methoxybenzoic acid.

The rôle of the terminal substituent is at its most critical in simple systems for which the unsubstituted compound is not liquid crystalline; in these cases, a terminal substituent can create liquid crystal formation. However, if the rigid core part of the molecule is sufficiently extensive, as in *p*-pentaphenyl, no terminal substituent is necessary, and if one were introduced, it would simply enhance the liquid crystal thermal stability.

In this section, information relating to the role of terminal groups and their relative efficiencies in promoting liquid crystal properties is drawn from systematic studies which have been made of a number of systems in which the nature of the terminal substituent has been systematically varied over quite a wide range. These studies are of two kinds and involve examination of homologous series of mesogenic compounds in which the terminal substituent, $Me(CH_2)_x$— or $Me(CH_2)_xO$—, is lengthened by successive increases in the number of methylene units, and studies of the influence of a range of terminal substituents on the liquid crystal properties of specific systems.

Studies of Homologous Series of Mesogenic Compounds

When the liquid crystal transition temperatures, e.g., N–I, S–I, S–N, S–Ch, Ch–I, smectic–smectic, for a homologous series are plotted against the number of carbons in the n-alkyl group, smooth curves may be drawn through points for like or related transitions. This was established [21] for about seventy extended series (C_1–C_{10}, C_{12}, C_{16}, C_{18}) or parts of series (at least four homologues). All series did not give the same type of temperature *v.* chain length plot. The type depended upon whether the series exhibited only smectic properties or both smectic and nematic properties, etc., and also, differences in the shapes of the curves were found for series exhibiting the same type(s) of liquid crystal. The situation has been fully reviewed [21], and it is necessary only to summarize the main points and to add that more recent studies [9], [36], [59], [60], [135] have substantiated the general principles.

A common type of liquid crystal transition temperature *v.* chain length plot is given in Fig. 4.1.2(*a*) for the 4′-n-alkoxybiphenyl-4-carboxylic acids [46]. As for many series, the melting points (C–S and C–N) do not show regular trends. However, the N–I temperatures alternate typically, the points fitting two falling curves, the upper for even and the lower for odd numbers of carbon atoms in the n-alkyl chain. For a system with the n-alkyl group attached *direct* to the ring, the reverse situation would arise, because the oxygen of the ether link is equivalent stereochemically to a methylene unit. The situation for n-alkyl esters (Ar·CO·OR) is also

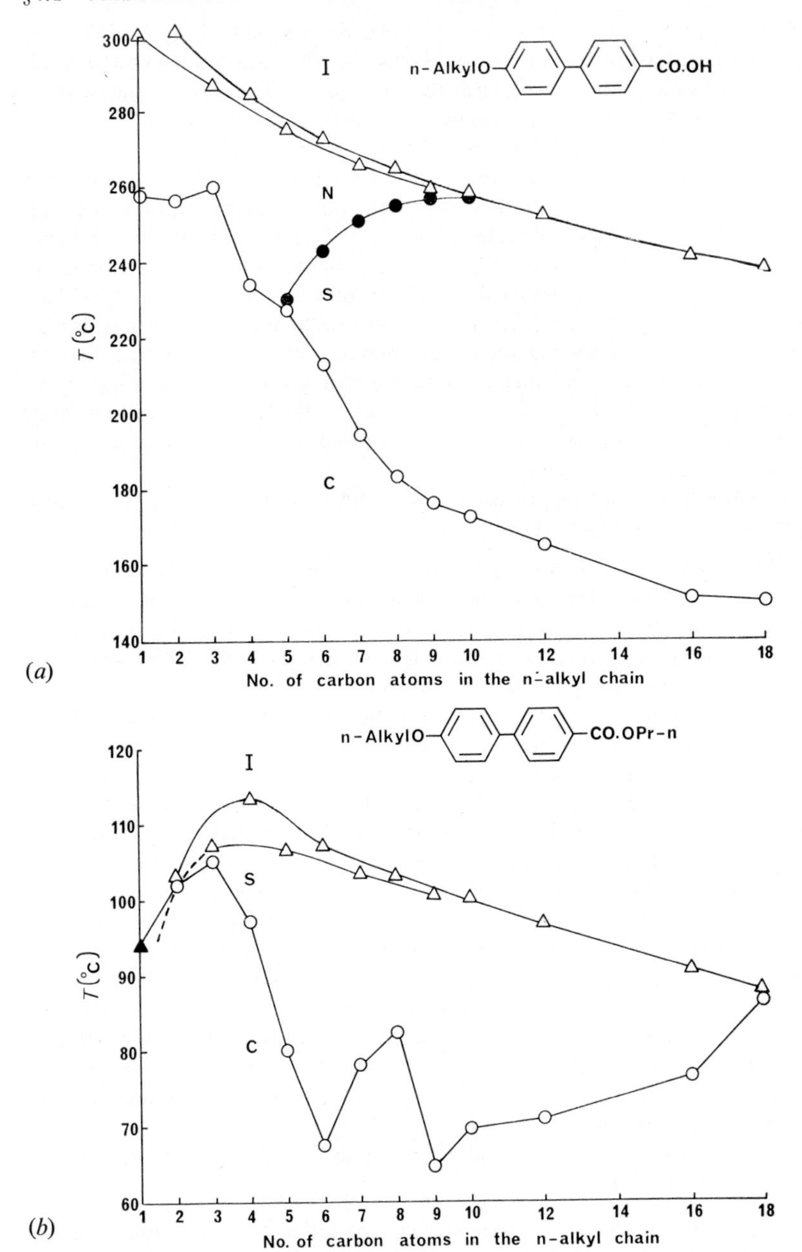

FIG. 4.1.2. Transition temperature *v.* chain length plots for two homologous series of mesogenic compounds

(*a*) 4′-n-alkoxybiphenyl-4-carboxylic acids

(*b*) n-propyl 4′-n-alkoxybiphenyl-4-carboxylates

Key: △ N–I or S–I ○ C–N or C–S
● S–N ▲ C–I (smectic liquid crystal not observed).

opposite to that for n-alkyl ethers (Ar·OR). Returning to Fig. 4.1.2(*a*), the odd–even alternation becomes less marked as the series is ascended, and the two curves actually merge later in many series. The S–N temperatures do not alternate, and lie on a single curve which rises steeply at first; the curve then levels out and merges with the falling N–I curves. After this point in a homologous series, no nematic properties are observed, and the S–I temperatures for the higher members lie on extrapolations of the falling N–I curves. In some series [21], the S–N curve falls slightly before merging with the N–I curves, and for the 4-*p*-n-alkoxy-3-chlorobenzoic acids [21] and 2-*p*-n-alkoxybenzylidene-aminofluorenones [21, 53], the S–N curve has not merged with the N–I curves at the highest homologue studied (C_{18}). The tendency for purely nematic behaviour to give way to predominantly or purely smectic behaviour in the higher homologues is quite general. The stage in a series at which smectic properties are first observed is of little significance, being dependent upon frequently irregular trends of the melting points.

The following variations on this type of transition temperature *v*. chain length plot have been observed [21].

1. The alternating N–I temperatures may fit two curves which either rise to maxima before falling or rise* progressively over an extended series.
2. The upper N–I curve may fall progressively as in Fig. 4.1.2(*a*), while the lower curve rises to a maximum before it falls and merges with the upper curve as the alternation diminishes.
3. The series may exhibit smectic–smectic transitions [21, 59], for which the temperatures change regularly. For the *NN'*-di-(*p*-n-alkoxybenzylidene)-1,4-phenylenediamines, the smectic–smectic temperatures for the complex range of polymorphic smectic states [36] do not give perfect fits on smooth curves, but their regularity is none the less striking. In the cases cited above, the temperatures for each related set of smectic–smectic transitions do not alternate and lie either on a curve which rises more or less steeply to a maximum before it falls or on a curve which falls progressively [36] along the series; however, recent results show that, for some series of compounds, alternation of S–S temperatures does occur [135].
5. The series may exhibit no nematic properties; the S–I temperatures then alternate and fit two curves. These may rise to early maxima as shown in Fig. 4.1.2(*b*) for the n-propyl 4'-n-alkoxybiphenyl-4-carboxylates [46]. Alternatively, both curves may rise progressively over the series, or one may rise to an early maximum while the other falls progressively over the series. In all cases the curves tend to merge as the alternation diminishes for the higher homologues.

No series which the writer has studied has failed to give smooth curve relationships between temperature and chain length, provided that careful

* In reports [122] of the liquid crystal properties of certain aldonitrones, the tendency of the N–I temperatures to rise as the homologous series was ascended was described as a unique or exceptional occurrence. This is not in fact so, for even as long ago as 1962, twelve series giving such trends for the N–I temperatures were known [130]; also [139, 149].

attention is paid to purity of the compounds. Although transition temperatures reported in the literature for some homologous series do depart from smooth curve relationships, the distribution of the temperatures is not random, and varying degrees of impurity of the homologues are probably responsible for the deviations. Attention is however drawn to the 2-*p*-n-alkoxybenzylideneaminofluorenones [21, 53] for which the temperature *v.* chain length plot remained anomalous after exhaustive purification of the anils; both N–I curves fell to shallow minima, then rose slightly to maxima and fell again. The series did not therefore depart from the principle of smooth curve relationships, but gave a type of N–I transition line which was unique until more recent evidence confirmed that N–I curves involving minima are obtained for esters of 4,4′-azoxyphenol [56] and for 4-*p*-n-alkoxybenzylideneaminoacetophenones [61].

In 1962, reliable data were available for only one series exhibiting cholesteric mesophases, i.e., the cholesteryl n-alkanoates [62]. The lower homologues gave only cholesteric mesophases, but from the heptanoate to the stearate, smectic and cholesteric mesophases were observed. The Ch–I temperatures alternated and gave two curves, one rising from acetate to butyrate and then falling progressively to the stearate, and the other falling from propionate to nonanoate. Taking the formate as the lowest homologue with an even number (zero) of carbon atoms in the *alkyl* chain, then the lower curve too rose to a maximum, at the propionate. The S–Ch temperatures rose to the laurate and then fell to the stearate, the curve approaching the Ch–I curve only very gradually over the last three homologues studied.

More recently, other homologous series of sterol derivatives have been examined, notably by Ennulat, Elser and Pohlmann. Some series were purely smectic in behaviour, e.g., derivatives of stigmasterol [63–67], but others did exhibit cholesteric properties [67–70]. Davis, Porter and Barrall [118] and Ennulat [69] have also re-examined the cholesteryl n-alkanoates, using differential scanning calorimetry to measure transition temperatures. Disagreement with the earlier data [62] was noted [69] for a few homologues (notably the pentanoate), and departures from smooth curve relationships between chain length and transition temperature arose despite the scrupulous attention paid to purity. Other series studied were the cholesteryl n-alkyl carbonates [70] and *S*-alkyl thiocarbonates [67, 123] 5α-cholestanyl n-alkyl carbonates [67, 68] and *S*-alkyl thiocarbonates [67], thiocholesteryl n-alkanoates [69, 129] and *S*-cholesteryl *O*-alkyl thiocarbonates [67]. In all cases, the transition temperatures deviated to some extent from *strict* smooth curve relationships with chain length, but the types of plot approximate to that for the cholesteryl n-alkanoates [21, 62], i.e., the Ch–I temperatures alternate and give curves which fall progressively or possibly in one series rise first to early maxima, and the S–Ch temperatures lie on a single curve which does not merge with the Ch–I curve even for extended chains. Eight members of the series of cholesteryl ω-phenylalkanoates were also studied [69]; the Ch–I temperatures showed very pronounced alternation giving two smooth curves, one falling steeply and the other rising steeply, and both levelling out for the higher homologues. This behaviour (pronounced alternation and smooth transition curves) has now been confirmed for extended series of 5α-cholestanyl ω-phenylalkanoates [124] and of thiocholesteryl ω-phenylalkanoates [69, 124], and it is concluded that the pronounced alternations must arise from the different spatial arrangements of the ω-phenyl substituent for odd and even homologues, cf., the similar explanation proposed (p. 141) by Gray and Harrison [125] to explain the high alternation of N–I temperatures for ω-phenylalkyl 4-*p*-substituted benzylideneaminocinnamates. In the case of the thiocholesteryl ω-phenylalkanoates, a notable feature was the alternation of the S–Ch temperatures for the higher homologues. This unusual behaviour for a homologous series is again reminiscent of recent results reported by Gray and Harrison [125] for two series of n-alkyl esters—see p. 121 and references [135, 139].

Knapp and Nicholas [71, 72] observed cholesteric liquid crystals which exhibited colours for certain esters (XXVII) obtained by reduction of 24-methylenecycloartanyl n-alkanoates; all had odd numbers of carbon atoms in the alkyl chains. The colourless liquid crystals given by the lower

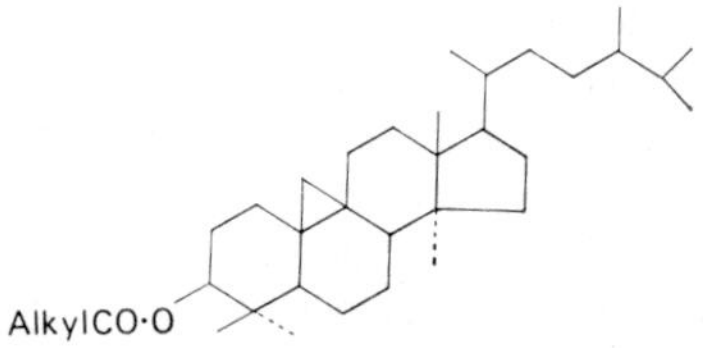

AlkylCO·O

24-Methylcycloartanyl n-alkanoate

XXVII

homologues were thought to be smectic, but it seems possible that these too were cholesteric, scattering light outside the visible range. Assuming this is so, the Ch–I temperatures show an appreciable scatter on plotting against alkyl chain length, but the best curve rises to an early maximum, then falls and levels off for the higher homologues, in a manner similar to the curve for the cholesteryl n–alkanoates [62].

Deviations from strict smooth curve relationships do therefore occur for several series giving cholesteric mesophases, despite scrupulous attention paid to purity. Pohlmann and Elser [68] and Elser and Ennulat [123] conclude however that, because of particular difficulties involved with sterol systems, undetected impurities may still be responsible for the deviations.

It is difficult to explain the observed trends in liquid crystal thermal stability with increasing length of the n-alkyl chain, particularly the regular changes in liquid crystal—amorphous isotropic liquid transition temperatures between odd and even carbon number homologues, if the alkyl group is free to adopt *completely* random conformations through flexing of the chain and rotation about carbon-carbon bonds. Interesting results showing that pendant (terminal) alkyl groups in nematic liquid crystals cannot adopt a *single* elongated conformation have however recently been made by Young *et al.* [122] for a homologous series of aldonitrones. It is possible however that the parallel alignment of rigid core parts of the molecules constrains movement within the alkyl chains to some extent and that, although the chains may not adopt a single or unique conformation, the conformations that are adopted are more or less extended in nature. Discussions of smooth curve relationships [21, 73] are usually based therefore on a relatively static picture of the system, assuming only some tendency of longer n-alkyl chains to bend. The observations requiring explanation are complex and the arguments are qualitative; these are given in detail elsewhere [21, 73] and only the major points need now be summarized.

S–N or S–Ch temperatures

Assuming that the alkyl chain adopts a zig-zag conformation [73] with the major axis of the chain statistically normal to the smectic planes, *each* methylene unit will increase by the same amount the lateral intermolecular attractions on which smectic thermal stability predominantly depends (see p. 127). S–N and S–Ch temperatures should not therefore alternate, and should increase as the chain is extended; this is the case for many homologous series. To explain the subsequent fall for some series, it is postulated [73] that the greater thickness of the region of "fluid" alkyl

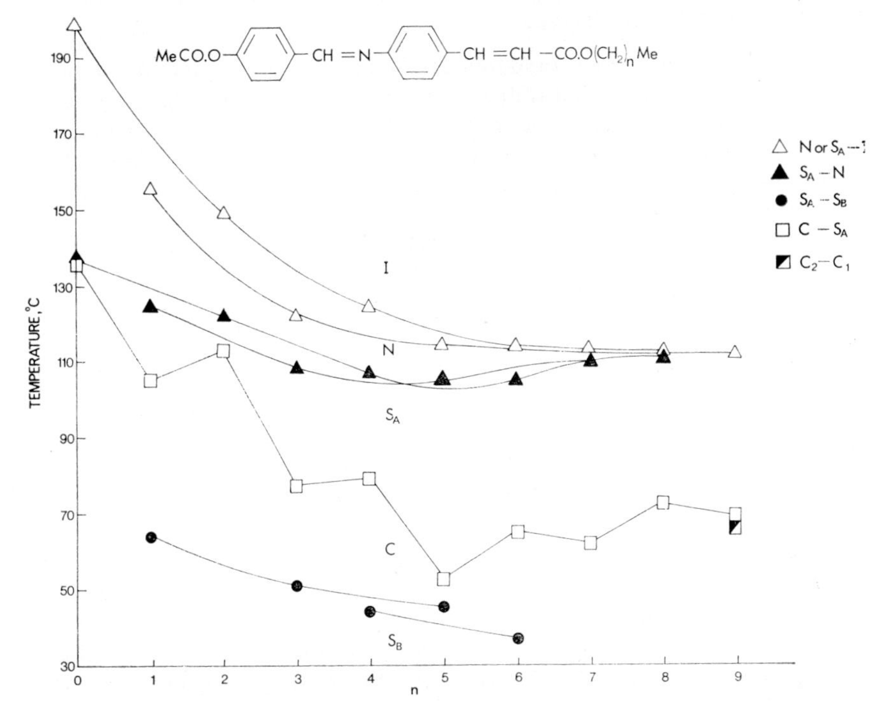

FIG. 4.1.3. Transition temperature *v.* chain length plot for the homologous series of n-alkyl 4-*p*-acetoxybenzylideneaminocinnamates.

chains in the higher homologues causes a more ready partial interpenetration of the layers, so diminishing smectic thermal stability. A combination of these two possibilities may explain the overall shapes of S–N and S–Ch curves. It should however be pointed out that alternation of S–N temperatures has recently been observed [125] in the homologous series of n-alkyl 4-*p*-methoxybenzylideneamino- and 4-*p*-acetoxybenzylideneamino-cinnamates. In both series, S_AN transitions are observed for a large number of the homologues (Fig. 4.1.3), and not only do the temperatures alternate, but also the shapes of the S_A–N temperature *v.* chain length curves are much more complex than any hitherto found for other homologous series. This behaviour has not been explained. As pointed out earlier, alternation of S–Ch temperatures has also been observed recently [124] for thiocholesteryl ω-phenylalkanoates—see p.120. See also references [135, 139].

N–I or Ch–I temperatures

As the n-alkyl chain is extended, one might expect that

(*a*) the longer molecules will be more difficult to disalign,
(*b*) the total molecular polarizability will increase,
(*c*) the frequency with which readily polarizable parts, e.g., core parts, of the molecules lie side by side [74] will decrease, and
(*d*) the terminal separation of readily polarizable parts, e.g., core parts, of the molecules will increase.

Factors (*a*) and (*b*) should increase and factors (*c*) and (*d*) should decrease the N–I and Ch–I transition temperatures. If the attractive forces between the core parts of the molecules are low, then factors (*a*) and (*b*) may predominate. In support of this view, when N–I transition temperatures for a series are relatively low, a rising N–I transition line is obtained as the series is ascended, and *vice versa*.

The alternation of N–I or Ch–I transition temperatures can be explained assuming that shorter alkyl chains extend strictly along an axis determined by the tetrahedral bond angles (Fig. 4.1.4). The nature of the contact

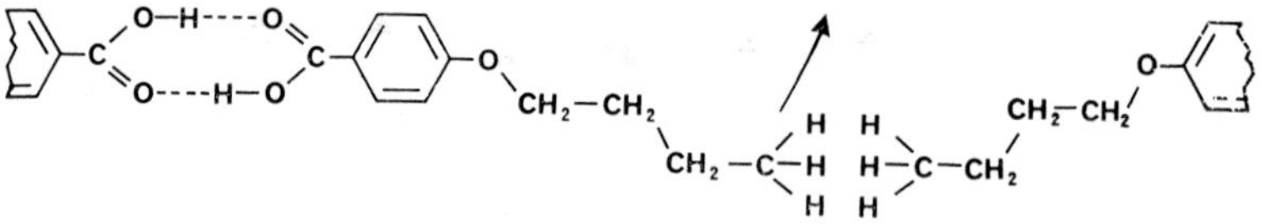

FIG. 4.1.4. End to end relationship between molecules with terminal substituents consisting of n-alkyl chains.

between terminal methyl groups of molecules arranged end to end will then differ for even and odd homologues. This would affect the terminal interactions between molecules and explain the alternation of the transition temperatures. With higher homologues, the longer chain may be constrained (see arrow in Fig. 4.1.4) to lie in line with the major axis of the core; the end to end contact would then ultimately be the same for odd and even homologues and the alternation would diminish with increasing tendency of the alkyl chain to bend.

An alternative explanation is as follows. With reference again to Fig. 4.1.4, with shorter alkyl chains, the contribution of each new carbon–carbon bond to the polarizability of the molecule in the direction of its long axis is different (greater) on passing from odd to even homologue compared with from even to odd homologue. The terminal intermolecular interactions should therefore alternate as the series is ascended. However, if longer n-alkyl chains bend and come into line with the molecular long axis, the contribution of each new carbon–carbon bond to the polarizability of the molecule in the direction of the long axis is the same, irrespective of whether the carbon number is odd or even. Alternation of the terminal intermolecular attractions and of the N–I temperatures should therefore again diminish with increasing tendency of the alkyl chain to bend. See also [146].

S–I temperatures

These too should alternate, as the transition again marks the breakdown of the total order of the system; the arguments used for N–I temperatures should therefore apply.

Other Studies of the Influence of Terminal Substituents on Liquid Crystal Behaviour

The effects of terminal substituents on the nematic properties of *mixtures* (see Chap. 4.2) were studied by Dave *et al.* [75–80]. They obtained from a series of binary mixtures an order of terminal group efficiency in promoting mixed liquid crystal formation of the nematic type. The order was:

$$NO_2 > MeO > NMe_2 > Me = Cl > Br > H$$

However, plots of S–N or S–I transition temperatures *v.* composition for binary mixtures give curves [81, 82], not straight lines, and an order for the effect of different terminal groups in promoting smectic phase formation cannot be obtained from mixed liquid crystal systems.

Information about the effects of terminal substituents on liquid crystal properties can however be obtained by studying a range of pure *mesogenic* compounds differing only in the nature of their terminal substituents. The order of the thermal stabilities, i.e., upper transition temperatures of the liquid crystals will then be the order of efficiency of the terminal groups in producing smectic, nematic or cholesteric liquid crystals in that system. The main conclusions which have been reached from such studies [9, 21, 73] are now summarized; in all cases substitution is of the type ring–X.

Nematic group efficiency order

Four systems have been examined—4-*p*-substituted-benzylideneamino-4′-n-octyloxy- and -4′-methoxy-biphenyls (1) and (2), racemic 2-methylbutyl 4-*p*-substituted-benzylideneaminocinnamates (3) and 4′-substituted-biphenyl-4-carboxylic acids (4)—giving the following nematic group efficiency orders for terminal substituents:

(1) Ph > NH·CO·Me > CN > OMe > NO_2 > NMe_2 > Me > H;
(2) NH·CO·Me > OMe > NO_2 > Cl > Br > NMe_2 > Me > F > H;
(3) CN > OMe > NO_2 > Cl > Br > H;
(4) Ph > CN > OMe > Cl > Me > F > H

Strictly comparable data have not been obtained from the four systems; for example, a terminal nitro-group in system (4) gives a high melting, non-mesogen [49], and fluoro- and bromo-substituents in system (1) give purely smectogenic compounds. However, the four orders are closely similar, and the following average group efficiency order for nematic liquid crystals may be constructed from the results:

Ph > NH·CO·Me > CN > OMe > NO_2 > Cl > Br
> NMe_2 > Me > F > H

This order may be used to give guidance as to the possible behaviour of a compound within a series, but it cannot be applied rigorously. For example, substituents such as Cl, Br and F, which are relatively low in the nematic order, may give purely smectogenic compounds if the molecular system in which they occur has smectogenic tendencies, i.e., there may be an early termination [9] of the nematic properties of the series. There are also certain irregularities in behaviour. For instance, although NMe_2 is lower in the order than Cl or Br, nematic properties are observed with this substituent in system (1), but not with chloro- or bromo-substituents; for system (3), the reverse is true [9]. The influence of these substituents on the thermal stability of smectic mesophases relative to that of nematic mesophases is therefore variable (see also p. 125), and the average order is most satisfactorily applied to highly nematogenic systems.

Considering only the substituents occurring in the order obtained by Dave *et al.*, the present results give for these the order:

OMe > NO_2 > Cl > Br > NMe_2 > Me > H

The two orders differ in detail, probably because the two methods of approach are giving information about somewhat different aspects of the same problem. Both orders do however show that any terminal substituent is superior to hydrogen in promoting nematic properties. This and the fact that the best substituents are highly polarizable (Ph, NH·CO·Me, NO_2, CN, OMe) may be explained in terms of the increased difficulty [73] of dis-aligning longer molecules between which stronger forces of attraction may operate, cf., points (*a*) and (*b*) on p. 122.

Cholesteric group efficiency order

By examining [9] the optically active esters corresponding to system (3) and *p*-substituted cholesteryl benzoates, an order of terminal group efficiency for cholesteric liquid crystals is obtained. This order is the same as the nematic order, and this yet again emphasizes the close relationship of cholesteric and nematic liquid crystals.

Smectic group efficiency order

The effects of terminal substituents on smectic thermal stability must be judged on rather limited data. Smectic properties are given by many compounds belonging to systems (1) and (3) [9, 73], and the following orders of smectic terminal group efficiency are obtained for the systems indicated:

(1) NH·CO·Me > Ph > Br > Cl > F > NMe_2 > Me > H > NO_2 > OMe > CN*

(3) Ph† > NMe_2 > Me > OMe > H

The highly polarizable Ph group and the NH·CO·Me substituent, which can hydrogen bond and has a strong dipole which will operate across the long molecular axis, both strongly enhance smectic and nematic thermal stabilities. Otherwise the order is totally different to the nematic/cholesteric order. Thus, OMe, CN and NO_2 groups, which lie high in the nematic order, are low in the smectic order, e.g., in system (1), all three groups are inferior to a terminal hydrogen in enhancing smectic thermal stability. It is tempting to relate this to the fact that nitro- and cyano-groups have strong dipoles which lie along the long molecular axis. Such dipoles can certainly be envisaged [73] as giving repulsions between molecules which lie parallel to one another, i.e., side by side, and *perpendicular* to the layer planes of a *smectic* liquid crystal. Yet, in system (1), chloro- and bromo-substituents with similar but weaker dipoles are quite high in the smectic order. The OMe group also presents problems. The dipole moment associated with this group should operate at an angle across the long molecular axis and could enhance smectic properties [9, 73]. The fact that the OMe group is low in the smectic order suggests that other factors, e.g., rotation about the ring-O bond, must be involved and that further systems should be studied to reach a fuller understanding of the role of terminal substituents in relation to smectic thermal stabilities. The variable position of OMe relative to hydrogen in the two smectic orders suggests however that no single smectic terminal group efficiency order is applicable to a range of systems.

The Effects of Lateral Substituents on Liquid Crystal Behaviour

Simple Cases of Lateral Monosubstitution

The effects of introducing lateral substituents into the elongated molecules of mesogenic compounds are now considered. This type of structural change has been made systematically for a few types of compound by

* No smectic phase appears on cooling the nematic phase to 130°C [83].

† It has been reported [11] that the optically *active* ester of system (3) with a terminal Ph group gives both a smectic and a *nematic* phase. This could not be correct, and it was assumed [9] that the compound must be smectogenic and *cholesterogenic*; in fact the active and racemic 2-methylbutyl esters are purely smectogenic [83]. This can now be understood, because system (3) is quite smectogenic, and it has been shown that the terminal phenyl group—see system (1)—strongly favours both smectic and nematic properties, smectic thermal stabilities being increased rather more than the nematic.

introducing a range of lateral substituents into the aromatic rings of core parts of the molecules, e.g., into the 3′-positions of 4′-n-alkoxybiphenyl-4-carboxylic acids giving the compounds (XXVIII) [46, 47, 84].

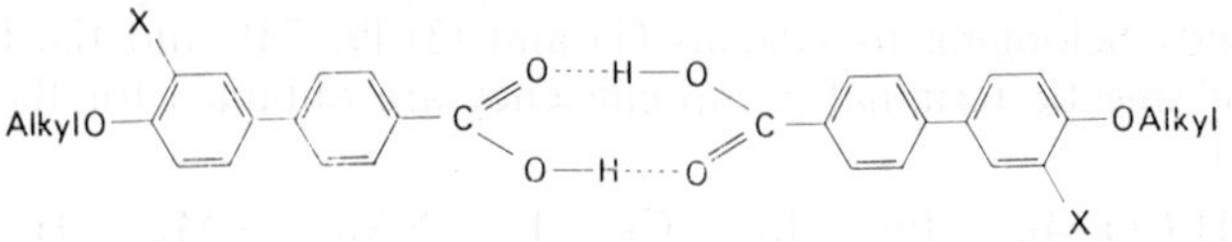

X

AlkylO— —C(=O···H—O)(O—H···O=)C— —OAlkyl

X

XXVIII

The substituent X will have two effects:

(*a*) The long molecular axes may be forced apart by the substituent, reducing intermolecular forces of attraction and thus lowering liquid crystal thermal stability. Since lateral interactions will be most affected, the thermal stability of the smectic mesophase should decrease more than that of the nematic mesophase.

(*b*) The change from a ring–H to a ring–X bond will increase the molecular polarizability and possibly also the molecular dipolarity. This should increase lateral intermolecular attractions, enhancing liquid crystal thermal stability, particularly that of the smectic mesophase.

Opposing effects are therefore operating, but experimental evidence [85] shows that effect (*a*) always predominates when the substituent X increases separation of the long molecular axes by the amount by which the ring–X bond length + the van der Waals radius of X exceeds the ring–H bond length + the van der Waals radius of H (see p. 129 for cases in which this condition does not apply). The thermal stabilities of both the smectic and nematic mesophases are therefore decreased, but the thermal stability of the smectic mesophase is not necessarily affected more than that of the nematic mesophase. This depends upon the change in dipole moment which results from introducing the substituent.

These effects have been demonstrated clearly [47] for system (XXVIII) with alkyl = n-octyl (C_8H_{17}) and a range of substituents (X = F, Cl, Br, I, Me and NO_2). Similar results have been obtained with other systems but are less instructive, because the larger substituents such as Br, I and NO_2 frequently eliminate liquid crystal formation if the liquid crystal thermal stability of the parent system is not sufficiently high, e.g., in 3-substituted 4-n-alkoxybenzoic acids [45]. Results for system (XXVIII) [47] are therefore used to illustrate the extent of the effects.

TABLE 4.1.1. *Liquid crystal transition temperatures for 3′-substituted 4′-n-octyloxybiphenyl-4-carboxylic acids*

3′-Substituent	H	F	Cl	Br	I	Me	NO_2
S–N (°C)	255	254·5	225	214	192·5	213	214*
N–I (°C)	264·5	255·5	233	224	214	237	214*

* S–I transition temperature

Decreases in smectic and nematic thermal stability accompany substitution of X for H in every instance. If the substituents are placed in order of their effectiveness in decreasing liquid crystal thermal stability, a different order is obtained for smectic and nematic mesophases. This reflects the fact that smectic thermal stability does not necessarily diminish more than nematic thermal stability on introduction of the substituent. In obtaining the orders given below, the S–I temperature for X = NO_2 is compared with the N–I temperatures for the other compounds. It is reasonable to do this, however, bearing in mind that the S–I curve for a homologous series is a continuation of the N–I curve for lower homologues, i.e., the S–N and N–I transitions become coincident.

Smectic order $H < F < Cl < Br = NO_2 < Me < I$
Nematic order $H < F < Me < Cl < Br < I = NO_2$

The reasons for the different orders become clear on plotting the relevant transition temperatures against the molecular breadths or diameters (defined as the diameters of the narrowest cylinders through which the dimerized molecules will pass, flexible parts such as alkyl chains being assumed not to interfere). Fig. 4.1.5(*a*) shows that the N–I temperatures and the S–I temperature for X = NO_2 fit a good curve (nearly a straight line), indicating that nematic thermal stability decreases regularly as the broadening effect or size of the substituent increases. It is significant that the ring–X bonds involved cover a range of dipolarities. Since this does not affect the temperature *v.* breadth relationship, dipolar interactions cannot be important as far as nematic thermal stability is concerned. The forces responsible for maintaining order in the nematic state must therefore be dispersion forces which will decrease rapidly with increasing separation of the long molecular axes. This effect will of course be counteracted to some extent by the increases in molecular polarizability accompanying substitution, but a near rectilinear relation between N–I transition temperature and molecular breadth would be interfered with only if the molecular polarizabilities changed irregularly with increasing size of the substituent. In fact the polarizabilities of the C–X units rise quite regularly with increase in size of the substituent as required by the relationship illustrated in Fig. 4.1.5(*a*).

Figure 4.1.5(*b*) shows that quite a different situation exists when the S–N or S–I temperatures for system (XXVIII) are plotted against molecular breadth. There is in fact an appreciable scatter of the points. However, a steeply falling curve may be drawn through the points for X = F, Cl, Br and I. Lying below this line are the points for X = H and X = Me, and above the line is the point for X = NO_2. Now the dipole moments of ring—halogen bonds are all about the same, whereas C–H and C–Me bonds are of much lower dipolarity, and the C–NO_2 bond is of much greater dipolarity. Dipole moments are therefore important as far as relative thermal stabilities of smectic mesophases are concerned, and a substituent's size effect on smectic thermal stability is decreased if the C–X bond is highly dipolar, and increased if the C–X bond is weakly dipolar.

The reasons for the two different orders are now clear, nematic thermal

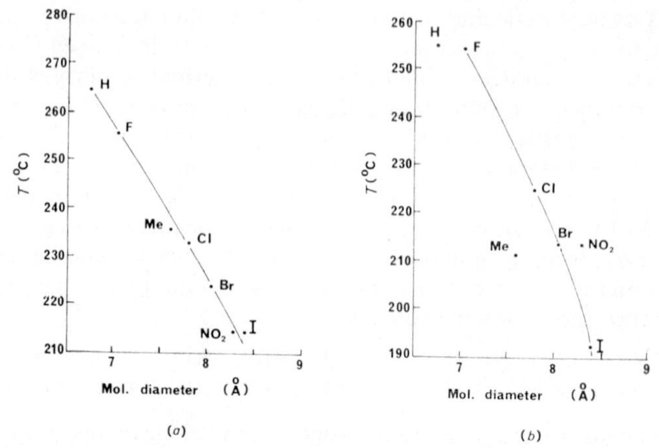

Fig. 4.1.5. Plots of transition temperatures *v.* molecular diameters for 3′-substituted 4′-n-octyloxybiphenyl-4-carboxylic acids
(*a*) N–I transition temperatures (except for NO_2 [S–I])
(*b*) S–N or S–I transition temperatures.

stability being influenced primarily by breadth effects, and smectic thermal stability by a combination of breadth and dipolar effects. We can also understand why, as shown in Table 4.1.2, the effect of a substituent on smectic thermal stability is not always greater than that on nematic thermal stability. This depends upon whether replacement of H by X increases the dipole moment by a large or a small amount, and whether the change in dipole strongly counteracts the breadth increase accompanying substitution. Referring to Table 4.1.2, the change H to Me involves a small change

TABLE 4.1.2. *Decreases in liquid crystal thermal stability accompanying substitution in the 3′-position of 4′-n-octyloxybiphenyl-4-carboxylic acid.*

3′-Substituent	F	Cl	Br	I	Me	NO_2
Decrease in smectic thermal stability (°C)	0·5	30	41	62·5	42	41
Decrease in nematic thermal stability (°C)	9	31·5	40·5	50·5	27·5	50·5

in dipole, and the smectic thermal stability suffers more than that of the nematic mesophase, whereas the change H to F involves a small breadth increase but a considerable increase in dipole, and the smectic thermal stability is affected to the smaller extent. However, with the larger halegeno-substituents, the greater breadth increases militate against the dipolar effect on smectic thermal stability; after chloro-, an inversion occurs, and the smectic thermal stability suffers more than the nematic thermal stability.

Since the effect of a change in substituent on smectic thermal stability is greater than that on nematic thermal stability when ring—substituent bonds of like dipolarity are involved, e.g., H to Me, F to Cl, Cl to I, it may be concluded that the effect of breadth increase alone is greater for smectic mesophases, i.e., allowing for the difference in scale, the gradient of the line in Fig. 4.1.5(*b*) is greater than that in Fig. 4.1.5(*a*). This is because the smectic order depends on dipole–dipole and induced interactions, as well as on dispersion forces, and all three interactions will decrease rapidly with increasing separation of the long molecular axes.

The above conclusions are supported indirectly by data [51, 52] for 5-substituted 6-n-alkoxy-2-naphthoic acids (XXIX). The dotted lines represent the perimeter of the dimerized parent molecule, i.e., the walls of the narrowest cyclinder through which the molecule will pass. It is clear

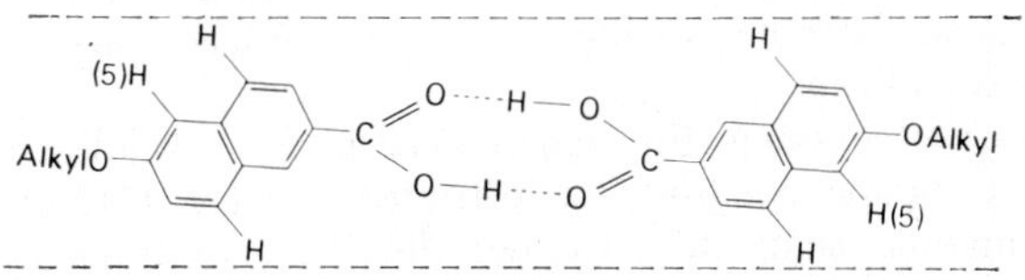

XXIX

that a "gap" exists between the surface of the hydrogen in the 5-position and the dotted line, i.e., a 5-substituent of considerable size, e.g., chloro-, may be introduced without broadening the molecule. Further, no 5-substituent in system (XXIX) will force the molecular axes apart as much as will the same substituent in the 3′-position of system (XXVIII).

Thus, effect (*a*) (page 126) is greatly reduced, and unless the 5-substituent is too large, it will simply fill in a "gap" between neighbouring molecules. Therefore, by increasing molecular polarizability and dipolarity, a 5-substituent could increase both nematic and smectic thermal stabilities. If the ring-substituent bond is dipolar, e.g., with a halogeno-substituent the increase in smectic stability should be greater than that (if any) in

TABLE 4.1.3. *Liquid crystal transition temperatures for 5-substituted 6-n-decyloxy-2-naphthoic acids*

5-Substituent	H	Cl	Br	I	NO_2
S–N (°C)	147	186·5	182·5	(164·5)*	(166·5)†
N–I (°C)	181	192·5	189·5	178·5	(166·5)†

* Monotropic transition temperatures are in parenthesis

† S–I transition temperature; "nematic" thermal stability assumed equal to that of the smectic mesophase.

nematic thermal stability. The results in Table 4.1.3 show that these effects are observed in practice.

Only for the largest substituents, i.e., iodo- and nitro-, are the nematic thermal stabilities reduced, whilst all four substituents enhance the smectic thermal stabilities. The importance of dipoles associated with lateral substituents is well illustrated by these results.

Other instances in which lateral substituents do not exert their full breadth increasing effect, and act by filling "gaps" between molecules are known, e.g., the 4-*p*-n-alkoxybenzylideneamino-3′-substituted biphenyls [73, 86].

It is more usual however for a lateral substituent to broaden the molecule and decrease both smectic and nematic thermal stabilities. For a given substituent, the extent of the decrease depends upon the system into which it is introduced. It would appear [47] that a halogeno-substituent gives smaller decreases in both smectic and nematic thermal stabilities when it is introduced into a longer molecule, and that on changing from a long molecule to a shorter one, the ratio of the decreases in smectic or nematic thermal stabilities becomes smaller as the substituent's size in increased from fluoro- to chloro-.

The number of systems for which experimental studies allow such comparisons to be made is however restricted, and for the 3-chloro-*trans*-*p*-n-alkoxycinnamic acids [87], smaller decreases in both smectic and nematic thermal stabilities have been observed than for the corresponding biphenyl acids (XXVIII; X = Cl), which involve longer molecules. However, in the 3-substituted cinnamic acids, a situation similar to that for the 5-substituted 6-n-alkoxy-2-naphthoic acids could arise [86].

Lateral Substituents exerting Steric Effects on the Molecules of Liquid Crystalline Compounds

Studies have also been made of the effects of lateral substitution of a type which, in addition to broadening the molecule, increases its thickness by imposing a steric effect on the system, i.e., by causing a twisting about one of the bonds, so that parts of the molecule are rotated out of the plane of the remainder of the molecule.

Biphenyl derivatives have again provided convenient systems for study. The presence of a 2- or 2′-substituent in biphenyl system (XXX) rotates the rings about the 1,1′-bond, so that the aromatic nuclei are no longer coplanar, as they are in the solid state of a biphenyl compound without a 2- or 2′-substituent.

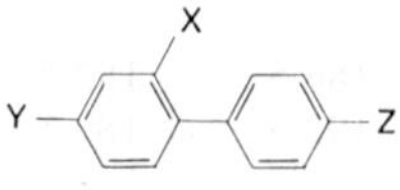

X

Y — [biphenyl] — Z

XXX

Substituent X will broaden the molecule in the usual way, but will also increase the molecular thickness, because the ring carrying substituent Z will lie with its plane out of the plane of the page. Three systems of this type have been studied systematically—the 2- and 2′-substituted 4-*p*-n-alkoxybenzylideneaminobiphenyls (XXXI and XXXII) [10] and the

2-substituted 4,4′-di-(*p*-n-alkoxybenzylideneamino)biphenyls (XXXIII) [88].

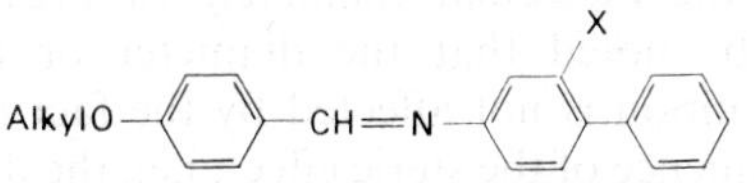

XXXI

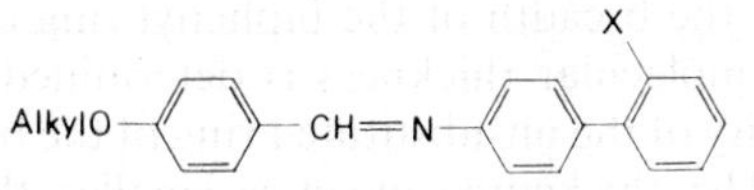

XXXII

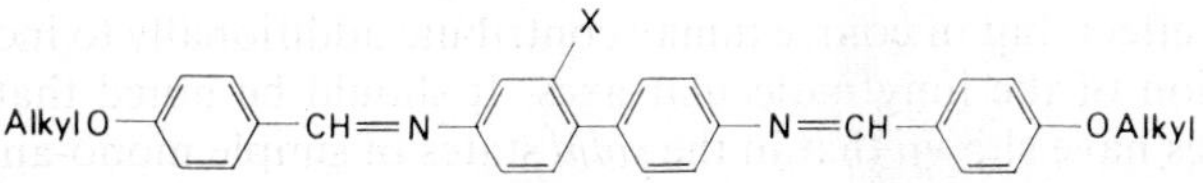

XXXIII

Comparison of the liquid crystal transition temperatures for such compounds with those for the analogous unsubstituted compounds confirms that 2- or 2′- substituents very markedly decrease liquid crystal thermal stabilities. The effects are much greater than would arise if the substituent merely broadened the molecule as in 3′-substituted 4′-n-alkoxybiphenyl-4-carboxylic acids (XXVIII). A combination of breadth and steric effects therefore greatly reduces the thermal stability of the ordered arrangement of molecules in the liquid crystal, and as expected, smectic thermal stability is affected more than nematic thermal stability. In fact smectic properties are not shown by systems (XXXI) and (XXXII) when X is bromo- or iodo-. However, smectic and nematic mesophases occur for system (XXXIII) with a range of substituents (X = F, Cl, Br, I, Me and NO_2) and most information about the effect of this type of substitution has come from studies of this system.

Results for system (XXXIII) are given in Table 4.1.4 and illustrate the magnitude of the decreases in liquid crystal thermal stability for a range of substituents differing in size and dipolarity. The large decreases with bromo- and iodo- substituents show why the simpler mono-anils (XXXI and XXXII) with 2- or 2′-bromo- or -iodo-substituents do not give smectic properties.

TABLE 4.1.4. *Decreases in smectic* and nematic† thermal stabilities from unsubstituted to 2-substituted 4,4′-di-(p-n-alkoxybenzylidene amino)biphenyls*

2-Substituent	Decrease in liquid crystal thermal stability (°C) Smectic	Nematic
F	58·5	29·4
Cl	157·5	84·1
Br	188	99·2
I	220·5	122·6
Me	144·5	85·4
NO_2	156	104·6

* Decreases for the n-decyl ethers only.
† Decreases are average values for the n-octyl–n-decyl ethers.

More information can be obtained from these 2-substituted systems (XXXIII) if the S–N and N–I temperatures and those for the unsubstituted compound are plotted against the molecular diameters or breadths as defined previously. It should be noted that the diameter or breadth calculated according to this definition is not affected by the fact that the molecule is thickened as a consequence of the steric effect, i.e., the diameter of the cylinder is determined by the breadth of the biphenyl ring carrying the substituent X, whereas the molecular thickness is determined by the extent of rotation from coplanarity of the unsubstituted ring of the biphenyl nucleus about the 1,1′-bond. The thickening effect is smaller than the broadening effect, but of course it may contribute additionally to increasing the separation of the long molecular axes. It should be noted that recent X-ray studies have shown that in the *solid* states of simple mono-anils such as benzylideneaniline [133] and its *p,p′*-disubstituted derivatives [134], the phenyl rings are not coplanar. It is not known however whether this twisting of the molecule applies in the liquid crystalline states and whether all anils are similarly affected. This twisting was not taken into account when the effects of lateral substitution on the mesomorphic properties of the mono- and di-anils were first discussed [88] along the lines which now follow. Clearly, however, if twisting about the anil linkages occurs, the molecular diameters will be affected. If however the degree of this deviation from planarity is the same for all anils, then the general situation with respect to the biphenyl ring system should not be altered. Our interest here lies in the difference between the mesomorphic properties of the unsubstituted and substituted biphenyl systems resulting from the additional twisting effect of the substituent in the biphenyl ring system.

Figure 4.1.6(*a*) shows that the N–I temperatures for the 2-substituted di-anils lie on an approximately straight line, implying a close relationship between nematic thermal stability and molecular diameter as found in the case of the acids (XXVIII)—see Fig. 4.1.5(*a*). However, as the 2-substituent increases in size, the increasing mean interplanar angle for the two rings of the biphenyl nucleus must cause the N–I temperatures for the 2-substituted di-anils (XXXIII) to fall more rapidly than if the breadth increasing effect of the substituent alone were operating, because, in addition, the molecules must now be pushed apart by the increase in molecular thickness. Since no deviations from a rectilinear relationship between transition temperature and breadth are observed, the mean interplanar angle must increase proportionately with increasing size of the 2-substituent, and calculations of the interplanar angles based on standard bond lengths, bond angles and van der Waals radii confirm [88] that this is the case. Therefore the increasing interplanar angles simply give the line in Fig. 4.1.6(*a*) a steeper gradient than that in Fig. 4.1.5(*a*) for system (XXVIII) in which breadth increasing effects alone are operating.

For the parent di-anils (XXXIII; X = H—page 131), the minimum geometrically possible interplanar angle calculated [88] on the basis of a van der Waals radius of 1·0 Å for hydrogen is in fact 20·1°. Although the mean interplanar angle for biphenyl itself is about 45° in solution [89], the molecule is known to be planar in the crystal [90, 91], and it is presumed

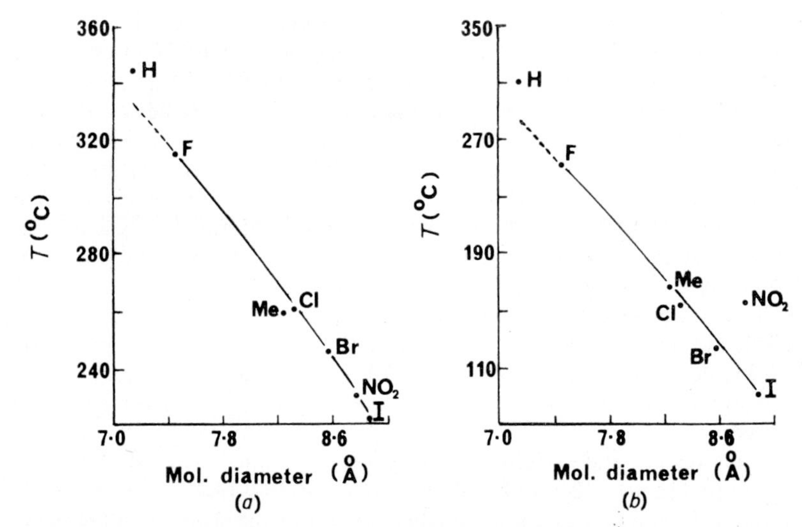

Fig. 4.1.6. Plots of transition temperatures *v.* molecular diameters for 2-substituted 4,4′-di-(*a*-n-alkoxybenzylideneamino) biphenyls
(*a*) average N–I transition temperatures for the n-octyl–n-decyl ethers
(*b*) S–N transition temperatures (n-decyl ethers only).

that crystal forces constrain the molecule to be planar. 2-Substituted derivatives of biphenyl are not however planar in the crystalline state, and even with the small fluoro- substituent, crystal forces are unable to overcome the steric effect within the molecule [92–94]. Now a difference between Fig. 4.1.5(*a*) and Fig. 4.1.6(*a*) is that the point for X = H in Fig. 4.1.6(*a*) lies about 10°C above the line. The large decrease in N–I temperature on passing from X = H to X = F implies that the increase in interplanar angle from parent di-anil to 2-fluoro-di-anil is larger than would be expected. This would be explained if *residual* crystal forces in the ordered nematic melt were without effect on di-anils carrying 2-fluoro- or larger 2-substituents, but were able to constrain the biphenyl ring system of the parent di-anil to adopt a conformation with a mean interplanar angle less than the calculated angle of 20·1°.

Turning now to Fig. 4.1.6(*b*) and comparing this with Fig. 4.1.5(*b*), similarities and differences are obvious. Drawing a virtual straight line through the points for the four halogeno- substituents, the point for X = NO_2 lies above the line. As in the case of 3′-nitro-4′-n-octyloxybiphenyl-4-carboxylic acid (XXVIII), the dipolar effect markedly reduces the effect of the nitro-group on *smectic* thermal stability. The point for X = H also lies well above the line, whereas in Fig. 4.1.5(*b*), this point lies below the line, as would be expected because of the low dipolarity of the C – H bond. The smectic thermal stability therefore falls more on passing to the 2-fluoro- derivative than would be expected if the interplanar angle for the biphenyl nucleus of the unsubstituted di-anil were 20·1°. The effect is more pronounced than for the nematic state, and implies an even larger increase in interplanar angle on 2-fluorination. This would be explained if the stronger *residual* crystal forces in the more ordered smectic state were tending to impose a planar conformation on the biphenyl ring system, as in crystals of biphenyl [90, 91] and its derivatives [95]. The remaining anomaly is that the point for X = Me lies on the line (Fig. 4.1.6(*b*)) and not below it, as would be expected, i.e., the interplanar angle for the smectic state of the 2-methyl di-anil is low in relation to the van der Waals radius (2·0 Å) associated with a spherical model for the Me group. The methyl group is however polyatomic, and not strictly a sphere, and residual crystal forces in the smectic state may arrange the molecule so that the 2′-hydrogen is accommodated in one of the H-C-H angles of the 2-methyl group. This would lower the interplanar angle and enhance the smectic thermal stability for the 2-methyl di-anil. Since the point for X = Me lies close to the curve in Fig. 4.1.6(*a*), the weaker residual crystal forces in the nematic state do not apparently cause this arrangement to be adopted.

Similar results and conclusions have been obtained from studies of a more limited range of 2- and 2′-substituted mono-anils (XXXI and XXXII).

Finally, in support of these conclusions about the 2- or 2′-substituted systems (XXXI, XXXII and XXXIII), much smaller decreases in nematic and particularly smectic thermal stability are given by the corresponding compounds with substituents X in the 3- or 3′- position. Such substituents cannot impose a steric effect on the biphenyl ring system*.

EFFECTS OF LATERAL POLYSUBSTITUTION ON LIQUID CRYSTAL THERMAL STABILITIES

Studies of 2-substituted di-anils (XXXIII) were extended [96] to di-, tri- and tetra-substituted analogues (XXXIV) with the following substituents in the positions indicated: 2,2′-, 2,3′-, 2,6- and 3,3′-dichloro, 2,2′-, 2,5-, 2,6- and 3,3′-dibromo-, 2,2′-, 2,3′-, 2,5- and 3,3′-dimethyl, 2,2′,6-trichloro- and 2,2′,6,6′-tetrachloro-.

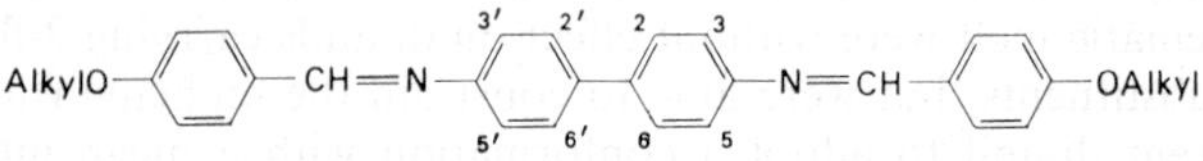

XXXIV

Very large decreases in nematic thermal stability were observed, and smectic properties occurred only with the 3,3′-dichloro- and -dibromo-compounds, in which the biphenyl ring system is not sterically affected. On comparing average results, e.g., for the n-heptyl to n-decyl or n-octyl to n-decyl ethers, the decreases in nematic thermal stability on 2,5-disubstitution were found to be close to the sums of the decreases accompanying 2-monosubstitution and 3-(equivalent to 5-)monosubstitution. This approximate additivity of substituent effect is understandable, because the substituents lie on opposite sides of the molecule. The approximation to additivity was also found, however, with the 2,3′- and 3,3′-disubstituted

* The use of steric effects in producing low melting nematogens is of related interest [147].

compounds, and in the latter case this applied to both smectic and nematic thermal stabilities. This suggests that these molecules adopt *trans*-conformations—effectively 2,5′- and 3,5′-, respectively, according to the numbering in XXXIV. This approximate additivity of the effects of two substituents on liquid crystal thermal stabilities is interesting, as is the fact that information about molecular conformation can be obtained from a knowledge of liquid crystal thermal stabilities. See also [148].

A further example is provided by the 2,2′-disubstituted di-anils (XXXIV). The decreases in nematic thermal stability were much greater than would be expected on an additivity basis if the molecule adopted a *trans*-conformation. The decreases could be explained, however, if the molecules adopted a *cis*-conformation having a considerably greater interplanar angle. This seemingly unlikely possibility is however supported by independent studies involving ultraviolet spectroscopy [97, 101], electron diffraction [98–100] and measurement of dipole moments [102] of 2,2′-disubstituted biphenyls, and X-ray analysis of crystalline derivatives of 2,2′-disubstituted biphenyls [103–105].

The largest decreases in nematic thermal stability were obtained for the 2,2′,6-trichloro- di-anils; the average decrease for the n-heptyl to n-decyl ethers was 222·1°C (from 345°C to 122·9°C). An even larger decrease would be expected for 2,2′,6,6′-tetrachlorination. The only compound prepared in this series was that with alkyl = n-heptyl, and the di-anil did not form liquid crystals. The crystals melted at 111·5°C and the isotropic liquid vitrified at 89°C. The decrease in thermal stability is therefore > 256°C. Such a large decrease in nematic thermal stability is not unexpected, because the effects of substitution are at a maximum in this compound—the interplanar angle will be at least as large as that in the 2,2′-dichloro- di-anils, and each of the four chloro-substituents will exert its full effect in broadening or thickening the molecule, increasing the separation of the long axes of neighbouring molecules.

Other studies of polysubstitution have been made by Goldmacher and Barton [58] and their results confirm [9] the conclusions reached about substituent effects from 5-substituted 6-n-alkoxy-2-naphthoic acids (XXIX) and also sterically hindered biphenyls (XXXIII and XXXIV). Their results are illustrated by the behaviour of the amide (XXXV) [58].

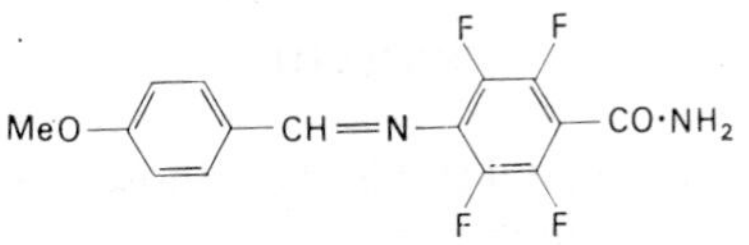

F F
MeO—C₆H₄—CH=N—C₆F₄—CO·NH₂
F F

XXXV

Two of the fluoro-substituents, one on each side of the molecule, will increase molecular breadth and decrease liquid crystal thermal stability. The other two fluoro- substituents will fill in "gaps" between molecules and contribute to the dipolarity of the compound. The unfluorinated amide is purely smectogenic (S–I, 195°C); the tetra-fluoro- amide (XXXV) is also purely smectogenic (S–I, 268°C). Tetrafluorination has therefore enhanced the smectic thermal stability by 73°C, and this provides a striking example of behaviour similar to that of the 5-substituted 6-n-alkoxy-2-naphthoic acids.

Such an example stresses the importance of considering carefully all relevant factors when assessing the effects of lateral substitution of a mesogenic compound.

Liquid Crystalline Compounds containing other Aromatic Ring Systems

Derivatives of fluorene and fluorenone

The molecules of fluorene (XXXVI) and fluorenone (XXXVII) are of comparable thickness but broader than that of biphenyl. Unlike 2-substituted biphenyl molecules however, the molecule of fluorene is planar [89], [106–109], and in addition, the ring–X and ring–Y bonds of 2,7-disubstituted fluorenes are not collinear, because the molecule is bent. It is assumed that the same situation obtains in the molecules of fluorenone and its derivatives.

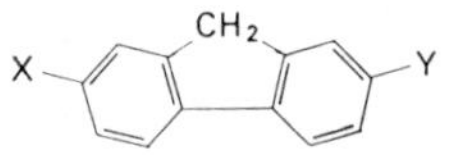

XXXVI (X, CH_2, Y)   XXXVII (X, C=O, Y)

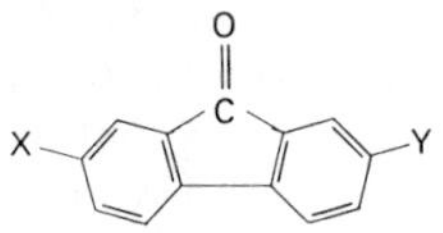

The relative liquid crystal thermal stabilities of derivatives of biphenyl, fluorene and fluorenone depend on the compounds involved, and have been fully discussed elsewhere [53, 54, 110, 149].

Derivatives of phenanthrene

Anils of the type (XXXVIII) prepared from 2-aminophenanthrene give nematic mesophases which are more thermally stable and smectic meso-

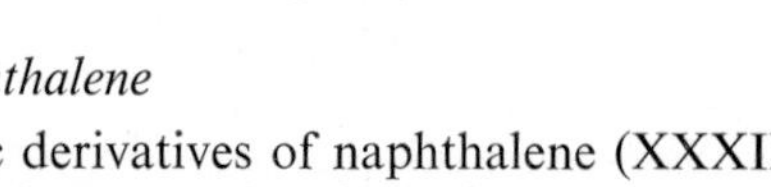

N=CH— —OAlkyl

XXXVIII

phases which are less thermally stable than those from the corresponding 4-*p*-n-alkoxybenzylideneaminobiphenyls.

Derivatives of naphthalene

Many mesogenic derivatives of naphthalene (XXXIX) are known [11]; usually these are 2,6-, 1,5- or 1,4-disubstituted systems. With 2,6-disubstituted naphthalenes the molecule is narrow and approximately linear. With 1,5-disubstituted naphthalenes, the axes of the substituents are not collinear, and the elongated molecule is kinked. With 1,4-disubstituted

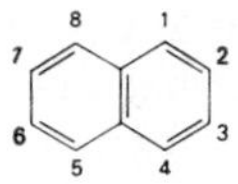

XXXIX

naphthalenes, the axes of the substituents are collinear, but the second ring protrudes from the molecule like a large substituent. On this basis, the predicted order of liquid crystal thermal stability might be:

$$2,6\text{-} \gg 1,5\text{-} > 1,4\text{-}$$

For example, the 2,6-, 1,5- and 1,4-di-(*p*-methoxybenzylideneamino)-naphthalenes all form nematic mesophases and the N–I temperatures are 356°C, 282°C and 263°C, respectively [50], confirming the above order.

LIQUID CRYSTALLINE COMPOUNDS WITH OTHER KINDS OF LATERAL SUBSTITUTION

The effects of introducing lateral substituents into positions other than those in aromatic rings of mesogenic compounds have not yet been studied systematically. Two instances are however worth noting.

Alkylation of a central —CH=X— group

The effect of replacing the hydrogen of such a group by a methyl or ethyl group has been examined, e.g., in anils (—CR=N) prepared from the amine and the methyl or ethyl ketone, and in α- or β-alkyl derivatives of suitably substituted cinnamic acids or cinnamates (—CH=CR— or —CR=CH—). The broadening effect of the alkyl group in all cases reduces liquid crystal thermal stabilities, and smectic mesophases are affected more than nematic.

Branching of terminal alkyl groups

Most commonly, the terminal alkyl groups found in mesogenic compounds are unbranched, because the flexible chain is then at its narrowest in the extended form. Examples of mesogenic compounds with branched terminal alkyl groups are of course known [11]. Regarding the shorter branch as a lateral substituent in a n-alkyl chain, this should increase intermolecular separation and give less thermally stable liquid crystals. Available results confirm that branched chain compounds usually do have lower liquid crystal transition temperatures. In this connection systematic studies have been made [83, 125, 135] on certain alkyl 4-*p*-substituted-benzylideneaminocinnamates (XL). The work is still incomplete, but a considerable number of branched chain esters have been prepared and *all* have lower liquid crystal thermal stabilities than the open chain esters.

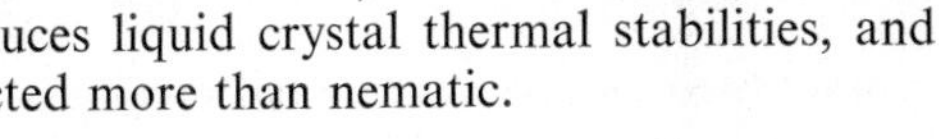

X—C₆H₄—CH=N—C₆H₄—CH=CH—CO·OAlkyl

XL

Some of the results are given in Table 4.1.5 and show that when branching occurs at the first carbon of the chain, the effect on liquid crystal thermal stability is greatest. Branching at any point appears to have a greater effect on nematic than on smectic thermal stability, and movement of the point of branching away from the first carbon of the chain towards the end of the chain causes the transition temperatures to rise again. Further experimental work is needed, however, to confirm the generality of these observations, see reference [135].

TABLE 4.1.5. *Effects of branching of the alkyl chain on the liquid crystal thermal stabilities (S–I or N–I) of esters of type (XL)*

	Transition temperature (°C)†		
	(S–I)	(N–I)	(N–I)
Alkyl	X = Ph	X = CN	X = NO_2
—$CH_2{\cdot}CH_2{\cdot}CH_2{\cdot}Me$	202	132·5	112·5
—$CHMe{\cdot}CH_2{\cdot}CH_2{\cdot}Me$	185	<63	—*
—$CH_2{\cdot}CHMe{\cdot}CH_2{\cdot}Me$	194·5	108	84
—$CH_2{\cdot}CH_2{\cdot}CHMe_2$	200·5	107	88·5

* Not liquid crystalline
† Determined by the capillary melting point method.

Features of Molecular Structure which favour the Formation of Smectic or Nematic/Cholesteric Mesophases

COMPOUNDS FORMING SMECTIC/NEMATIC MESOPHASES

As discussed previously [21], certain guide lines exist regarding features of molecular structure which favour the formation of smectic or nematic mesophases, but the predominance of one or other type of liquid crystal is difficult to explain in many cases. In fact, recent experimental results show that subtle changes in molecular structure can strongly influence the situation. Assuming however that a compound has a suitable molecular structure for liquid crystal formation, certain conditions must be satisfied if the smectic mesophase is to arise:

(*a*) the molecules must be predisposed to form a layer crystal lattice;
(*b*) the terminal intermolecular attractions in the crystal must be relatively weak so that the melting point of the compound is not too high;
(*c*) when the crystal has begun to melt, the lateral intermolecular attractions must be adequate to retain the layer arrangement in the smectic state.

The features of molecular architecture which bear upon condition (*a*) cannot be defined with any certainty. Consequently, no matter how suitable a structure appears, if the crystal preceding the mesophase adopts a non-layer lattice, the mesophase will not be smectic. This uncertainty is always present. Assuming however that the crystal has a layer lattice, satisfaction of (*b*) and (*c*) requires a suitably high ratio of lateral to terminal intermolecular attractions (dispersion forces, induced dipole interactions and dipole–dipole interactions—see p. 129). Herein lies the difficulty, because we must rely on qualitative judgments as to how this ratio is affected by given changes in molecular structure, stereochemistry, etc.

However, it is fairly obvious that dipole moments acting across the long molecular axis may strongly favour smectic behaviour, and for this reason, a compound with a *terminal* function such as ring–CO·OR, ring–CH=CH·CO·OR, ring–CO·NH_2 or ring–O·CF_3 is strongly inclined to be smectogenic. Ionised groups too favour lateral interactions, and salts of carboxylic acids (R·CO_2^- M^+) and amine salts (R·NH_3^+ X^-) are predominantly smectogenic systems [111]. The importance of cross dipoles and lateral interactions has also been emphasized when considering lateral substituent effects (p. 129). Smectic liquid crystals are very sensitive to increases in intermolecular separation, but when the ring—substituent bond is highly dipolar, the decrease in smectic thermal stability is smaller than expected, cf., also the enhanced smectic thermal stabilities of 5-substituted 6-n-alkoxy-2-naphthoic acids [51, 52].

Therefore, a narrow, rod-like molecule with dipoles acting across the long axis should favour smectic phase formation. However, it is often difficult to assess the significance of the number, strength and disposition of the dipoles associated with a given structure. Neighbouring dipoles in a molecule may interact, and there may also be electronic interactions between widely separated dipolar centres in a conjugated molecule which affect the dipole situation [53].

A classical example of the difficulties involved in considering dipole moments is provided by *p*-azoxyanisole (XLI); the molecule appears to

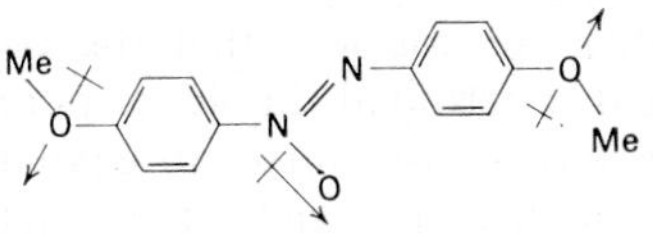

Me O N N O O Me

XLI

have three cross dipoles, but the compound is purely nematogenic. The crystal of *p*-azoxyanisole does not in fact have a layer lattice [24]. This is not, however, a sufficient explanation, for, as already discussed (p. 124), terminal ring–OMe groups always strongly *enhance* the nematic thermal stabilities and in some cases diminish the smectic thermal stabilities relative to the system with a terminal ring–H bond. The reasons remain uncertain as to why the terminal dipoles of ring–OMe groups contribute less to lateral intermolecular attractions than would be thought at first sight. It is possible of course that the ring–OMe dipole may have quite a different orientation to that indicated in (XLI), particularly if rotation about the ring–O bond is relatively unrestricted, or in systems in which the —OMe group may interact conjugatively with the rest of the molecule. Thus, it is known that liquid crystal thermal stabilities change differently on introducing a terminal —OMe group at the end of a n-alkyl chain

[83, 112]. The situation is admittedly different in many respects, and on passing from an ester with a terminal —CH_2·Me group to one with a terminal —CH_2·CH_2·OMe group, both smectic and nematic thermal stabilities are reduced. However, the significant point is that, in this case in which conjugative interactions involving the —OMe group cannot occur, the smectic thermal stability is affected less than the nematic. This example draws attention to the dangers of applying information about a terminal ring–X substituent to another molecule with the same substituent in a different terminal environment.

Problems connected with the effects of terminal ring–X substituents on the relative thermal stabilities of smectic and nematic phases occur with groups other than —OMe. As already indicated (p. 125), nematic and smectic terminal group efficiency orders are different, and there are variations in the relative effects of substituents from one system to another. Whilst certain terminal ring–X substituents favour nematic and particularly smectic thermal stabilities, e.g., —NH·CO·Me, —O·CO·Me and —Ph, others, eg., —NO_2 and —CN, markedly enhance nematic and diminish smectic thermal stabilities. Both nitro- and cyano- groups have

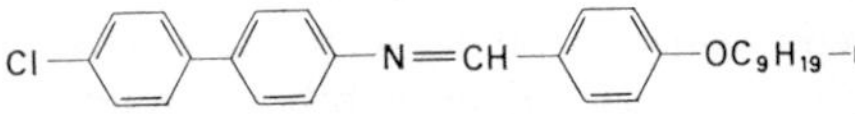

Cl—⟨benzene⟩—⟨benzene⟩—N=CH—⟨benzene⟩—OC_9H_{19}—n

XLII

strong dipoles which act out from the end of the molecule along its major axis, and although such dipoles may well destabilize a smectic molecular arrangement, there is no evidence that permanent dipoles play an important rôle in stabilizing nematic mesophases (p. 127). It is also difficult to interpret the effects of cyano- and nitro- groups in terms of permanent dipoles when we consider that the compound (XLII) [113], with a similar but weaker terminal dipole, is purely smectogenic.

Cyano- and nitro- groups are of course more polarizable than the chloro-substituent, and presumably this and other factors which cannot be assessed qualitatively are responsible for the enhanced thermal stabilities of the nematic mesophases that arise when such terminal ring–X substituents are present.

Even when as many relevant factors as possible are considered, examples such as the above emphasize the great caution which must be exercised in predicting the type of liquid crystal which may predominate in a system.

It is however quite clear that long terminal n-alkyl chains strongly favour formation of smectic mesophases. If the lower homologues of a series of esters, ethers or alkyl aromatic compounds are purely nematogenic, the middle members will form smectic and nematic mesophases, and the long chain homologues will be predominantly or purely smectogenic; many such examples are known. With a fairly smectogenic system such as (XLIII), when R = racemic 2-methylbutyl, only when X = OMe does the ester show nematic in addition to smectic properties. When X = OEt or a longer n-alkyloxy group, the compounds are purely smectogenic. Conversely, the very powerful effect of the cyano- group in promoting nematic mesophase formation is highlighted by the fact that the compound (XLIII) [125] with X = CN and R = n-octyl is still a purely nematogenic system.

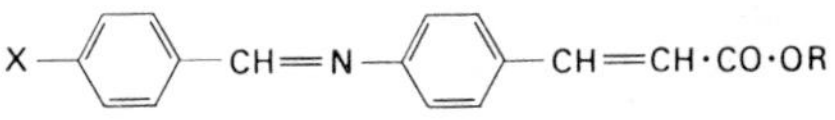

X— —CH=N— —CH=CH·CO·OR

XLIII

Interesting results have been obtained [125, 135] with system (XLIII) showing that very subtle factors may affect the balance between smectic and nematic mesophase formation. The situation is summarized by the data in Table 4.1.6.

TABLE 4.1.6. *Types of liquid crystal formed by some esters (XLIII) of 4-p-substituted benzylideneaminocinnamic acids*

		Liquid crystals formed by esters having different values of n in R				
X	R	0	1	2	3	4
Ph	$—(CH_2)_nPh$	S,N	S	S,N	S	S,N
MeCO·O	$—(CH_2)_nCHMe_2$	S	S,N	S	S,N	—*
CN	$—(CH_2)_nCHMe_2$	All nematic—quite pronounced alternation of N–I temperatures —upper curve where n is *odd*				—*
CN	$—(CH_2)_nPh$	All nematic—the N–I temperatures alternate				
NO_2	$—(CH_2)_nPh$	very markedly—upper curve where n is *even*				

* Not examined.

Irrespective of X, there is a marked alternation in the nematogenic tendencies for both types of ester as n is increased. This manifests itself either by a high alternation of the N–I temperatures, or, in the case of X = Ph— and R = $—(CH_2)_nPh$, and X = MeCO·O— and R = $—(CH_2)_nCHMe_2$, by an alternation between purely smectic, and smectic together with nematic mesophase formation by the compounds. Note that when smectic properties occur, X is either Ph- or MeCO·O-, i.e., ring–X substituents which favour nematic mesophases less than smectic. Only a close study of molecular models reveals a possible explanation of the results in Table 4.1.6. Assuming again that the methylene chain adopts a zig-zag conformation perpetuating the major axis of the rest of the molecule, when n is odd for $—(CH_2)_nPh$ and when n is even for $—(CH_2)_nCHMe_2$, the terminal Ph and Me groups, respectively, project beyond a line defining the perimeter of the rest of the molecule. In fact, the long molecular axes bend over slightly at their extremities. Although a smectic layer comprised of such bent molecules may be quite stable, the side spacing of the molecules in the nematic state to which this layer would give rise will be greater than that in a nematic state comprised of simple rods. The tendency to form such a nematic state comprised of such bent molecules could be so low that a direct S–I transition occurs on heating. This is in fact in accord with the data in Table 4.1.6. Only when the molecules contain terminal ring–CN or ring–NO_2 groups, which strongly favour nematic behaviour, do the esters with R = —$(CH_2)_n$Ph, when n is odd, and R = —$(CH_2)_n$$CHMe_2$, when n is even, give nematic phases. In these cases, the N–I temperatures are much lower than those for the related esters for which n is even and odd, respectively. The marked alternations in transition temperatures (N–I) and liquid crystal type are therefore tentatively explained. These results clearly emphasize how finely the balance may be poised between smectic and nematic phase formation in a compound, and demonstrate well the complex interplay of different molecular parameters determining which liquid crystal predominates in a system.

Having drawn attention to some of the problems involved in assessing the type of liquid crystal which a particular molecular structure may favour, those guiding principles which can be applied with some certainty are now summarized.

1. Salts of carboxylic acids and amine salts are predominantly smectogenic compounds.
2. Terminal ring–CO·OR, ring–CH = CH·CO·OR, ring–CO·NH_2 and ring–O·CF_3 groups strongly favour smectic mesophase formation.
3. Increasing the length of a terminally situated n-alkyl chain favours smectic mesophase formation.
4. Ring–X terminal substituents increase N–I transition temperatures.
5. Ring–X terminal substituents such as —Ph, —NH·CO·Me and —O·CO·Me favour smectic more than nematic mesophase formation.
6. Ring–X terminal substituents such as —CN, —NO_2 and —OMe strongly enhance nematic thermal stabilities and often diminish smectic thermal stabilities.
7. Lateral substituents which broaden a molecule markedly reduce the tendency of a compound to form liquid crystals. Smectic liquid crystals are decreased in thermal stability more than nematic liquid crystals if the substituent is either weakly dipolar, e.g., Me, or large and only moderately dipolar, e.g., Br, I.
8. Lateral substituents which do not broaden the molecule enhance liquid crystal thermal stabilities; the thermal stability of the smectic phase is particularly enhanced if the substituent is dipolar.
9. Steric effects within a molecule markedly reduce the potential of a compound to form liquid crystals; smectic thermal stabilities are affected more than nematic.

Compounds Forming Cholesteric Mesophases

Any compound consisting of geometrically anisotropic molecules that are disposed to arrange themselves parallel to one another without the ends of the molecules lying in planes may exhibit nematic properties when the crystal is heated. If such a compound possesses a centre of asymmetry in the molecule and is optically active, then the liquid crystal formed will be cholesteric, not nematic, i.e., the optical isomers are cholesterogenic.

It is therefore fortuitous that for many years most of the compounds giving cholesteric properties were derivatives of cholesterol. Cholesterol is obtained from natural sources and is optically active, and its esters and other derivatives have the molecular requirements to give a molecular organization which is nematic in type. The result is that the compounds give cholesteric liquid crystals. Had Reinitzer's cholesteryl benzoate been racemic, the first liquid crystal he discovered would have been nematic. The predominance of derivatives of cholesterol amongst compounds exhibiting cholesteric properties is simply a reflection of the fact that cholesterol is a readily available optically active compound giving derivatives which form liquid crystals. The realization that optical activity of the compound is the primary requirement has led to the preparation of many more compounds which exhibit cholesteric properties but are not derived from cholesterol or other sterols. The predominance of derivatives of cholesterol amongst recorded cholesteric compounds should therefore gradually decrease.

The close relationship between the nematic and the cholesteric states means that those factors which determine whether a system gives a smectic or nematic mesophase will also determine whether an optically active system gives a smectic or cholesteric mesophase. Thus structural features which favour nematic behaviour will favour cholesteric behaviour in an optically active system. Of course, if the optically active compound has structural features which strongly favour smectic behaviour, the compound will simply be smectic, but the mesophase may be a twisted smectic [136].

In assessing whether a compound will be cholesterogenic, we do not therefore have to look for subtle features relating the shape of the molecule to that of a derivative of cholesterol or some other sterol. We simply need to know that the compound is optically active and that the molecular geometry and structure are conducive to an organization of the molecules which is nematic in type.

Consequently, in considering in the previous section those features of molecular structure which favour nematic mesophase formation, we have at the same time considered the features of molecular structure which favour cholesteric phase formation in an optically active compound.

However, since the involvement of mesogenic substances derived from sterols in biological processes and membrane functions is a subject of considerable interest, it seems worthwhile reviewing the situation with regard to those derivatives of sterols which are known to form liquid crystals, i.e., to consider how the smectic or cholesteric tendencies of mesogenic derivatives of sterols vary in relation to the structure of the sterol skeleton.

Derivatives of sterols forming liquid crystals

The number of derivatives of pure sterols which have been examined and found to form liquid crystals is limited and comparisons are possible over only a restricted range of compounds.

There is no doubt that cholesterol is a favourable sterol from which to derive mesogenic compounds. Cholesteryl halides and cholesteryl esters of various kinds, e.g., cholesteryl n-alkanoates, cholesteryl ω-phenylalkanoates [145], cholesteryl *O*-alkyl carbonates [145] and α,ω-polymethylene bischolesteryl carbonates exhibit cholesteric properties. The change from an *O*-alkyl carbonate to an *S*-alkyl thiocarbonate of cholesterol appears to increase the Ch–I temperature more than the S–Ch temperature, so widening the temperature range of the cholesteric phase, and derivatives of thiocholesterol also give cholesteric liquid crystals [67, 129]. For the series of cholesteryl n-alkanoates [62], the onset of smectic properties appears, at the earliest, at the heptanoate,* and even for cholesteryl stearate purely smectic behaviour is not observed. Other homologous series of cholesteryl esters behave similarly, and we can infer that derivatives of cholesterol do have a marked tendency for their asymmetric molecules to arrange themselves in an essentially nematic-type of order, resulting in cholesteric behaviour.

Wiegand [114] first studied the influence of modifying the sterol system on liquid crystal behaviour, and found that only derivatives of cholesterol and cholestanol gave cholesteric properties. Other sterol derivatives prepared from epicholesterol (a 3α-sterol) and coprostanol (a sterol in which rings A and B are *cis*) did not form liquid crystals. As pointed out by Knapp and Nicholas [72], these sterols are "spatially kinked", and the absence of liquid crystals is hardly surprising. Based upon experimental evidence, Wiegand concluded that liquid crystal behaviour was found only with derivatives of sterols in which the substituent is in the 3β-position and rings A and B are quasi-planar. This quasi-planar arrangement arises in derivatives of cholestan-3β-ol, because rings A and B are *trans*. For derivatives of cholesterol, a 3β-sterol with a double bond in the 5,6-position (see structure (XLIV)), we cannot describe rings A

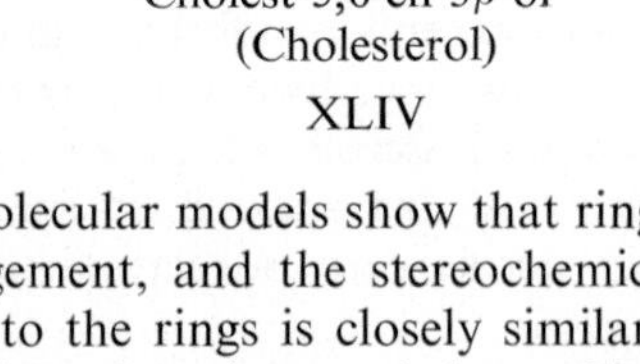

Cholest-5,6-en-3β-ol
(Cholesterol)
XLIV

and B as *trans*, but molecular models show that rings A and B again have a quasi-planar arrangement, and the stereochemical distribution of the substituents attached to the rings is closely similar to that in cholestan-3β-ol. He concluded that when the sterols have this quasi-planar arrangement of rings A and B, the 3β-substituent effectively extends the long axis of the molecule. This stereochemical situation would of course favour

* The monotropic smectic phase was detected by optical means [62]; more recent studies [118] by differential scanning calorimetry have indicated that smectic properties may not in fact be observable in the odd homologues of the series until cholesteryl nonanoate.

liquid crystal formation in general, and not the cholesteric mesophase in particular. Wiegand's diagnosis appears to have been accepted and subsequent studies of mesogenic derivatives have been confined to 3β-sterols. This does not mean that all 3β-sterols give cholesteric liquid crystals, e.g., derivatives of stigmasterol (24-ethylcholest-5,6; 22,23-dien-3β-ol) give exclusively smectic mesophases [64].

On passing from a cholesteryl to a cholestanyl ester, the double bond in the 5,6-position is saturated. This should decrease the polarizability and increase the flexibility of the sterol system, and both factors should lower the cholesteric thermal stability. As discussed elsewhere [21], this is the case for four of the five examples for which comparisons were possible.

Confining attention to 3β-sterols, Wiegand explored the consequences of moving the double bond from the 5,6-position in cholesterol (XLIV) to the 7,8-, 8,9-, 8,14-, and 14,15-positions. In the first three cases, the benzoates were cholesterogenic. The Ch–I temperatures were about the same as for cholesteryl benzoate in the first two instances, but 38°C lower in the third instance; in the fourth case, the benzoate did not form a liquid crystal.

Benzoates of cholestadienols with the two double bonds in various positions were also studied. The thermal stability of the cholesteric phase is obviously sensitive to the position of a single double bond, if this is moved out of ring B, and the introduction of a second double bond does not always give the increase in Ch–I temperature which might be expected. Small increases occur if the double bonds are in conjugation in ring B, but location of one of the two double bonds at the 14,15-position destroys liquid crystallinity, cf., cholest-14,15-en-3β-yl benzoate. Knapp and Nicholas [72] have pointed out that molecular models show that the orientation of the alkyl side chain of a sterol relative to the rest of the molecule is quite different when a double bond is in the 14,15-position compared with the orientation of the chain in sterols which do not have such a double bond, e.g., in cholestan-3β-ol, cholest-5,6-en-3β-ol (XLIV) or cholest-5,6;7,8-dien-3β-ol.

The absence of a liquid crystalline phase for cholest-14,15;24,25-dien-3β-yl benzoate is therefore best attributed to the adverse influence of the double bond in the 14,15-position and not to the double bond in the 24,25-position, despite the fact that there is a large decrease in the Ch–I temperature from cholesteryl benzoate to cholest-8,9;24,25-dien-3β-yl benzoate. In support of this, Knapp and Nicholas [71, 72] showed that both cycloartanyl palmitate (XLV) and the corresponding cycloartenyl ester with a double bond in the 24,25-position are cholesterogenic.

However, double bonds in other positions in the sterol side chain can have important effects. Esters of 24-methylenecycloartenol (XLVI) either give smectic phases or are non-liquid crystalline, but on reduction of the double bond giving 24-methylcyloartanyl esters, cholesteric properties are observed. It appears that either the unsaturated function in the side chain enhances lateral intermolecular attractions to such an extent that smectic mesophases predominate, or that reduction of the double bond gives a more flexible side chain which adopts a conformation more

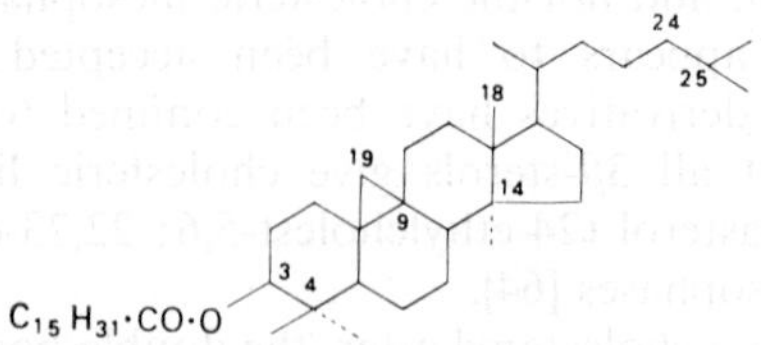

9,19-Cycloanostan-3β-yl palmitate (Cycloartanyl palmitate)

XLV

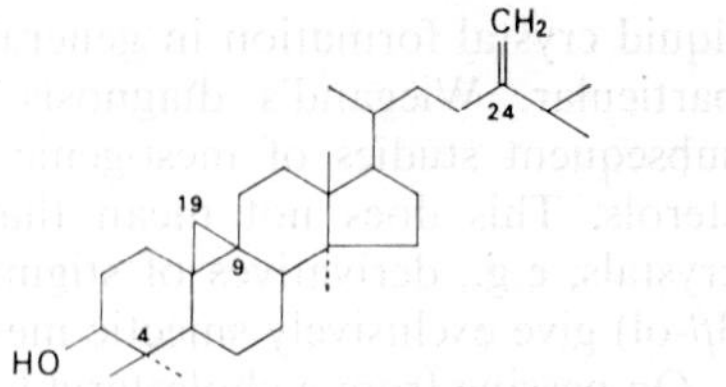

24-Methylenecycloartenol

XLVI

suitable for formation of a liquid crystal with a nematic-type packing. The result was confirmed by studying derivatives of cycloeucalenol which differs from 24-methylenecycloartenol (XLVI) in having only one methyl group in the 4-position. Cycloeucalenyl palmitate did not form a liquid crystal (C–I, 60–62°C), but reduction of the double bond at the 24-position yielded 4α,14α,24(?)-trimethyl-9,19-cyclocholestan-3β-yl palmitate, a cholesterogenic compound (C–I, 62–64°C; I–Ch, 52°C).

It is worth noting that according to the results of Knapp and Nicholas [72], the second methyl group at the 4β-position in 24-methylcycloartanyl palmitate gives an increase of 12°C in the I–Ch transition temperature compared with that of the reduction product of cycloeucalenyl palmitate with only one methyl group at the 4α-position. This appears to be an instance of lateral substitution enhancing cholesteric thermal stability, although in this case the methyl groups project above and below the mean planes of the molecules.

The sensitivity of cholesterogenic tendencies of sterol derivatives is also shown by the behaviour of esters of stigmasterol which differs from cholesterol only in having a double bond at the 22,23-position and an ethyl substituent at the 24-position. These modifications to the sterol side chain eliminate cholesteric properties in stigmasteryl n-alkanoates, stigmasteryl alkyl carbonates and α,ω-polymethylene bis-stigmasteryl carbonates [63–66], and the compounds are smectogenic. This is also the case for β-sitosteryl alkyl carbonates and β-sitosteryl n-alkanoates in which the double bond at the 22,23-position in the analogous stigmasteryl esters is reduced [64], [115]. Only monotropic smectic phases were observed, even with β-sitosteryl n-octanoate. Similarly, ergosteryl n-alkanoates* have recently been shown to be purely smectogenic [121]; ergosterol is 24α-methylcholest-5,6;7,8; 22,23-trien-3β-ol.

As mentioned before, too few mesogenic derivatives of sterols are known to permit any *general* conclusions to be reached from the above review of the factual position, and further studies in this area must be awaited with interest. In this connection, some interesting and elegant studies of the molecular structural dependence of cholesteric liquid crystals were reported recently by Pohlmann, Elser and Boyd [126, 127, 128] in papers presented at the Third International Conference on Liquid

* It should be noted that the S–I temperatures for this series alternate typically as the series is ascended and lie on two smooth curves.

Crystals, Berlin (1970). These studies involved the examination of a considerable number of sterol 3β-alkanoates in which systematic alterations in the nature of the 17β-side chain had been made. The effects of introducing a double bond in ring A—see structure (XLIV)—were also investigated. Interesting information concerning the effects of these changes on the tendencies of the systems to give cholesteric mesophases was presented, but the effects are subtle, and it is still too early to understand the co-ordinating principles behind them. See also references [150, 151].

REFERENCES

[1] FRIEDEL, G. *Annls Phys.* **18,** 273 (1922).
[2] ROBINSON, C. and WARD, J. C. *Nature* **180,** 1183 (1957).
[3] ROBINSON, C., WARD, J. C. and BEEVERS, R. B. *Discuss. Faraday Soc.* **25,** 29 (1958).
[4] SACKMANN, E., MEIBOOM, S. and SNYDER, L. C. *J. Am. chem. Soc.* **89,** 5981 (1967).
[5] CANO, R. *Bull. Soc. fr. Minér. Cristallogr.* **90,** 333 (1967).
[6] CANO, R. *Bull. Soc. fr. Minér. Cristallogr.* **91,** 20 (1968).
[7] LECLERCQ, M., BILLARD, J. and JACQUES, J. *C. r. hebd. Séanc. Acad. Sci., Paris* **266,** 654 (1968).
[8] LECLERCQ, M., BILLARD, J. and JACQUES, J. *Molec. Crystals Liqu. Crystals* **8,** 367 (1969).
[9] GRAY, G. W. *Molec. Crystals Liqu. Crystals* **7,** 127 (1969).
[10] BYRON, D. J., GRAY, G. W., IBBOTSON, A. and WORRALL, B. M. *J. chem. Soc.* 2246 (1963).
[11] KAST, W. In *Landolt-Börnstein*, 6th edn., Springer, Berlin (1960), Vol. II, Part 2a, p. 266.
[12] WALTER, R. *Ber. dt. chem. Ges.* **59,** 962 (1926).
[13] KNIGHT, G. A. and SHAW, B. D. *J. chem. Soc.* 682 (1938).
[14] Gray, G. W. *J. chem. Soc.* 552 (1958).
[15] Kreide, W. *Phys. Z.* **14,** 979 (1913).
[16] VORLÄNDER, D. *Z. phys. Chem.* **126,** 449 (1927).
[17] VORLÄNDER, D. *Ber. dt. chem. Ges.* **54,** 2261 (1921).
[18] VORLÄNDER, D. *Trans. Faraday Soc.* **29,** 910 (1933).
[19] URBAN, R. *Dissertation, Halle* (1921).
[20] MEYER, F. and DAHLEM, K. *Justus Leibigs Annln Chem.* **326,** 334 (1903).
[21] GRAY, G. W. *Molecular Structure and the Properties of Liquid Crystals*, Academic Press, London and New York (1962).
[22] GRAY, G. W. *Molec. Crystals* **2,** 189 (1966).
[23] REINITZER, F. *Mh. Chem.* **9,** 421 (1888).
[24] BERNAL, J. D. and CROWFOOT, D. *Trans. Faraday Soc.* **29,** 1032 (1933).
[25] ARNOLD, H., DEMUS, D. and SACKMANN, H. *Z. phys. Chem.* **222,** 15 (1963).

[26] ARNOLD, H. and SACKMANN, H. *Z. Elektrochem., Ber. Bunsenges. phys. Chem.* **63,** 1171 (1959); *Z. phys. Chem.* **213,** 137, 145 and 262 (1960).
[27] DEMUS, D. and SACKMANN, H. *Z. phys. Chem.* **222,** 127 (1963).
[28] SACKMANN, H. and DEMUS, D. *Z. phys. Chem.* **222,** 143 (1963), **224,** 177 (1963) and **227,** 1 (1964).
[29] SACKMANN, H. and DEMUS, D. *Molec. Crystals* **2,** 81 (1966); *Fortschr. chem. Forsch.* **12** (2), 349 (1969).
[30] ARNOLD, H. *et. al. Molec. Crystals* **2,** 63 (1966); *Z. phys. Chem.* **231,** 407 (1966), **234,** 401 (1967), **239,** 283 (1968) and **240,** 177 and 185 (1969).
[31] SACKMANN, H., DIELE, S. and BRAND, P. Paper presented at the International Crystallographic Conference, Stony Brook, New York, August (1969); *Molec. Crystals Liqu. Crystals* **16**, 105 and **17,** 163 (1972).
[32] DEMUS, D., SACKMANN, H., KUNICKE, G., PELZL, G. and SALFFNER, R. *Z. Naturf.* **23a,** 76 (1968).
[33] DEMUS, D., KUNICKE, G., NEELSEN, J. and SACKMANN, H. *Z. Naturf.* **23a,** 84 (1968).
[34] HERRMANN, K. *Trans. Faraday Soc.* **29,** 972 (1933).
[35] SAUPE, A. *Angew. Chem., Internat. Edn.* **7,** 97 (1968); *Molec. Crystals Liqu. Crystals,* **7,** 59 (1969).
[36] ARORA, S. L., TAYLOR, T. R., FERGASON, J. L. and SAUPE, A. *J. Am. chem. Soc.* **91,** 3671 (1969).
[37] CHISTYAKOV, I. G., SCHABISCHEV, L. S., JARENOV, R. I. and GUSAKOVA, L. A. *Molec. Crystals Liqu. Crystals* **7,** 279 (1969).
[38] MARKAU, K. and MAIER, W. *Chem. Ber.* **95,** 889 (1962).
[39] KRENTSEL, B. A., AMERIK, VU. B. and KONSTANTINOV, I. I. *Dokl. Akad. Nauk SSSR* **165,** 1097 (1965).
[40] FARINHA-MARTINS, A. *C. r. hebd. Séanc. Acad. Sci., Paris* **B268,** 1731 (1969).
[41] FARINHA-MARTINS, A. *J. Phys. Radium, Paris* **30,** C4, No. 11–12, 83 (1969).
[42] GRAY, G. W. and JONES, B. *J. chem. Soc.* 4179 (1953).
[43] BROWN, G. H. and SHAW, W. G. *Chem. Rev.* **57,** 1049 (1957).
[44] BROWN, G. H. *Analyt. Chem.* **41,** No. 13, 26A (1969).
[45] GRAY, G. W. and JONES, B. *J. chem. Soc.* 2556 (1954).
[46] GRAY, G. W., HARTLEY, J. B. and JONES, B. *J. chem. Soc.* 1412 (1955).
[47] GRAY, G. W. and WORRALL, B. M. *J. chem. Soc.* 1545 (1959).
[48] CULLING, P., GRAY, G. W. and LEWIS, D. *J. chem. Soc.* 2699 (1960).
[49] BYRON, D. J., GRAY, G. W. and WILSON, R. C. *J. chem. Soc.* (C), 840 (1966).
[50] WIEGAND Ch. *Z. Naturf.* **9b,** 516 (1954).
[51] GRAY, G. W. and JONES, B. *J. chem. Soc.* 683 (1954).
[52] GRAY, G. W. and JONES, B. *J. chem. Soc.* 236 (1955).
[53] GRAY, G. W., HARTLEY, J. B., IBBOTSON, A. and JONES, B. *J. chem. Soc.* 4359 (1955).
[54] GRAY, G. W. and IBBOTSON, A. *J. chem. Soc.* 3228 (1957).

[55] WEYGAND, C. and LANZENDORF, W. *J. prakt. Chem.* **151,** 221 (1938).
[56] KELKER, K. and SCHEURLE, B. *J. Phys. Radium, Paris* **30,** C4, No. 11–12, 104 (1969).
[57] EABORN, C. and HARTSHORNE, N. H. *J. chem. Soc.* 549 (1955).
[58] GOLDMACHER, J. and BARTON, L. A. *J. org. Chem.* **32,** 476 (1967).
[59] DAVE, J. S. and PATEL, P. R. in *Liquid Crystals* (edited by G. H. Brown, G. J. Dienes and M. M. Labes), Gordon and Breach, New York (1967), p. 363.
[60] DAVE, J. S. and PATEL, P. R. in *Liquid Crystals* (edited by G. H. Brown, G. J. Dienes and M. M. Labes), Gordon and Breach, New York (1967), p. 375.
[61] CASTELLANO, J. A., GOLDMACHER, J. E., BARTON, L. A. and KANE, J. S. *J. org. Chem.* **33,** 3501 (1968).
[62] GRAY, G. W. *J. chem. Soc.* 3733 (1956).
[63] POHLMANN, J. L. W. *Molec. Crystals* **2,** 15 (1966).
[64] POHLMANN, J. L. W. *Molec. Crystals Liqu. Crystals* **8,** 417 (1969).
[65] MAIDACHENKO, G. G. and CHISTYAKOV, I. G. *Zh. obshch. Khim.* **37,** 1730 (1967).
[66] MAIDACHENKO, G. G. and CHISTYAKOV, I. G. *J. gen. Chem. U.S.S.R.* **37,** 1649 (1967).
[67] ELSER, W. *Molec. Crystals Liqu. Crystals* **8,** 219 (1969).
[68] POHLMANN, J. L. W. and ELSER, W. *Molec. Crystals Liqu. Crystals* **8,** 427 (1969).
[69] ENNULAT, R. D. *Molec. Crystals Liqu. Crystals* **8,** 247 (1969).
[70] ELSER, W. in *Liquid Crystals* (edited by G. H. Brown, G. J. Dienes and M. M. Labes), Gordon and Breach, New York, (1967), p. 261.
[71] KNAPP, F. F. and NICHOLAS, H. J. *J. org. Chem.* **33,** 3995 (1968).
[72] KNAPP, F. F. and NICHOLAS, H. J. *J. org. Chem.* **34,** 3328 (1969).
[73] GRAY, G. W. *Molec. Crystals* **1,** 333 (1966).
[74] GALIGNÉ, J. L. *J. Phys. Radium, Paris* **30,** C4, No. 11–12, 4 (1969).
[75] DAVE, J. S. and DEWAR, M. J. S. *J. chem. Soc.* 4616 (1954).
[76] DAVE, J. S. and DEWAR, M. J. S. *J. chem. Soc.* 4305 (1955).
[77] DAVE, J. S. and LOHAR, J. M. *Proc. natn. Acad. Sci. India* **29A,** 35 (1960).
[78] DAVE, J. S. and LOHAR, J. M. *Proc. natn. Acad. Sci. India* **29A,** 260 (1960).
[79] DAVE, J. S. and LOHAR, J. M. *Proc. natn. Acad. Sci. India* **32A,** 105 (1962).
[80] DAVE, J. S. and VASANTH, K. L. in *Liquid Crystals* (edited by G. H. Brown, G. J. Dienes and M. M. Labes), Gordon and Breach, New York (1967), p. 385.
[81] DAVE, J. S., PATEL, P. R. and VASANTH, K. L. *Ind. J. Chem.* **4,** 505 (1966).
[82] DAVE, J. S., PATEL, P. R. and VASANTH, K. L. *Molec. Crystals Liqu. Crystals* **8,** 93 (1969).
[83] GRAY, G. W. and HARRISON, K. J. previously unpublished results.
[84] GRAY, G. W., JONES, B. and MARSON, F. *J. chem. Soc.* 393 (1957).

[85] GRAY, G. W. *Molecular Structure and the Properties of Liquid Crystals*, Academic Press, London and New York (1962), p. 240 *et seq.*

[86] GRAY, G. W. *Molecular Structure and the Properties of Liquid Crystals*, Academic Press, London and New York (1962), p. 247.

[87] GRAY, G. W., JONES, B. and MARSON, F. *J. chem. Soc.* 1417 (1956).

[88] BRANCH, S. J., BYRON, D. J., GRAY, G. W., IBBOTSON, A. and WORRALL, B. M. *J. chem. Soc.* 3279 (1964).

[89] BRAUDE, E. A. and FORBES, W. F. *J. chem. Soc.* 3776 (1955).

[90] DHAR, J. *Indian J. Phys.* **7,** 43 (1932).

[91] DHAR, J. *Proc. natn. Inst. Sci. India* **15,** 11 (1949).

[92] GRAY, G. W., SUTHERLAND, H. H. and YOUNG, D. W. *J. chem. Soc.* 4208 (1965).

[93] HOY, G. T. and SUTHERLAND, H. H. *Z. Kristallogr. Kristallgeom.* **123,** 319 (1966).

[94] SUTHERLAND, H. H., TOLLIN, P. and YOUNG, D. W. *Acta Cryst.* **B24,** 161 (1968).

[95] FIRAG, M. S. and KADAR, N. A. *J. Chem. U.A.R.* **3,** 1 (1960).

[96] BYRON, D. J., GRAY, G. W. and WORRALL, B. M. *J. chem. Soc.* 3706 (1965).

[97] BEAVEN, G. H. in *Steric Effects in Conjugated Systems* (edited by G. W. Gray), Butterworths, London (1958), p. 22.

[98] BASTIANSEN, O. *Acta chem. scand.* **3,** 408 (1949).

[99] BASTIANSEN, O. *Acta chem. scand.* **4,** 926 (1950).

[100] BASTIANSEN, O. and SMEDVIK, L. *Acta chem. scand.* **8,** 1593 (1954).

[101] BEAVEN, G. H. and HALL, D. M. *J. chem. Soc.* 4637 (1956).

[102] LITTLEJOHN, A. C. and SMITH, J. W. *J. chem. Soc.* 2552 (1954).

[103] SMARE, D. L. *Acta Cryst.* **1,** 150 (1948).

[104] FOWWEATHER, F. and HARGREAVES, A. *Acta Cryst.* **3,** 81 (1950).

[105] FOWWEATHER, F. *Acta Cryst.* **5,** 820 (1952).

[106] BURNS, D. M. and IBALL, J. *Nature, Lond.* **173,** 635 (1954).

[107] BURNS, D. M. and IBALL, J. *Proc. R. Soc.* **A227,** 200 (1955).

[108] BORTNER, M. H. and BROWN, G. M. *Acta Cryst.* **7,** 139 (1954).

[109] BROCKWAY, L. O. and KARLE, J. *J. Am. chem. Soc.* **66,** 1974 (1944).

[110] GRAY, G. W. *Molecular Structure and the Properties of Liquid Crystals*, Academic Press, London and New York (1962), p. 241 *et seq.*

[111] KAST, W. in *Landolt-Börnstein*, 6th edn., Springer, Berlin (1960), Vol. II, Part 2a, p. 288–289.

[112] WEYGAND, C., GABLER, R. and BIRCON, N. *J. prakt. Chem.* **158,** 266 (1941).

[113] GRAY, G. W. *Molecular Structure and the Properties of Liquid Crystals*, Academic Press, London and New York (1962), p. 172.

[114] WIEGAND, Ch. *Z. Naturf.* **4b,** 249 (1949).

[115] KUKSIS, A. and BEVERIDGE, J. M. R. *J. org. Chem.* **25,** 1209 (1960).

[116] JAEGER, F. M. *Recl. Trav. chim. Pays-Bas Belg.* **26,** 334 (1907).

[117] DEWAR, M. J. S. and GOLDBERG, R. S. *J. Am. chem. Soc.* **92,** 1582 (1970).

[118] Davis, G. J., Porter, R. S. and Barrall, E. M. *Molec. Crystals Liqu. Crystals* **10,** 1 (1970).
[119] Arora, S. L., Fergason, J. L. and Saupe, A. *Molec. Crystals Liqu. Crystals* **10,** 243 (1970).
[120] de Vries, A. *Molec. Crystals Liqu. Crystals* **10,** 31 (1970).
[121] Knapp, F. F. and Nicholas, H. J. *Molec. Crystals Liqu. Crystals* **10,** 173 (1970).
[122] Young, W. R. *Molec. Crystals Liqu. Crystals* **10,** 237 (1970); Young, W. R., Haller, I. and Aviram, A. *I.B.M. J. Res. Develop.* **15,** No. 1, 41 (1971).
[123] Elser, W. and Ennulat, R. D. *J. phys. Chem.* **74,** 1545 (1970).
[124] Elser, W., Pohlmann, J. L. W. and Boyd, P. R., Proc. 3rd Int. Conf. on Liquid Crystals, Berlin (1970); *Molec. Crystals Liqu. Crystals* **15,** 175 (1971).
[125] Gray, G. W. and Harrison, K. J. *Molec. Crystals Liqu. Crystals* **13,** 37 (1971).
[126] Pohlmann, J. L. W., Elser, W. and Boyd, P. R. *Molec. Crystals Liqu. Crystals* **13,** 243 (1971).
[127] Elser, W., Pohlmann, J. L. W. and Boyd, P. R. *Molec. Crystals Liqu. Crystals* **13,** 255 (1971).
[128] Pohlmann, J. L. W., Elser, W. and Boyd, P. R. *Molec. Crystals Liqu. Crystals* **13,** 271 (1971).
[129] Elser, W., Pohlmann, J. L. W. and Boyd, P. R. *Molec. Crystals Liqu. Crystals* **11,** 279 (1970).
[130] Gray, G. W. *Molecular Structure and the Properties of Liquid Crystals*, Academic Press, London and New York (1962), p. 200–205.
[131] Sackmann, H. and Demus, D., information presented, Colloque sur les Cristaux Liquides, Pont à Mousson, France, July 1971.
[132] Fergason, J. L., Conference Lecture, Colloque sur les Cristaux Liquides, Pont à Mousson, France, July 1971.
[133] Bürgi, H. B. and Dunitz, J. D. *Helv. chim. Acta* **53,** 1747 (1970).
[134] van der Veen, J. and Grobben, A. H. *Molec. Crystals Liqu. Crystals* **15,** 239 (1971).
[135] Gray, G. W. and Harrison, K. J. *Symposium of the Faraday Society* No. 5, 54 (1971); Coates, D., Gray, G. W. and Harrison, K. J. *Molec. Crystals Liqu. Crystals* **22,** 99 (1973).
[136] Helfrich, W. and Chan, S. OH. *Molec. Crystals Liqu. Crystals* **14,** 289 (1971).
[137] de Vries, A. and Fishel, D. L. *Molec. Crystals Liqu. Crystals* **16,** 311, (1972).
[138] Levelut, A.-M. and Lambert, M. *C. r. hebd. Séanc. Acad. Sci., Paris* **B272,** 1018 (1971).
[139] Dietrich, H. J. and Steiger, E. L. *Molec. Crystals Liqu. Crystals* **16,** 263 (1972).
[140] Gray, G. W., Plenary Lecture on Recent Advances in Preparations of Liquid Crystals presented at the 4th Int. Conf. on Liquid Crystals, Kent, Ohio, U.S.A. (1972), *Molec. Crystals Liqu. Crystals* **21,** 161 (1973) and references, cited therein.

[141] GRAY, G. W., HARRISON, K. J. and NASH, J. A. *Electron. Lett.* **9,** 130 (1973).

[142] SCHUBERT, H. and DEHNE, H. *Z. Chem.* **7,** 241 (1972).

[143] GRAY, G. W. and NASH, J. A. *Molec. Crystals Liqu. Crystals* (1973), in press; SCHUBERT, H. *Wiss. Z. Univ. Halle,* XIX '70 M, H. 5, S. 1; CHAMPA, R. A. *Molec Crystals Liqu. Crystals* **19,** 233 (1973).

[144] HIRATA, H., WAXMAN, S. N., TEUCHER, I. and LABES M. M. *Molec. Crystals Liqu. Crystals* **20,** 343 (1973).

[145] ELSER, W., POHLMANN, J. L. W. and BOYD, P. R. *Molec. Crystals Liqu. Crystals* **20,** 77 and 87 (1973).

[146] DE JEU, W. H., VAN DER VEEN, J. and GOOSSENS, W. J. A. *Solid St. Commun.* (1973), in press.

[147] YOUNG, W. R., AVIRAM, A. and COX, J. *Angew. Chem.* **83,** 399 (1971); YOUNG, W. R., HALLER, I. and AVIRAM, A. *Molec. Crystals Liqu. Crystals* **15,** 311 (1972).

[148] VAN METER, J. P. and KLANDERMAN, B. H., Proc. 4th Int. Conf. on Liquid Crystals, Kent, Ohio (1972), to be published.

[149] GRAY, G. W., HARRISON, K. J. and NASH, J. A. Paper presented at Am. Chem. Soc. Symp. on Ordered Fluids and Liquid Crystals, Chicago (1973).

[150] ELSER, W. *Cholesteric Liquid Crystals*, Adv. in Liquid Crystal Research, Academic Press, New York, to be published.

[151] ATALLAH, A. M. and NICHOLAS, H. J. *Molec. Crystals Liqu. Crystals* **17,** 1 (1972); **18,** 321 and 333 (1972); **19,** 217 (1973).

4.2 INFLUENCE OF MOLECULAR STRUCTURE AND COMPOSITION ON THE LIQUID CRYSTALS FORMED BY MIXTURES OF NON-AMPHIPHILIC COMPOUNDS

J. S. Dave and *R. A. Vora*

Just as the melting points of compounds are depressed by the addition of other substances, so also are the liquid crystal transition temperatures of mesogens lowered by the presence of foreign substances. When a mesogen is mixed with another mesogen or a non-mesogen, the solid-mesophase, mesophase-mesophase and mesophase-amorphous liquid transition temperatures may be depressed; the degree of depression will depend upon the concentration of the added component in the mixture. If both the components of a range of mixtures give enantiotropic mesophases by themselves, the mixtures will also give enantiotropic mesophases with transition temperatures depending on the temperature ranges over which the mesophases of the two separate components are thermally stable. Binary mixtures of a mesogen with other non-mesogenic compounds also exhibit mesophases, which may be enantiotropic or monotropic. Such mixtures form mixed liquid crystals over a range of temperature and concentration, depending upon the nature of the added non-mesogenic component. A non-mesogenic compound mixed with an enantiotropic mesogenic compound, however, will give an enantiotropic mesophase within a certain range of composition, even though this may be small. Liquid crystallinity in binary mixtures of non-mesogenic compounds is also known. The phenomenon of mixed liquid crystals may be called mixed mesomorphism. Such binary mixtures exhibiting mixed mesomorphism can therefore be of three types: (1) where both the components are mesogens, (2) where one component is a mesogen and the other is a non-mesogen, (3) where both the individual components are non-mesogens. Let us first of all mention some of the work involving such systems carried out prior to 1939.

Schenck [1] first showed that the amorphous liquid–nematic transition temperature for *p*-azoxyanisole was lowered by the addition of other substances. He also calculated the molecular depression of the transition temperature in the case of *p*-azoxyanisole and from this he calculated the heat of transition. Later, Kast [2] showed that various samples of *p*-azoxyanisole containing different amounts of impurities gave different values for the critical frequency of the applied electric field at which the orientation of the mesomorphic phase by the field disappeared, and Svedberg [3] reported that the anisotropy of the nematic phase of *p*-azoxyanisole decreased with addition of amorphous solvents.

Meantime, Lehmann [4] studied binary mixtures of substances, one or both of which exhibited mesomorphism, and measured the transition temperatures; later, a detailed qualitative study of the lowering of the crystal–mesophase transition temperature caused by mixing several pairs of mesogenic substances was carried out by Vorländer and Ost [5]. In some cases the crystal-mesophase transition point of the mixture was calculated empirically and compared with the observed value.

Prins [6] and de Kock [7] studied binary systems in which one or both components individually could form liquid crystals. Binary mixtures of *p*-azoxyphenetole and *p*-azoxyanisole, both of which give nematics, form a continuous series of mixed liquid crystals, the clearing point curve falling rectilinearly from the nematic–amorphous liquid transition point of the former to that of the latter component. However, when *p*-azoxyanisole and *p*-methoxycinnamic acid, both of which again give nematics, were mixed, their nematic–amorphous liquid transition points as well as the melting points (crystal-nematic) were depressed and the upper

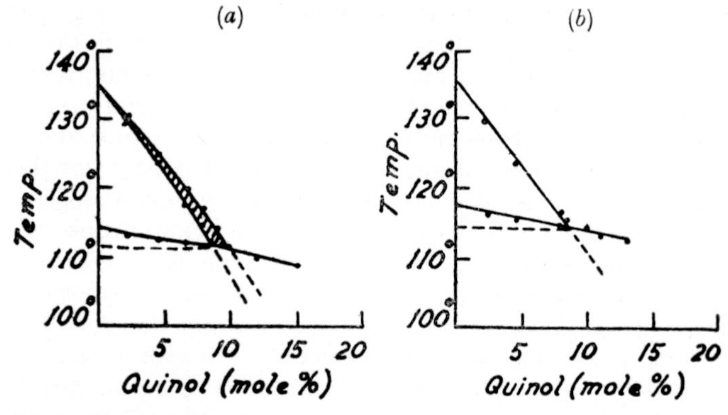

FIG. 4.2.1. The phase diagram for the system *p*-azoxyanisole-quinol.
(*a*) according to de Kock [7]
(*b*) according to Dave and Dewar [8]

transition temperature *v.* concentration curve exhibited a rounded minimum. They discussed the general problem in terms of the phase rule. They concluded that the transition from anisotropic to amorphous isotropic liquid should not in general occur sharply for two-component systems, but that there should, as a rule, be a range of temperature over which two liquid phases (of different composition) can coexist. The phase behaviour of one such system, *p*-azoxyanisole–quinol, studied by de Kock is illustrated in Fig. 4.2.1(*a*), taken from de Kock's paper. Here quinol does not form a liquid crystal on fusion, whereas *p*-azoxyanisole does; in the shaded area of the diagram, according to de Kock, two distinct liquids, one isotropic and the other anisotropic, coexist.

In the early 1950's, Dave and Dewar [8] re-investigated this particular problem by studying the behaviours of various binary mixtures of different compositions, each mixture comprising the nematogen, *p*-azoxyanisole, and a non-mesogenic compound. In all the cases studied, the liquid phase was apparently a homogeneous single phase which was or was not anisotropic. They did not find any evidence for two conjugate liquid phases, one isotropic and the other anisotropic coexisting over a range of temperature for a binary mixture of given composition. Moreover, the breaks in the cooling curves for binary mixtures of given composition were similar in form to the break observed for the pure mesogen, *p*-azoxyanisole. Dave and Dewar then very carefully repeated the experiment of de Kock with *p*-azoxyanisole and quinol. The phase diagram obtained by Dave and Dewar [8] is given in Fig. 4.2.1(*b*) and does not show any indication of the two-phase liquid system observed by de Kock.

The transitions were followed by de Kock mainly by observing the change in appearance of the liquid from cloudy to clear without adequate temperature control. It is, therefore, not surprising that the transitions were not quite sharp. Dave and Dewar concluded that de Kock's observations were based on pre-conceived ideas rather than on direct evidence, and noted that no results were quoted for the critical range of composition corresponding to the horizontal break in the melting line in Fig. 4.2.1(*a*). Dave and Dewar, however, pointed out that in mixed systems of this kind it is thermodynamically impossible for a normal first order phase transition to occur. The possibilities are that either (1) the transition from anisotropic liquid to amorphous isotropic liquid is not a first order transition, in which case, the two phases would necessarily have identical compositions or (2) the two phase region is too small to be detected by normal methods of investigation due to the small heats of transition involved at mesophase-amorphous liquid transitions. Dave and Dewar, could not distinguish by their results between the two possibilities and suggested that the same considerations will apply generally for anisotropic liquids containing two components, such liquids being in general true solutions and not mixtures of phases.

Returning again to earlier studies, Bogojawlensky and Winogradow [9] and later Walter [10] studied mixtures of isomorphous substances and reported the formation of mixed liquid crystals from pairs of substances of which both or only one could individually form nematics. They pointed out that the transition lines separating the anisotropic and amorphous liquid regions in the phase diagrams were linear, implying that the transition from mesophase to amorphous liquid is sharp and that a well defined boundary separates the anisotropic and amorphous liquids, i.e., that there is no intermediate two phase zone. Perfectly homogeneous mixed liquid crystals were formed, analogous to a similar series of ideal solid solutions of two components, and the mesomorphic states of such mixtures were perfectly homogeneous. Since the transition lines were also straight in cases where one component, but not the other, formed a nematic, they deduced that the second non-mesogenic component must have a potential or a latent ability to give the nematic form, but that this cannot normally be observed because the temperature of the transition from the amorphous liquid to the nematic lies below the normal melting point; this virtual transition temperature could however be determined by extrapolation. Such latent transitions were actually observed in some cases in supercooled melts.

Dave and Dewar, however, pointed out that transition lines are not invariably straight, cf. the binary system *p*-azoxyanisole and *p*-methoxy-cinnamic acid mentioned earlier. They obtained different values for the virtual transition temperature of 4,4′-dichloroazoxybenzene depending upon whether the second component used was *p*-azoxyanisole or *p*-azoxyphenetole. This observation can be reconciled with the views of Bogojawlensky and Winogradow only if the parts of the transition lines which cannot be obtained experimentally bend appreciably, and in that case reliable extrapolation is not possible. Reliable extrapolation of the transition curve would it seemed be possible only in the case of binary systems comprising components of similar molecular shape and structure (see pp. 157–8) and for which the course of the extrapolated transition line could be checked over a reasonable distance by the observed monotropic mesomorphic behaviour of the supercooled melt [11].

Dewar and Goldberg [12] in an attempt to examine the role of *p*-phenylene groups in the formation of nematic liquid crystals studied binary mixtures of isomorphous substances where one or both of the components were mesogens. The transition lines are either linear or slightly curved. They obtained the same value for the latent or virtual transition temperature for *p*-phenylene-di-4-methoxybicyclo[2,2,2]octane-1-carboxylate by extrapolation of the transition lines as that obtained for mixtures of the ester with two different nematic liquid crystals. They, however, mentioned that this extrapolation procedure is known to be unreliable, because the transition lines may show unexpected bends [8]. The extrapolation in this particular case is however supported by the concordant results from the two different systems.

Schroeder and Schroeder [13] obtained latent transition temperature values for *N*-*p*-nitrobenzylidene-*p*-anisidine and *N*-*p*-nitrobenzylidene-*p*-toluidine by extrapolation of the transition lines for mixtures with 4,4′-di-n-hexyloxyazoxybenzene (a substance giving a nematic mesophase and a monotropic smectic mesophase); they stated that their extrapolated values compared well with those that they obtained for these substances by extrapolating transition lines for binary mixtures studied by Dave and Dewar [8]. They mentioned however that the phase transitions in the case of the mixtures studied by them occurred over a temperature range, whether the transition was smectic-nematic or nematic-amorphous liquid, and that this range became larger with increase in the concentration of the non-mesogenic component except at a maximum in the transition curve. These authors discussed the maxima observed for certain of the nematic-amorphous liquid and smectic-nematic transition lines in terms of complexing between the components of the mixtures and therefore used the higher transition temperatures for the purposes of extrapolation. It should be pointed out here that Dave *et al.* [14, 15] in their study of mixed systems giving more than one type of mesophase and various other workers who have studied such systems have all reported that the transitions are sharp; there is no evidence for a two phase region existing over a range of temperature for a binary mixture of given composition.

Furthermore, Dave and Lohar [16], Dave and Vasanth [17], Demus [18],

Castellano, McCaffrey and Goldmacher [19] and Pohl and Steinsträsser [20] have reported binary systems of isomorphous components which individually give liquid crystals. The transition lines were almost rectilinear and the transitions were sharp. Continuous series of mixed liquid crystals were formed over the entire range of concentration.

Castellano, McCaffrey and Goldmacher [19] used various mixtures of nematogenic compounds to obtain materials with lower melting points and wide nematic ranges. Binary mixtures of nematogenic compounds with subtle differences in molecular structure did not exhibit minima in the nematic-amorphous liquid transition temperature curves with changing molar composition, although eutectic points for the crystal-nematic transition temperatures were obtained; the nematic-amorphous transition temperatures formed a smooth curve over the entire range of molar composition. These authors obtained a further lowering in the crystal-nematic transition temperature by preparing ternary mixtures.

Pohl and Steinsträsser [20] have observed that binary mixtures of *p,p'*-disubstituted benzylideneanilines with each other or with structurally related compounds always show the same eutectic composition of two moles of low melting and one mole of higher melting component; some of these eutectic compositions had melting points below 0°C and were characterized by particularly wide temperature ranges for the mesophases —up to 120°C. On the basis of differential thermal analysis, as well as X-ray diffraction and N.M.R. measurements, they proposed that the considerably enlarged liquid crystal temperature range was attributable to a nematic secondary structure consisting of freely mobile ribbons of differing length each one consisting of three molecular chains; in each chain the molecules adopt a head to tail sequence, which is reversed in adjacent chains.

According to Dave and Dewar, any substance with elongated molecules would form a liquid crystal if it could be obtained in liquid form at a sufficiently low temperature. If the molecules are anisodimensional (elongated) or if the force fields round them are not spherically symmetrical, the internal energy of the liquid will be less if the molecules are all oriented parallel to one another in an appropriate mutual relationship. In other words a mesophase must have a lower energy than the normal amorphous liquid, and this implies that the normal liquid should undergo transition to the mesomorphic form if it can be cooled sufficiently without crystallisation taking place. Whether the mesomorphic form can be observed in practice will of course depend on the melting point of the solid; in most cases the transition will in fact be virtual, i.e., it will lie well below the melting point, and the mesophase will not be observed. In a mixture of two different substances both with elongated molecules, liquid crystal formation will depend on two factors: first, the ability of the molecules to pack into a single liquid crystal "lattice" and, secondly, the mean orientational cohesive energy density of the mixture, i.e., the decrease in energy on orientation within the liquid crystal. If the two components are of similar size and shape, the steric factors will be about the same for mixtures of all compositions; the orientational cohesive energy density will vary more or less linearly with composition, and therefore, the transition line in the phase diagram should be more or less straight. If, however, the molecules of the two components differ appreciably, there will probably be more difficulty in packing them together. The transition temperature should then be less than that predicted for the "ideal" behaviour considered. The transition line in the phase diagram should then be concave upwards, and when the components differ markedly, it should show a minimum. Such a behaviour has been observed by de Kock [7] for *p*-azoxyanisole: *p*-methoxycinnamic acid, by Dave and Dewar [8] for *p*-azoxyanisole: *p*-methoxybenzoic acid and by Dave and Lohar [21] for *p*-azoxyanisole: *p*-ethoxybenzoic acid.

Dave and Dewar [8] studied a number of systems in which the nematogen *p*-azoxyanisole was mixed with a second component—a non-mesogen—of the type:

$$p\text{-A-}C_6H_4\text{-B-}C_6H_4\text{-C-}p$$

where B is azo, azoxy or —CH:N—; all these molecules, with given groups A and C, should be similar in shape and size. The transition lines were almost linear. The main differences in the phase diagrams lay in the slopes of the anisotropic liquid-amorphous liquid transition curves. They proposed that the slope of the transition line should, in such cases, provide an indication of the virtual transition temperature of the second non-mesogenic component, the virtual transition temperature being higher, the lower the slope value. Therefore, the lower the slope value, the greater will be the tendency of the non-mesogenic component to form a liquid crystal and *vice versa*. In Table 4.2.1 are given the transition line slopes for the mixtures of non-mesogens with *p*-azoxyanisole studied by Dave and Dewar [8].

Dave and Dewar reached the following interesting conclusions from these results. Slopes are similar for isomeric pairs of Schiff's bases, e.g., No. 16 and 17 in the Table; this is not surprising, since such pairs of

TABLE 4.2.1. *Transition line slopes (in °C per 10% change in molar composition) for compounds of the type* p-*A-C_6H_4-B-C_6H_4-C*-p *admixed with* p-*azoxyanisole.*

System No.	A	B	C	Slope	System No.	A	B	C	Slope
1	H	CH:N	H	29·5	12	OMe	CH:N	Br	11·0
2	H	N:N	H	28·0	13	Cl	N:NO	Cl	10·0
3	Me	CH:N	H	26·0	14	OMe	CH:N	Cl	9·5
4	H	N:NO	H	25·0	15	Cl	CH:N	OMe	9·0
5	H	CH:N	NMe_2	23·0	16	Me	CH:N	OMe	9·0
6	H	CH:N	OMe	22·0	17	OMe	CH:N	Me	8·5
7	Me	CH:N	Br	16·0	18	NO_2	CH:N	Me	8·0
8	Me	CH:N	Me	14·5	19	OMe	CH:N	NMe_2	7·5
9	Cl	CH:N	Cl	14·5	20	OMe	CH:N	OMe	4·0
10	Cl	CH:N	Me	14·5	21	NO_2	CH:N	OMe	2·5
11	Me	CH:N	NMe_2	13·0					

molecules are very similar. Secondly, two polar terminal groups are necessary for a low value of the slope of the transition line. Thirdly, the effects of the terminal groups are approximately additive; Dave and Dewar were able to ascribe "group slope" values for the terminal groups, such that the slope of the transition line for a given compound could be expressed as the sum of the group slopes for the terminal substituents present. From this they deduced an order of group efficiency in promoting nematic behaviour:

$$\underset{0{\cdot}5}{NO_2} > \underset{2{\cdot}0}{OMe} > \underset{5{\cdot}6}{Me_2N} > \underset{7{\cdot}2}{Me} = \underset{7{\cdot}2}{Cl} > \underset{8{\cdot}9}{Br} > \underset{19{\cdot}0}{H}$$

Fourthly, Me and Cl substituents have similar effects; this may be a consequence of their similar size. Fifthly, the presence of an NO_2 group has a very large effect. From this study of mixtures involving similar non-mesogenic compounds differing in their central groups, they deduced that the effect of the central group is either negligible or small.

Dave and Lohar [22] later studied more such binary systems comprising in this case non-mesogenic Schiff's bases, having a wider range of different terminal groups, mixed with *p*-azoxyanisole, and obtained similar results from which they deduced the following order of group efficiency in promoting nematic properties:

$$\underset{0{\cdot}5}{NO_2} > \underset{1{\cdot}0}{OEt} > \underset{2{\cdot}0}{OMe} > \underset{2{\cdot}4}{OCO{\cdot}Et} > \underset{3{\cdot}0}{OCO{\cdot}Me} > \underset{5{\cdot}5}{NMe_2} > \underset{7{\cdot}2}{Me}$$

$$= \underset{7{\cdot}0}{Cl} > \underset{9{\cdot}0}{Br} > \underset{12{\cdot}5}{I} > \underset{14{\cdot}3}{OH} > \underset{19{\cdot}0}{H}$$

Dave and Vasanth [11, 23] extended this study to a range of other binary systems involving *p*-acetoxybenzylidene-*p*-phenetidine (which forms a nematic) and Schiff's bases which do not themselves form liquid crystals. The order of group efficiency was:

$$\underset{0{\cdot}5}{NO_2} > \underset{1{\cdot}0}{OEt} > \underset{2{\cdot}0}{OMe} > \underset{2{\cdot}5}{OPr} > \underset{5{\cdot}6}{NMe_2} > \underset{7{\cdot}2}{Me} = \underset{7{\cdot}2}{Cl} > \underset{8{\cdot}4}{F}$$

$$= \underset{8{\cdot}5}{Br} > \underset{11{\cdot}0}{I} > \underset{17{\cdot}2}{NEt_2}$$

The group efficiency order is therefore the same whether the mesogen is *p*-azoxyanisole or *p*-acetoxybenzylidene-*p*-phenetidine. From all the above results, the combined group efficiency order is:

$$\underset{0{\cdot}5}{NO_2} > \underset{1{\cdot}0}{OEt} > \underset{2{\cdot}0}{OMe} > \underset{2{\cdot}4}{OCO{\cdot}Et} = \underset{2{\cdot}5}{OPr} > \underset{3{\cdot}0}{OCO{\cdot}Me} > \underset{5{\cdot}6}{NMe_2} > \underset{7{\cdot}2}{Me}$$

$$= \underset{7{\cdot}2}{Cl} > \underset{8{\cdot}4}{F} = \underset{8{\cdot}5}{Br} > \underset{11{\cdot}0}{I} > \underset{14{\cdot}3}{OH} > \underset{17{\cdot}2}{NEt_2} > \underset{19{\cdot}0}{H}$$

Dave and Vasanth [11] also studied binary systems comprising *p*-n-butoxybenzoic acid (a nematogen) with non-mesogenic *p*-substituted benzoic acids.

They obtained the following order of group efficiency:

$$\underset{1{\cdot}0}{OMe} > \underset{1{\cdot}6}{CH_3} = \underset{1{\cdot}6}{Cl} > \underset{1{\cdot}8}{Br} > \underset{2{\cdot}3}{I} > \underset{7{\cdot}2}{OH}$$

Apparently therefore the efficiency order remains the same, but the magnititudes of the group slope values in this case are low compared with those obtained in the study of Schiff's bases with *p*-acetoxybenzylidene -*p*-phenetidine [11, 23] or *p*-azoxyanisole [8, 22]. They ascribed this to a possible closer packing of the molecules of the benzoic acids (III) than that of the molecules of the Schiff's bases (I) with one another or with *p*-azoxyanisole (II).

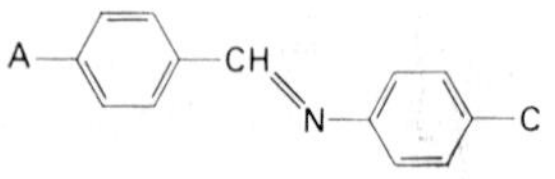

I

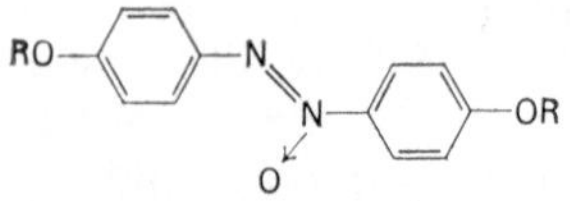

II

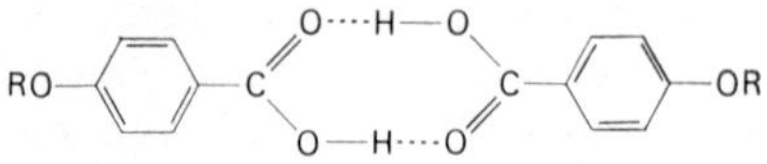

III

Dave and Lohar [21] and Dave and Vasanth [11] also studied binary mixtures of non-mesogenic Schiff's bases with *p*-methoxycinnamic acid and *p*-n-butoxybenzoic acid, both nematogens. They observed that mixed liquid crystals were formed, but the slopes of the transition lines differed markedly from those obtained with the mixed liquid crystals formed by the same Schiff's bases either with *p*-azoxyanisole or *p*-acetoxybenzylidene-*p*-phenetidine, and no consistent order of group efficiency could be obtained. See for example the results in Table 4.2.2. They ascribed this behaviour to difficulties in packing together of the molecules of the acids and the Schiff's bases.

TABLE 4.2.2. *Transition line slopes (in °C per 10% change in molar composition) for compounds of the type* p-*A*-C_6H_4-*CH*=*N*-C_6H_4-*C*-p *admixed with* p-*azoxyanisole (I) or* p-*methoxycinnamic acid (II). Dave and Lohar* [21]

A	C	Slope I	Slope II	A	C	Slope I	Slope II
OMe	OEt	3·0	11·5	OH	OEt	16·0	21·0
NO_2	OEt	1·5	5·5	OMe	OMe	4·0	12·5
NMe_2	OEt	6·5	20·5	OH	H	29·0	23·0
Cl	OEt	8·0	11·0	OH	Me	21·5	20·5
H	OEt	20·0	18·0	OH	OMe	17·0	20·5
				OMe	I	14·5	16·0

Dave and Dewar [8] have related the efficiency order to the order of decreasing polarity of the end groups regardless of sign. Gray [24], in his study of pure mesogens, has obtained orders of group efficiency in promoting nematic liquid crystal formation by substitution of various groups in the *p*-positions of different aromatic systems. Below are given the nematic group efficiency orders which he obtained with some of the substituents listed in the extended group efficiency order of Dave *et al.*, for mixed mesomorphic systems; these orders may be compared with the extended order of Dave *et al.* and also with smectic and cholesteric group efficiency orders which have been obtained.

Nematic: NO_2 > OEt > OMe > n-PrO > NMe_2 > Me
= Cl > F = Br > H
(Dave *et al.*—extended order.)
MeO > NO_2 > n-PrO > Cl > Br > NMe_2 > Me > H
(4-*p*-substituted-benzylidineamino-4′-methoxybiphenyls. Gray [24].)
MeO > n-PrO > NO_2 > NMe_2 > Me > H
(4-*p*-substituted-benzylidineamino-4′-n-octyloxybiphenyls. Gray [24].)
OEt > OMe > Cl > n-PrO > Me > F > H
(4′-substituted-biphenyl-4-carboxylic acids. Gray [24].)

Smectic: Br > Cl > F > NMe_2 > n-PrO > Me > H > NO_2 > OMe
(4-*p*-substituted-benzylidineamino-4′-n-oxtyloxybiphenyls. Gray [24].)

Cholesteric: OMe > NO_2 > Br = Cl > I > Me > H
(*p*-substituted-benzoates of cholesterol. Dave and Vora [25])

The nematic orders are only broadly similar, and the smectic order is quite different. The cholesteric order is more similar to the nematic order than to the smectic one, agreeing with the view that the cholesteric mesophase resembles the nematic mesophase more closely than the smectic mesophase (see also p. 171). If the magnitude of the group dipole is considered as the main factor in promoting liquid crystallinity, these orders should be closely similar; evidently, since they are not, efficiency orders must reflect the cumulative effect of many factors like dipole interactions, the overall polarizability of the system, the sizes of the groups, the shapes of the molecules and even hydrogen bonding and conjugation [26]. Gray [26] has pointed out that the group efficiency orders for nematogenic systems will not therefore hold good for smectogenic systems whose thermal stabilities are governed by quite different factors.

In a mixed liquid crystal, consisting of a mesogen and a non-mesogen, the system is presumably composed of oriented domains and these should be formed readily when the molecules of the non-mesogen are similar in shape and size to the molecules of the mesogen. The greater the dissimilarity of the two molecules, the steeper will be the slope of the anisotropic liquid–amorphous liquid transition line. This might be expected, for the disturbing effects of a solute on the mesomorphic phase should be greater the smaller the tendency of the solute molecules themselves to give mesophases. Theoretically, it would be expected that there might arise a condition when the effect of disturbance by even a small concentration of a non-mesogenic solute consisting of highly dissimilarly shaped molecules would be a maximum and wholly prevent liquid crystal formation. Thus, Gray [26] has suggested that if the non-mesogenic component consisted of bulky spherical molecules, then the parallel distribution of the molecules of the mesomorphic system would be difficult to maintain; the mesophase-amorphous liquid transition temperature would probably then be depressed more than the melting point and liquid crystal formation would be prevented at quite low concentrations of the non-mesogenic solute.

Dave and Dewar [8], Dave and Lohar [27] and Dave and Vasanth [11] have in fact studied binary systems consisting of a mesomorphic substance and a compound such as α-naphthylamine; β-naphthylamine; α-naphthol; β-naphthol; 4,4′-dimethoxy-3,3′-diacetylbiphenyl; 6′,6‴-dihydroxy-4,4″-dimethoxy-3′,3‴-bichalconyl; 6′,6‴-dihydroxy-3′,3‴-bichalconyl; triphenylmethane; 2,2′,6,6′-tetramethoxybiphenyl; 7-benzyloxy-3-benzoylflavone, i.e., compounds with very low tendencies to form liquid crystals. The structures and numbering systems for 3′,3‴-bichalconyl and flavone are given below in (IV) and (V), respectively

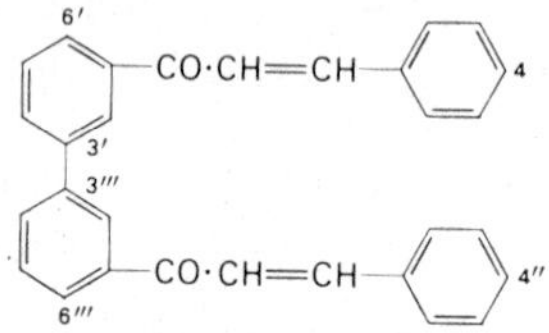

6′ CO·CH=CH 4
3′
3‴
CO·CH=CH 4″
6‴

IV

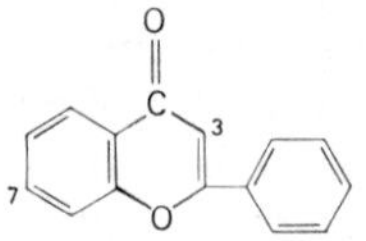

O
C 3
7 O

V

In all cases studied, mixed liquid crystals were formed over a certain range of low concentrations, irrespective of the size, shape or structure of the non-mesogenic solute molecule. However, the results support the view that, in the case of binary liquid crystal mixtures where one component is a non-liquid crystal, the extent of the composition range over which liquid crystallinity is observed depends on the degree of similarity of the molecules of the non-liquid crystalline and the liquid crystalline substances in shape, size, polarity of the end groups and polarizability. If the solute molecules are structurally compatible with the molecular arrangement of the mesophase, they should adjust to the parallel alignment of the mesophase and mixed liquid crystals should form over a wider composition range in the phase diagram; the most favourable condition for mixed liquid crystal formation would thus arise if the molecules of the non-mesogen are long and lath-shaped.

It was suggested by de Kock [7] that the depression of liquid crystal transition temperatures by solutes might be used as a method of determining molecular weights. Du Pont and Lozac'h [28] did use *p*-azoxyanisole as a cryoscopic solvent in the Rast method, but Dave and Dewar [8] have pointed out that this would be correct in principle only if the solute were insoluble in the liquid crystal, the two-component liquid being a mixture of two phases, one containing only the mesogen and the other being a normal liquid containing both components. They reasoned that since the birefringent two-component liquids consist of a single liquid crystalline phase, depression of transition temperatures cannot be used as a reliable means for molecular weight determination; indeed in the systems studied by them, the molar depression varied from compound to compound. Dave and Lohar [29] and Dave and Vasanth [11] have obtained more evidence supporting this viewpoint.

It has been stated earlier that mixtures of non-mesogenic substances often give rise to mixed mesomorphism, cf., page 153. Gaubert [30] obtained mesomorphic systems by mixing molten cholesterol with each of the following compounds—succinimide, tartaric, malic, maleic, malonic, succinic, anisic, cinnamic and lactic acids. He also studied liquid crystal mixtures obtained by heating ergosteryl acetate, propionate or butyrate with glycollic acid, glycerol or orcin and on melting certain cholesterol and ergosterol derivatives with urea [31]. Kravchenko and Pastukhova [32] have reported liquid crystallinity in mixtures of non-mesogenic compounds such as indene, naphthalene and isoquinoline with each other and with coumarone. Mlodzeevskii [33] also observed that mixtures of cholesterol and cetyl alcohol and of cholesterol and glycerol give rise to mesomorphic phases and this early work has been supported by the observations of Chistyakov and Usol'tseva working with the same systems [35].

Lawrence [34] too has recently reported mixed liquid crystal formation in anhydrous systems containing cholesterol and a C_{12}, C_{13}, C_{14}, C_{16} or C_{18} alkanol with the formation of a 2:1 alkanol:cholesterol complex.

Vorländer and Gahren [36] and Bogojawlensky and Winogradow [9] have also reported the formation of mixed liquid crystals from pairs of substances which do not form liquid crystals by themselves, and Walter [10] obtained enantiotropic mixed liquid crystals within certain limits of concentration using a mixture of *p*-methoxybenzoic acid and α-anisal-propionic (*p*-methoxy-α-methylcinnamic) acid which themselves are non-mesomorphic. He determined the latent nematic-amorphous liquid transition temperatures for each of these acids from the behaviour of their mixtures separately with *p*-methoxycinnamic acid; this was done by extrapolation of the clearing point curves. He found that the curve which joined the latent amorphous liquid-nematic transition points of the two non-liquid crystalline compounds and passed through those nematic-amorphous liquid transition temperatures which could be experimentally determined was a straight line.

Tammann's [37] view that formation of liquid crystals from a mixture of substances that do not form liquid crystals is a matter of emulsion formation is disproved by the foregoing results.

Bennett and Jones [38] showed that the high melting *p*-methoxybenzoic acid and *p*-ethoxybenzoic acid do not form liquid crystals by themselves, but do form nematic liquid crystals when particular mixtures of the two are melted; they suggested that this was related to the lower melting points of the mixtures. Dave and Lohar [29] confirmed this result and showed that mixed liquid crystals are formed by mixtures of the two acids over a small range of temperature and composition. Walter [10] also reported that the system *p*-ethoxybenzoic acid in admixture with *p*-methoxycinnamic acid can give liquid crystals and he determined the latent nematic-amorphous liquid transition temperatures for the individual acids by extrapolation of the transition line. Dave and Lohar [29] repeated the work on the systems *p*-methoxybenzoic acid and *p*-ethoxybenzoic acid in admixture separately with *p*-methoxycinnamic acid and obtained the values for the latent nematic-amorphous liquid transition temperatures for the two acids. Dave and Vasanth [17] studied mixtures of these acids separately with another mesogen, *p*-n-butoxybenzoic acid, and determined the latent transition temperatures by extrapolation of the transition lines. The extrapolated values for the latent transition temperatures obtained in these studies are in close agreement; these are incorporated in Table 4.2.3 below:

TABLE 4.2.3. *Comparison of latent transition temperatures (LTT) for* p-*methoxybenzoic acid and* p-*ethoxybenzoic acid.*

Non-mesogenic component	LTT (°C) extrapolated using mixtures with			
	p-methoxycinnamic acid (Dave and Lohar [29])	*p*-methoxycinnamic acid (Walter [10])	*p*-ethoxy- or *p*-methoxybenzoic acid (Dave and Lohar [29])	*p*-n-butoxy benzoic acid (Dave and Vasanth [17])
p-Methoxybenzoic acid	157·5	155·6	156·5	155·5
p-Ethoxybenzoic acid	169·0	165·0	169·5	168·5

Walter has pointed out that substances which do not individually form mesophases but which as binary mixtures exhibit mixed mesomorphism, are generally very highly crystalline and are not readily supercooled from the amorphous melt; therefore, monotropic liquid crystals which might have been anticipated for them by reason of their chemical constitution remain latent. If two such chemically similar substances are mixed, it may happen that the melting point of the solid phase of the mixture, which is generally lower than the melting points of the individual components, falls below the mesophase-amorphous liquid transition temperature of the mixed liquid crystalline phase, a temperature which, with chemically related substances, should lie between the two latent mesophase-amorphous liquid transition temperatures of the pure components. Such a mixture exhibits enantiotropic liquid crystal properties. Based on his studies of mixtures of isomorphous substances, Walter suggested the equation $t_m = (t_1c_1 + t_2c_2)/(c_1 + c_2)$ for the calculation of the mesophase–amorphous liquid transition points, based on the law of mixtures, where t_m is the transition point of the mixed melt, t_1 and t_2 are the transition temperatures of the two components and c_1 and c_2 their concentrations.

Dave and Lohar [39] and Dave and Vasanth [17] studied other mixtures of substances, one or both of which exhibit monotropic nematic mesomorphism, and again obtained latent transition temperatures for the non-mesogenic component by extrapolation of the transition lines. Mixed liquid crystals were formed over a range of temperature and concentration, and were either enantiotropic or appeared in the supercooled region. Dave and Vasanth [17] have pointed out that polarity cannot be taken as the only pre-requisite for a binary mixture to exhibit mixed mesomorphism, for many systems with highly polar groups failed to exhibit mixed mesomorphism. The virtual or latent transition temperatures in such cases are much below the normal melting points and no liquid crystalline phase is obtained.

Mixtures Forming Smectic Mesophases

The binary systems reported so far have mainly consisted of components one or both of which may be either nematogenic or non-mesogenic. Arnold and Sackmann [40] have examined the miscibility of liquid crystal phases, and reported that the nematic phases of two different compounds are completely miscible and form a homogeneous nematic liquid crystalline phase. Smectic phases of two different compounds are likewise miscible, forming a homogeneous smectic mesomorphic phase. However, the smectic phase of one compound and the nematic phase of another compound, being distinct phases, are not of course miscible. Demus and

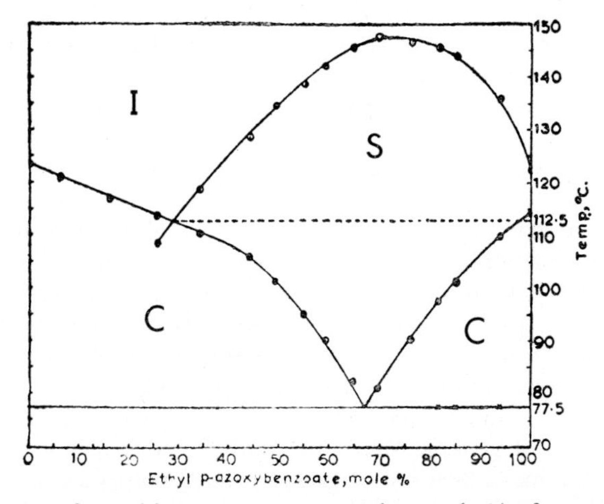

FIG. 4.2.2. Plot of transition temperatures against mole % of smectogen for the system ethyl *p*-azoxybenzoate: *p*-nitrobenzylidene-*p*-phenetidine—Dave, Patel and Vasanth [14]. C = crystal; S = smectic; I = amorphous isotropic liquid.

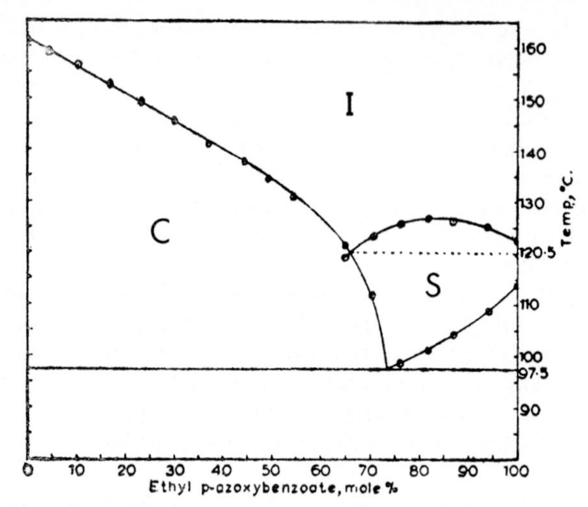

FIG. 4.2.3. Plot of transition temperatures against mole % of smectogen for the system ethyl *p*-azoxybenzoate: methyl 4-*p*-chlorobenzylideneaminobenzoate—Dave, Patel and Vasanth [14]. C = crystal; S = smectic; I = amorphous isotropic liquid.

Sackmann [41] have extended this work and shown that different polymorphic forms of smectic mesophases may be identified on the basis of miscibility behaviour and the nature of their microscopic textures.

However, no extensive studies have been made for mixed liquid crystals of the smectic type. Dave, Patel and Vasanth [14] studied binary systems comprising ethyl *p*-azoxybenzoate (a smectogen) and Schiff's bases (non-mesogenic solutes). A continuous series of mixed smectic liquid crystals was formed over a range of temperature and concentration. The smectic-amorphous liquid transition lines were usually depressed except in the case of nitro- and in some cases chloro-substituted Schiff's bases when the curve exhibited a maximum (Figs. 4.2.2 and 4.2.3).

The smectic phase consists of strata of molecules. Some of the molecules of the compound dissolved in such a phase may occupy positions between the molecules of the smectogen in each stratum and the rest of the solute molecules may form a molecular stratum which may pack itself between two strata of the smectic component, thus forming a homogeneous smectic melt. The tentative picture of such a melt is shown in Fig. 4.2.4(*a*) and (*b*) where X represents the molecules of ethyl *p*-azoxybenzoate and Y represents the molecules of the non-mesogen. Presumably, the stratum of molecules (Y) must also contain a few of the molecules (X), but for simplicity this has not been shown in Fig. 4.2.4(*b*). The degree of packing in this mode will depend upon the shape and size of the solute molecule and its functional groups. When the parallel molecules are arranged in layers with the mean position of each molecule perpendicular to the layer interfaces (Fig. 4.2.5), terminal dipole moments lying parallel to the major molecular axis of the solute molecules, as in the case of *p*-nitro- or *p*-chloro-compounds, would be *expected* to disturb the efficiency of the attraction between the dipoles of the terminal groups of the neighbouring molecules of the smectogen, causing *reduction* in lateral intermolecular attractions and consequently in the smectic thermal stability of the mixture compared with that of the pure smectic components. A possible

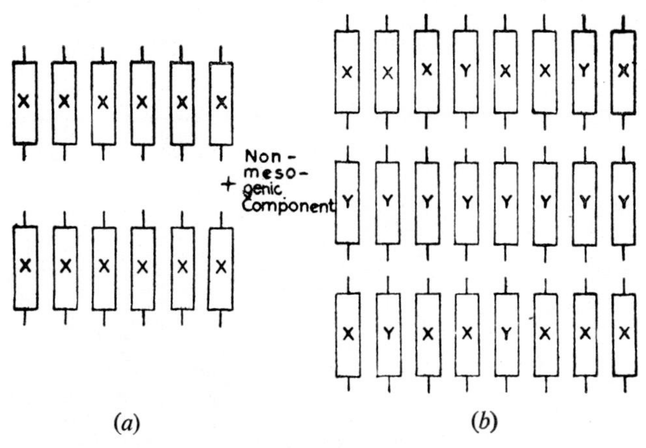

FIG. 4.2.4. Structure of (*a*) a smectic phase (pure) and (*b*) a smectic phase containing a non-mesogenic component in solution—Dave, Patel and Vasanth [14].

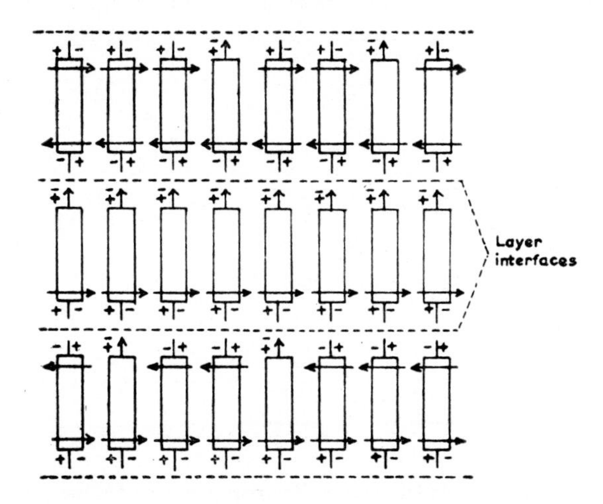

FIG. 4.2.5. Arrangement of parallel molecules in a mixture consisting of layers wherein each molecule is perpendicular to the layer interfaces—Dave, Patel and Vasanth [14].

explanation of the maxima in the transition lines, i.e. of the *enhanced* smectic thermal stability, is given on page 170.

The transition lines are almost all curved, and hence no order for the tendency of the terminal polar groups present in non-mesogenic components to promote mixed smectic mesomorphism can be derived.

Dave, Patel and Vasanth [14] studied binary mixtures of *p*-n-dodecyloxybenzylidene-*p*-n-butoxyaniline, a polymesogen exhibiting smectic (101–111°C) and nematic (111–111·5°C) phases, and other non-mesogenic Schiff's bases (Fig. 4.2.6). Similarly, Dave and Patel [15] examined binary mixtures of *p*-n-octyloxybenzoic acid, a compound giving smectic and

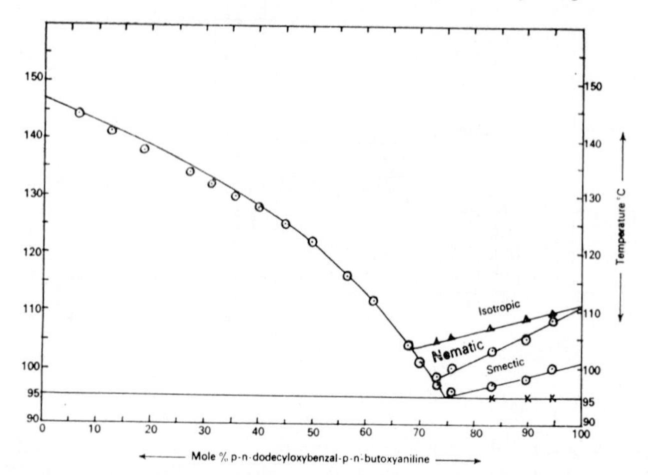

FIG. 4.2.6. System *p*-n-dodecyloxybenzylidene-*p*-n-butoxyaniline: *p*-anisylidene-*p*-anisidine—Dave, Patel and Vasanth [14].

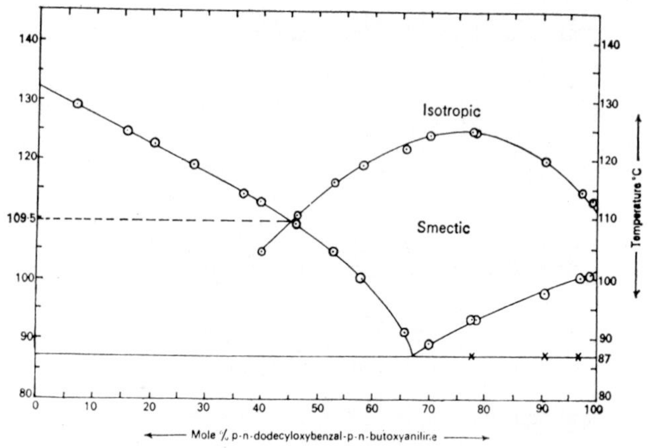

FIG. 4.2.7. System *p*-n-dodecyloxybenzylidene-*p*-n-butoxyaniline: *p*-nitrobenzylidene-*p*-chloroaniline—Dave, Patel and Vasanth [14].

nematic mesophases, and non-mesogenic *p*-substituted benzoic acids. In both these cases mixed mesomorphism of the smectic and nematic types was observed over a range of temperature and concentration. The smectic–nematic and nematic–amorphous liquid transition lines fell regularly, as usual, with increasing concentration of the non-mesogen, the range of temperature over which the nematic existed increasing at the cost of the smectic mesophase. However, the binary mixtures comprising Schiff's bases with a nitro-group showed a peculiar behaviour (Fig. 4.2.7); the nematic phase was completely eliminated, and the smectic phase showed an enhanced thermal stability, the smectic–amorphous liquid transition curve showing a maximum in the phase diagram.

Dave and Dewar [8] in their study of nematic mixed liquid crystals in binary mixtures of *p*-azoxyanisole and *p*-nitro-substituted Schiff's bases have reported that the presence of the nitro-group enhances the thermal stability of the mixed nematic mesophase; its group slope value is the lowest. They have ascribed this to the high polarity of the NO_2 group.

Schroeder and Schroeder [13] have also reported that mixtures of 4-nitro-substituted Schiff's bases and 4,4′-di-n-hexyloxyazoxybenzene (a polymesogen exhibiting a nematic and a monotropic smectic phase), exhibit smectic phases which are more thermally stable than that of the

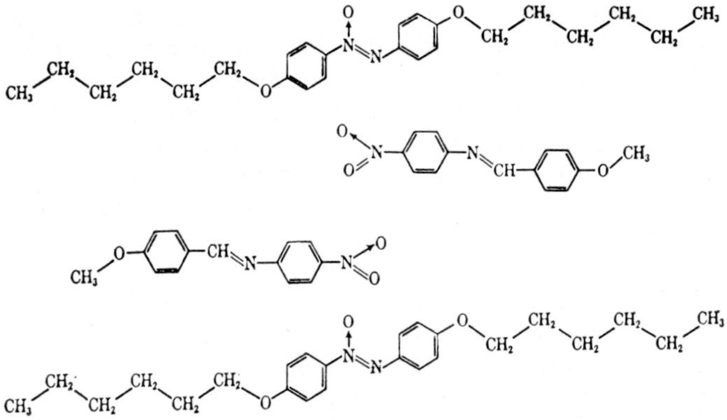

FIG. 4.2.8. Head to head arrangement of the nitro-substituted molecules in the layer structure of the mixed smectic mesophase—Schroeder and Schroeder [13]

pure azoxy-compound (see also page 166). They mention that a highly polar nitro-substituent is a most unlikely terminal group for a smectogen and that to their knowledge no compound with a nitro-group in the terminal position has ever been reported to be smectic. This does not seem to be a fair comment, because Gray [26] has studied the smectic properties of a number of compounds including Schiff's bases with NO_2 groups in the terminal as well as other positions. Schroeder and Schroeder have tried to explain the enhanced smectic thermal stability of their mixtures by postulating a head to head arrangement of the nitro-substituted molecules in the layer structure of the mixed smectic mesophase, with the nitro groups adjacent to one another in the middle of each layer (Fig. 4.2.8). They in fact draw attention to the analogy between this "dimer" of the nitro-Schiff's base and the dimers of carboxylic acids which often form liquid crystals. According to this postulate, the nitro-Schiff's bases themselves would be expected to exhibit mesomorphism and this is not the case. Moreover, a compound like *p*-nitrobenzoic acid or *p*-nitrocinnamic acid, on dimerization *via* the carboxyl group and then, according to this postulate *via* the nitro-group would comprise at least four benzene rings in the molecular arrangement; the acid might therefore be expected to exhibit mesomorphism, but this is also not the case. An explanation of the enhanced mesophase thermal stabilities due to such terminal groups has been given by Byron and Gray [26]. They explained the enhanced mesomorphic thermal stability of 4-*p*-n-nonyloxybenzylidineamino-4′-chlorobiphenyl, in which the C–Cl dipole acts parallel to the major molecular axis, by assuming that the molecules are tilted at an angle to the layer interfaces; by tilting the molecules away from the perpendicular arrangement, the separation of the positive and of the negative charges of the C–Cl dipoles does not necessarily increase, but the positive charge of one dipole can be brought nearer to the negative centre of a neighbouring dipole. At some suitable angle, the attractive forces will outweigh the repulsive forces and the nett energy of attraction will enhance the lateral attraction and the smectic properties. In the mixture of ethyl *p*-azoxybenzoate and nitro-substituted Schiff's bases mentioned earlier, the smectic component has two dipoles, due to two terminal groups acting at an angle to the main molecular axis, while one dipole due to one of the two terminal groups of the non-mesogenic component acts parallel and the other at an angle to the main molecular axis; in some cases both the dipoles of the two terminal groups of the non-mesogenic component act parallel to the main molecular axis. If the molecular arrangement of the mixture is considered as shown in Figs. 4.2.4 and 4.2.5, then a way to account for the enhanced mesomorphic properties of binary mixtures comprising non-mesogenic compounds possessing terminal groups of high dipole moment, e.g., C–NO_2 etc., is by suggesting that the long axes of the molecules in the mesomorphic melt lie tilted at an angle to the planes of the smectic strata; the nett energy of attraction might then effectively enhance the lateral attractions and the mesomorphic properties.

It is however rather difficult to explain the role of terminally situated groups and their dipole moments on the mesomorphic characteristics of mixtures as these systems are fairly complex, and any attempt to relate molecular structure to liquid crystalline properties of the mixtures is necessarily qualitative.

Mixtures of compounds of which at least one gives a cholesteric mesophase

Liquid crystalline mixtures have been studied in order to examine the effects on the microscopic textures of the various mesophases. Friedel [42] observed that addition of very small amounts of suitable cholesteryl derivatives to a nematogen such as *p*-azoxyanisole or *p*-azoxyphenetole imparted to the latter the characteristic properties of a cholesteric. From a study of this type of mixture, Friedel concluded that the addition of any substance having rotatory power, to a nematogen produces the cholesteric structure, the rotatory power being proportional to the amount of asymmetric substance added; as this amount approaches zero, there will be a gradual transition from cholesteric to nematic properties. He [43] considered that the cholesteric phase is really a nematic state.

Vorländer [44] had earlier studied the effect of optically active substances on liquid crystallinity by mixing them with optically inactive mesogens. He observed that purified, crystalline abietic acid transformed optically inactive nematic liquid crystals to a marked degree into cholesteric liquid crystals. He reported that racemic 2-methylbutyl 4-*p*-methoxybenzylideneamino-α-methylcinnamate gives a nematic liquid crystal, whereas active forms of this substance exhibit cholesteric properties. Luckhurst and Smith [45] have reported that addition of an optically active solute to a nematogen brings about changes in the structure of the mesophase which in fact becomes cholesteric. Friedel and Royer [46] observed equidistant planes in mixtures of cholesteryl derivatives with *p*-azoxyphenetole (as observed by Grandjean with active 2-methylbutyl 4-*p*-cyanobenzylideneaminocinnamate). Royer [47] also made interesting observations on mixtures of cholesteric compounds with nematic compounds.

Billard and co-workers [48] have recently reported that racemic nematogens on resolution yield optical isomers which exhibit a cholesteric mesophase. They have prepared mesogenic compounds namely (*a*) 4′-alkoxybiphenyl-4-carboxylic acids having an asymmetric centre in the ether chain and (*b*) 4-*p*-substituted arylideneaminocinnamic acids esterified with an alcohol containing an asymmetric centre. They observed nematic phases in the case of racemic derivatives and cholesteric phases in the case of the enantiomers; mixtures of two enantiomers in 50% composition exhibited a nematic mesophase.

Sackmann *et al.* [49] have similarly reported that when two cholesteric compounds with opposite optical rotatory powers are mixed in different proportions, a stage is reached at which that mixture exhibits a nematic mesophase. Chistyakov and Gusakova [50] have reported a new type of texture for a mixture of *p*-azoxyanisole and cholesteryl propionate, and the studies of optical rotatory power in mixtures of cholesteric and nematic substances by Cano *et al.* [51], Chatelain and Martin [52], Buckingham *et al.* [53] and Stegemeyer *et al.* [54] should also be mentioned.

Gray [55] too has reported that the racemic forms of some alkyl 4-*p*-substituted benzylidineaminocinnamates are nematic whereas their optical isomers exhibit cholesteric mesophases. He obtained the order of terminal group efficiency for the cholesteric phases of the above system and compared it with the corresponding group efficiency order for nematic phases reported earlier. He reported that the two efficiency orders were identical, thus providing convincing evidence of the close relationship between cholesteric and nematic liquid crystals. It should be emphasized here that the study of mixtures has been instrumental in arriving at this conclusion; the observations of Friedel [43] regarding the nature of the cholesteric phase thus stand vindicated after about 50 years.

Kinetic Studies Involving Liquid Crystals as Solvents

Reference should also be made here to studies of rates of reactions in liquid crystalline media. Such experiments depend on finding suitable mesogenic compounds to act as anisotropic solvents such that a reasonable concentration range exists over which the non-mesogenic solute or solutes will not destroy the mesomorphic properties of the anisotropic solvent or solvents. Svedberg [56] measured the rate of decomposition of picric acid, trinitroresorcinol and pyrogallol in *p*-azoxyphenetole at 140°C; he observed that the reaction rate diminished when the solution was subjected to the influence of a magnetic field. This type of effect was not observed in similar experiments with maleic acid as the dissolved substance. He studied the reaction between picric acid and *p*-azoxyphenetole and observed a sudden increase in the rate of reaction at 165°C, which is the nematic-amorphous liquid transition temperature of the mesomorphic compound. He attributed this to an increased rate of diffusion. He extended this type of experiment to other solutes and reported that similar effects were obtained when the anisotropy was destroyed by the addition of foreign substances such as biphenyl or naphthalene. Svedberg also studied the rate of diffusion of *m*-nitrophenol in a mixture of *p*-azoxyanisole and *p*-azoxyphenetole in a magnetic field and found that the rate of diffusion was increased by the application of a field parallel to the concentration gradient and retarded by a field at right angles to this. More recently, Bacon and Brown [57] have studied the rates of the Claisen rearrangement of *p*-substituted allyl phenyl ethers using the nematic phase of di-*p*-methoxyphenyl cyclohexane-*trans*-1,4-dicarboxylate as solvent. The rate of rearrangement was faster in the nematic liquid than in the isotropic solvent. They state that nematic liquids may serve as good solvents for studies of many molecular rearrangement and polymerization processes.

REFERENCES

[1] Schenck, R. *Z. phys. Chem.* **25,** 337 (1898), **29,** 546 (1899).
[2] Kast, W. *Z. Physik.* **76,** 19 (1932).
[3] Svedberg, T. *Kolloidzeitschrift*, **16,** 103 (1915).
[4] Lehmann, O. *Annln Phys.* (4), **21,** 181 (1906).
[5] Vorländer, D. and Ost, K. *Ber. dt. chem. Ges.* **71B,** 1688 (1938).
[6] Prins, A. *Z. phys. Chem.* **67,** 689 (1909).
[7] de Kock, A. C. *Z. phys. Chem.* **48,** 129 (1904).
[8] Dave, J. S. and Dewar, M. J. S. *J. chem. Soc.* 4617 (1954); 4305 (1955).
[9] Bogojawlensky, A. and Winogradow, N. *Z. phys. Chem.* **60,** 433 (1907); **64,** 229 (1908).
[10] Walter, R. *Ber. dt. chem. Ges.* **58B,** 2303 (1925).
[11] Dave, J. S. and Vasanth, K. L. Previously unpublished work.
[12] Dewar, M. J. S. and Goldberg, R. S. *J. Am. chem. Soc.* **92,** 1582 (1970).
[13] Schroeder, J. P. and Schroeder, D. C. *J. org. Chem.* **33,** 591 (1968).
[14] Dave, J. S., Patel, P. R. and Vasanth, K. L. *Ind. J. Chem.* **4,** 505 (1966); *Molec. Crystals Liqu. Crystals* **8,** 93 (1969).

[15] Dave, J. S. and Patel, P. R. *J. Ind. chem. Soc.* **47,** 815 (1970).
[16] Dave, J. S. and Lohar, J. M. *Proc. natn. Acad. Sci. India* **32A,** 105 (1962).
[17] Dave, J. S. and Vasanth, K. L. *Ind. J. Chem.* **7,** 498 (1969).
[18] Demus, D. *Z. Naturf.* **22A,** 285 (1967).
[19] Castellano, J. A., McCaffrey, M. T. and Goldmacher, J. E. *Molec. Crystals Liqu. Crystals* **12,** 345 (1971).
[20] Pohl, L. and Steinsträsser, R. Proc. 3rd Int. Conf. on Liquid Crystals, Berlin (1970); *Z. Naturf.* **26B,** 26 (1971).
[21] Dave, J. S. and Lohar, J. M. *Ind. J. Chem.* **4,** 386 (1966).
[22] Dave, J. S. and Lohar, J. M. *J. chem. Soc.* **(A),** 1473 (1967).
[23] Dave, J. S. and Vasanth, K. L. *Molec. Crystals* **2,** 125 (1966).
[24] Gray, G. W. in *Liquid Crystals* (edited by G. H. Brown, G. J. Dienes and M. M. Labes), Gordon and Breach, New York, (1967), p. 129.
[25] Dave, J. S. and Vora, R. A. *Ind. J. Chem.* **11,** 19 (1973).
[26] Gray, G. W. *Molecular Structure and the Properties of Liquid Crystals*, Academic Press, London and New York (1962).
[27] Dave, J. S. and Lohar, J. M. *Proc. natn. Acad. Sci. India* **29A,** 260 (1960).
[28] Du Pont, G. and Lozac'h, O. *C. r. hebd. Séanc. Acad. Sci., Paris* **221,** 751 (1945).
[29] Dave, J. S. and Lohar, J. M. Previously unpublished work.
[30] Gaubert, P. *C. r. hebd. Séanc. Acad. Sci., Paris* **156,** 149 (1913); *Bull. Soc. fr. Minér. Cristallogr.* **35,** 64 (1912).
[31] Gaubert, P. *C. r. hebd. Séanc. Acad. Sci., Paris* **147,** 498 (1908); **149,** 608 (1909).
[32] Kravchenko, V. M. and Pastukhova, I. S. *Zh. prikl. Khim., Mosk.* **25,** 313, 328, 343 (1952).
[33] Mlodzeevskii, A. *Z. Physik.* **20,** 317 (1923); *Z. phys. Chem.* **135,** 129 (1928).
[34] Lawrence, A. S. C. in *Liquid Crystals and Ordered Fluids* (edited by J. F. Johnson and R. S. Porter), Plenum Press, New York, (1970), p. 289.
[35] Chistyakov, I. G. and Usol'tseva, V. A. *Izv. vўssh. ucheb. Zaved.* **5,** 589 (1962); *Chem. Abstr.* **58,** 13239 (1963).
[36] Vorländer, D. and Gahren, A. *Ber. dt. chem. Ges.* **40,** 1966 (1907).
[37] Tammann, G. *Aggregatzustande,* Leopold Voss, Leipzig (1922), p. 286.
[38] Bennett, G. M. and Jones, B. *J. chem. Soc.* 420 (1939).
[39] Dave, J. S. and Lohar, J. M. *Chemy Ind.* 597 (1959); 494 (1960).
[40] Arnold, H. and Sackmann, H. *Z. phys. Chem.* **213,** 137 (1960); *Z. Electrochem* **63,** 1171 (1959).
[41] Demus, D. and Sackmann, H. in *Liquid Crystals* (edited by G. H. Brown, G. J. Dienes and M. M. Labes), Gordon & Breach, New York (1967) p. 341.
[42] Friedel, G. *C. r. hebd. Séanc. Acad. Sci., Paris* **176,** 475 (1923).
[43] Friedel, G. *Annls Phys.* **18,** 273 (1922).

[44] VORLÄNDER, D. and JANECKE, F. *Z. phys. Chem.* **85,** 691, 697 (1914); *Chem. Abstr.* **8,** 602, 608 (1914).

[45] LUCKHURST, G. R. and SMITH, H. J. Proc. 3rd Int. Conf. on Liquid Crystals, Berlin (1970).

[46] FRIEDEL, G. and ROYER, L. *C. r. hebd. Séanc. Acad. Sci., Paris* **173,** 1320 (1921).

[47] ROYER, L. *C. r. hebd. Séanc. Acad. Sci., Paris* **174,** 1182 (1922).

[48] LECLERCQ, M., BILLARD, J. and JACQUES, J. *C. r. hebd. Séanc. Acad. Sci., Paris, Ser. C.* **266,** 654 (1968); *Molec. Crystals Liqu. Crystals* **8,** 367 (1969).

[49] SACKMANN, E., MEIBOOM, S., SNYDER, L. C., MEIXNER, A. E. and DIETZ, R. E. *J. Am. chem. Soc.* **90,** 3567 (1968).

[50] CHISTYAKOV, I. G. and GUSAKOVA, L. A. *Kristallografiya* **14,** 153 (1969).

[51] CANO, R. and CHATELAIN, P. *C. r. hebd. Séanc. Acad. Sci., Paris* **251,** 1139 (1960); **259,** 352 (1964).

[52] CHATELAIN, P. and MARTIN, J. *C. r. hebd. Séanc. Acad. Sci., Paris, Ser. C.* **268,** 898 (1969).

[53] BUCKINGHAM, A. D., CEASAR, G. P. and DUNN, M. B. *Chem. Phys. Lett.* **3,** 540 (1969).

[54] STEGEMEYER, H. and MAINUSCH, K. J. Proc. 3rd Int. Conf. on Liquid Crystals, Berlin, (1970).

[55] GRAY, G. W. *Molec. Crystals Liqu. Crystals* **7,** 127 (1969).

[56] SVEDBERG, T. *Kolloidzeitschrift* **18,** 101 (1916); **20,** 73 (1917); **22,** 68 (1918); *Annln Phys.* **49,** 437 (1916).

[57] BACON, W. E. and BROWN, G. H. Proc. 3rd Int. Conf. on Liquid Crystals, Berlin (1970).

Additional References

HOMER, J. and DUDLEY, A. R. *J. chem. Soc.* 926 (1972).

STEGEMEYER, H. *et al. Naturwissen.* **58,** 599, 621 (1971); *Chem. Phys. Lett.* **6,** 5 (1970); **8,** 425 (1971); **16,** 38 (1972).

BERNHEIM, R. A. and SHUHLER, T. A. *J. phys. Chem.* **76,** 925 (1972).

STEINSTRÄSSER, R. *et al. Z. Naturf.* **26B,** 87 (1971); **27B,** 774 (1972); *Angew. Chem.* **11,** 633 (1972); *Tetr. Lett.* 1921 (1971).

FISHEL, D. L. and HSU, YING YEN. *J. chem. Soc.* (D), 1557 (1971).

CLADIS, P. E., RAULT, J. and BURGER, J. P. *Molec. Crystals Liqu. Crystals* **13,** 1 (1971).

4.3 CHOLESTERIC LIQUID CRYSTALS FORMED BY CERTAIN POLYPEPTIDES WITH ORGANIC SOLVENTS

E. T. Samulski and A. V. Tobolsky

Introduction

Some twenty years ago it was found that synthetic polypeptides, —(—NH·CHR·CO—)$_n$—, adopt a rigid rod-like helical conformation in solution in contrast to the random coil conformation exhibited by most other synthetic polymers. This observation with its implications in the study of protein structure stimulated a concentrated and sustained investigation of this class of polymers. In the early stages of this research with synthetic polypeptides, the mesomorphic behaviour of concentrated polypeptide solutions was discovered. Elliott and Ambrose observed the formation of a spontaneously birefringent liquid phase in a chloroform solution of poly-γ-benzyl-L-glutamate while preparing solid films for infrared studies [1]. Robinson investigated the birefringent solutions in detail pointing out the similarity between the supramolecular structure in the concentrated polypeptide solution and the structure found in thermotropic cholesteric liquid crystals [2–6]. The unusual properties and structure of the lyotropic polypeptide liquid crystal along with the influence that studies of this mesophase impose on structural considerations of the solid state of these polymers will be discussed in this section on non-amphiphilic liquid crystalline systems.

Throughout the text experimental results and observations refer primarily to one synthetic polypeptide, poly-γ-benzyl-L-glutamate (PBLG). This has been most extensively studied since it was one of the first readily available synthetic polypeptides and has good solubility in a wide range of solvents. However, other lyotropic polypeptide liquid crystals have been examined in less detail and the results confirm that the mesomorphic behaviour and properties discussed below are general phenomena characteristic of this class of polymers.

The Lyotropic Polypeptide Liquid Crystal

Formation

When a synthetic polypeptide exceeds a certain limiting concentration in a solvent, the solution separates into two liquid phases. The polypeptide-rich phase is spontaneously birefringent and separates from the more dilute phase initially in the form of spherical liquid droplets or "spherulites". If the two phase solution is cooled or the polypeptide concentration further increased, the droplets grow in size and coalesce forming a continuous birefringent fluid. This anisotropic solution is the lyotropic polypeptide liquid crystal. Table 4.3.1 contains a list of polypeptides which form the liquid crystal phase. The phase occurs in all solvents in which the polypeptides are soluble at sufficiently high concentrations. A birefringent phase with striking similarities to the polypeptide liquid crystal has been observed in concentrated aqueous solutions of certain nucleic acids; this will be referred to later in the text.

The PBLG liquid crystal forms in a number of solvents and mixed solvents including benzene, chloroform, *NN*-dimethylformamide (DMF), dioxane, dichloromethane, the isomers of dichloroethane and dichloroethene, dichloroacetic acid and *m*-cresol. In all circumstances in which the liquid crystalline phase is present PBLG is in the extended helical molecular conformation. (ORD measurements indicate that PBLG reverts

TABLE 4.3.1. *Polypeptides forming liquid crystals with suitable organic solvents*

Polypeptide —(—NH·CHR·CO—)$_n$—	R	Ref.	Abbreviation
Poly-L-glutamic acid	$—CH_2CH_2COOH$	27	PLGA
PLGA esters: γ-methyl	$—CH_2CH_2COOMe$	5	PMLG
γ-ethyl	$—CH_2CH_2COOCH_2Me$	6	PELG
γ-benzyl (L and D isomers)	$—CH_2CH_2COOCH_2C_6H_5$	2–6	PBLG and PBDG
Poly-β-benzyl-L-aspartate	$—CH_2COOCH_2C_6H_5$	5	PBLA
Poly-ε-carbobenzyloxy-L-lysine	$—(CH_2)_4NHCOOCH_2C_6H_5$	38	PCLL
A copolymer of equal parts PBLA and PDGA		5	

from random coil to helix at high polypeptide concentrations in the denaturing solvent dichloroacetic acid [7].) The polymer concentration, at which the birefringent phase initially appears and that at which the entire solution becomes a continuous birefringent fluid are designated the *A* point and the *B* point, respectively. The *A* and *B* points depend on the particular solvent used, but no correlation with solvent properties is apparent. On the other hand, in any particular solvent the *A* and *B* points are related to the degree of polymerization, $n =$ molecular weight/monomer weight, of the synthetic polypeptide. The *A* and *B* points occur at lower polymer concentrations the higher the value of *n*.

These qualitative observations alone suggest that the rod-like shape of the polypeptide molecule and not specific polymer–solvent interactions are responsible for the formation of the polypeptide liquid crystal. Theoretical investigations of the possibility that phase separation or self-ordering could occur in solutions of rod-like solutes were carried out by Onsagar [8] and Flory [9]. (Reference [10] will introduce interested readers to current literature on the theoretical aspects.) Flory applied the lattice model for polymer solutions to a system composed of non-interacting, impenetrable, rod-like solute particles and inert, spherical solvent molecules. The former occupy a continuous linear array of sites on the lattice while the latter occupy single lattice sites. This model treatment indicates that in dilute solutions the rod-like particles are free to adopt random orientations. However, as the concentration of the rod-like particles is increased, a point is eventually reached beyond which randomness of orientation for these particles is no longer compatible with the confinement to the number of lattice sites at their disposal. There is a larger number of arrangements or a higher configurational entropy for solutions in which the solute particles assume orientations locally correlated to one another (a semi-parallel arrangement of the rod-like solute particles) than for an isotropic solution (randomly oriented solute). For solute concentrations above the limiting concentration point, A, the theory predicts that the solution will separate into two phases, the solute-dilute phase being isotropic while the more concentrated phase has an anisotropic arrangement of the solute particles. If the solute particle has a large axis ratio, $X =$ length/diameter, the dependence of the solute volume fraction at the A point, C_A, is given approximately by

$$C_A = \frac{8}{X}(1 - 2/X) \qquad (4.3.1)$$

This phase separation is unique in that it occurs as a consequence of the rod-like shape of the solute particles; attractive forces between the particles are not required.

Experimental values for the A and B points were determined optically by noting the concentrations of polypeptide at which phase separation and coalescence occurred. It was also reported that the viscosity of the solution below the A point was about 2·5 times that of the more concentrated liquid crystal phase [5]. Subsequently, a study of shear viscosity for PBLG solutions with polypeptides of differing degrees of polymerization (axis ratio, X) as a function of polypeptide concentration has been shown to yield independent determinations of the A point [11]. In Fig. 4.3.1, the calculated

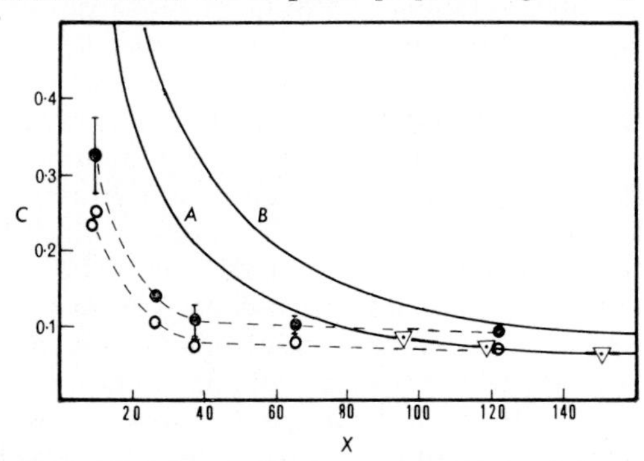

FIG. 4.3.1. Comparison of theoretical and experimental determinations of A and B points. $C =$ PBLG vol. fraction; $X = n \times 1{\cdot}5$ Å/15·5 Å [12b].
○ A points; ● B points. Optical measurements [6].
▽ Shear rate values for A points [11]
— Lattice theory [12].

dependence of the A and B points on axis ratio (from Flory's model) is shown with the experimentally determined values. The polymer concentration at which there is an abrupt change in shear viscosity is in excellent agreement with the value of C_A given by equation (4.3.1). In general, the agreement between theory and experiment is rather good considering the idealizations incorporated into the former and the lack of sensitivity in the latter. Flory suggested that better correlations between theory and experiment might be achieved if the model included mixing of the polypeptide side chains with the solvent [12]. In any case, the results do indicate that the predominant factor responsible for the formation of the polypeptide liquid crystal is geometrical in nature. Reference 45 is a current representative study of phase equilibria in the system PBLG–dimethyl formamide using NMR, hydrodynamic and polarizing microscopy measurements to locate the phase boundaries between isotropic and liquid crystal phases.

The supramolecular structure of the polypeptide liquid crystal, as is shown in the next section, is cholesteric. We have suggested—see also Robinson [6], p. 477—that the reason a cholesteric structure is present in these systems is also geometrical in nature—a consequence of the asymmetry of the helical molecular conformation of synthetic polypeptides. This topic is considered in more detail in the section Origin of the Cholesteric Structure.

Optical Properties

An examination of the polypeptide liquid crystal with a polarizing microscope reveals a number of remarkable features. The most prominent characteristic of the birefringent fluid is a network of equally spaced parallel lines somewhat reminiscent of a fingerprint. The unusual pattern permeates the entire macroscopic volume of the fluid and is visible in ordinary light. The periodicity in the pattern, i.e., the distance between the parallel lines, is remarkably large, regular and reproducible. Figure 4.3.2 is a photograph of a solution of PBLG in dioxane with the polymer concentration between the A and B points (two phase region). The distance between the alternating bright and dark lines is designated by the letter S. The magnitude of S is not sensitive to the molecular weight of the polypeptide, but does depend on the polymer concentration, the solvent and the temperature. S has a reproducible value for a given set of conditions and is not influenced by the shape of the vessel in which the solution is contained. S may be as large as 10^6 Å (100 microns) or have values too small for resolution with an optical microscope. S decreases as the polymer concentration is increased in the anisotropic phase, appearing to be proportional to $1/C^2$, where C is the volume fraction of polymer in the birefringent phase [5]. Between the A and B points, changes in the overall polymer concentration do not affect S since only the relative amounts of isotropic and anisotropic fluid are changed in the two phase region.

If the anisotropic fluid is confined in a thin (<1 mm) rectangular cell, a better defined pattern of lines forms. When the direction of viewing is parallel to the thin dimension of the cell, one observes that the equally spaced lines generally align parallel to the walls of the container, and further away from the walls, small uniform areas appear; these are surrounded by the parallel lines but within them no lines are observed. On standing, these uniform polygonal areas grow as two or more neighbouring areas unite. The appearance of the pattern suggests that the parallel lines are layers seen edge-on, while the uniform areas are layers lying parallel to the plane of the liquid-container interface (normal to the

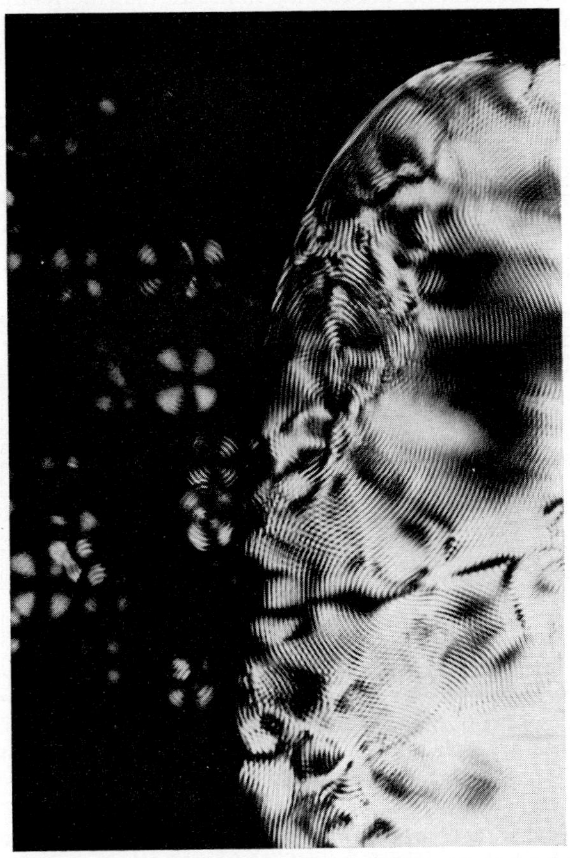

FIG. 4.3.2. Two phase solution of PBLG in dioxane; $S = 14$ microns. Polars are crossed and are oriented parallel to the photograph edges [6].

direction of viewing). A simple model which facilitates visualizing the optical pattern consists of close packed groups of nested transparent polyhedra of progressively increasing size. A schematic drawing of the model is given in Fig. 4.3.3. In the drawing, the parallel lines represent a set of equally spaced, parallel sides of a nested set of polyhedra seen edge on. The uniform areas are the sides of the smallest polyhedra in the set (normal to the direction of viewing) as seen through the transparent sides of the larger polyhedra in the set.

More quantitative investigations of the optical properties of the polypeptide liquid crystal are summarized below.

(*a*) When the birefringent liquid crystal is viewed with one polar (either polarizer or analyser), the parallel lines are invisible wherever the lines are perpendicular to the electric vector of the light.

(*b*) When seen between crossed polars, the alternating light and dark parallel lines are replaced by parallel bands of very bright birefringence colours. The variation in the colours indicates that the retardation oscillates through a regular series of local maxima and minima as one moves along a direction at right angles to the parallel bands. The colour change that occurs when a compensator is inserted shows that the direction of high refractive index is nearly parallel to the lines.

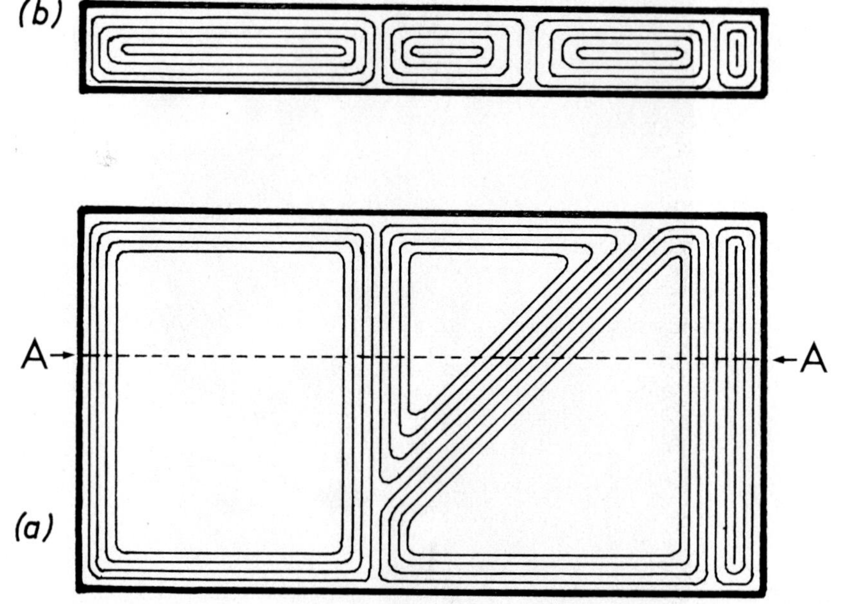

FIG. 4.3.3. A schematic drawing of an arrangement of transparent, close-packed, nested polyhedra which results in the appearance of the uniform areas [6]; (*b*) is a cross section through A–A in (*a*).

(*c*) In solutions with more regular patterns, bright colours exist in the uniform polygonal areas when they are observed with crossed polars. These colours cannot be extinguished with a quartz wedge and are reproducible depending on the thickness of the area. The colour changes in a continuous manner when one polar is rotated, and with monochromatic light, two extinction positions are found 180° apart. The extinction position varies considerably with the wavelength of light being used.

Observation (*c*) signifies the presence of a very high form optical rotation with a correspondingly high rotatory dispersion which accounts for the observed colours. Using the polarizing microscope as a polarimeter, form optical rotations of 20,000° to 150,000° mm^{-1} were measured in the uniform areas of the solution.*

The highest refractive index of the PBLG molecule is parallel to the helix axis [13]. This fact taken together with the sign of the birefringence in the liquid crystal pattern and the other optical features implies that the helix axes lie parallel, or nearly parallel, to the equally spaced retardation lines. Hence, a line of constant retardation is a plane seen edge on, in which the helix axes lie, and in this plane the helices assume a certain preferred orientation. The direction of preferred orientation changes in a continuous manner as one moves along at right angles to the retardation lines.

The optical characteristics listed above are also present in the birefringent "spherulites". The "spherulites" contain additional features which are described in detail by Robinson [6], important amongst these being the radial line of dislocation. Generally speaking, the appearance of the pattern of retardation lines in the "spherulites" suggests an arrangement which is similarly idealized by a model consisting of a nested set of concentric spheres. The polypeptide helices lie in planes tangential to the spheres and have a preferred orientation in each plane, this direction of preferred orientation changing continuously as one moves out along the radius of the "spherulite". Related structures are found in spherulites in the solid state of ordinary synthetic polymers [14].

All of the optical features of the polypeptide liquid crystal can be explained with a structure proposed by Robinson and described in the next section.

Supramolecular Structure

Stumpf discovered that a certain class of thermotropic liquid crystals, which he named cholesteric, has optical rotatory powers of the order of several thousand degrees per millimeter [15]. Other equally remarkable optical properties are associated with this type of liquid crystal. For example, when white light is incident on the surface of the liquid crystal, selective reflection takes place over a small region of the spectrum, the wavelength of maximum reflection varying with the angle of incidence in accordance with Bragg's Law. At normal incidence, the reflected light is circularly polarized; one circular component is totally reflected over a spectral range of some 200 Å, while the other passes through unchanged. Also, quite contrary to usual experience, the reflected wave has the same sense of circular polarization as that of the incident wave.

* The differences between the behaviour of an optically active molecule in the vicinity of an absorption and the form optical rotation to which we are referring should be emphasized. The former more familiar optical rotation occurs in isotropic solutions which contain optically active molecules and has its origin in the selective absorption of one circularly polarized component of the light (Cotton effect). The form optical rotation which we are discussing is caused by the supramolecular arrangement of the molecules in the liquid crystal and originates in the selective reflection of circularly polarized light by this arrangement. The form optical rotation is so large that the contribution from ordinary optical rotation coming from the asymmetric polypeptide helices may be neglected.

A model for the supramolecular structure in cholesteric liquid crystals was proposed by de Vries [16]. He assumed the structure to be composed of a large number of thin birefringent layers, with the principal axis of successive layers turned through a small angle. The liquid crystal molecules lie in the plane of the layer with their long axes approximately parallel. Robinson proposed a similar structure for the polypeptide liquid crystal [2]. In the structure, the polypeptide helix axes are essentially parallel to one another within any microscopic domain of the liquid crystal. In one of these domains it is possible to imagine a set of equally spaced parallel planes in which the helices lie. Helices in one plane have a slight angular displacement relative to those in adjacent planes. The displacement occurs about an axis of torsion, Z, which is normal to the planes. The incorporation of successive layers of appropriately oriented helices results in a helicoidal or cholesteric supramolecular structure. A schematic drawing

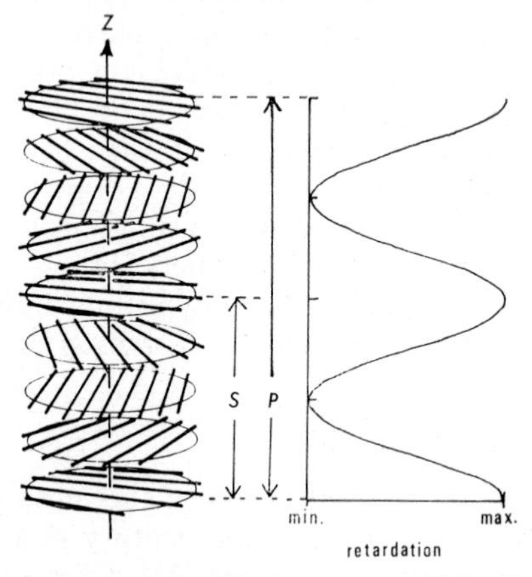

FIG. 4.3.4. Cholesteric supramolecular structure. Z is the optic axis (axis of torsion), P is the pitch and $S = \frac{1}{2}P$ is the visible periodicity. Polypeptide molecules are represented by the parallel lines.

of the cholesteric structure is shown in Fig. 4.3.4. In the drawing, the helical polypeptide molecules are represented by the parallel lines.

This cholesteric structure successfully explains all of the optical properties of polypeptide liquid crystals. Since the refractive index is different along the length of the polypeptide helix and along the helix diameter, the periodic oscillation of the retardation, e.g., the appearance of the equally spaced retardation lines, follows directly from the cholesteric structure as a consequence of viewing the structure at right angles to Z.

In addition to accounting for selective reflection of circularly polarized light at normal incidence without changing the sense of the polarization, the de Vries theory leads to the following formula for rotatory power ($\mathscr{R}$)

$$\mathscr{R} = -\frac{2\pi\alpha^2}{P8\lambda'^2(1 - \lambda'^2)} \tag{4.3.2}$$

where P is the pitch of the structure, $\alpha = (\varepsilon_1 + \varepsilon_2)/2\varepsilon$, ε_1 and ε_2 are the principal dielectric constants of the untwisted structure, $\varepsilon = \frac{1}{2}(\varepsilon_1 + \varepsilon_2)$, $\lambda' = \lambda/(\varepsilon)^{\frac{1}{2}}P$, and λ is the wavelength in vacuum. Chandrasekhar and Rao also arrived at a similar expression for the optical rotatory power using the Jones Calculus for optical systems and the de Vries model [17]. The rotatory power in radians per unit length is given by

$$\mathscr{R} = -\frac{\pi(\mu_1 - \mu_2)^2 P}{4\lambda^2} \tag{4.3.3}$$

where $(\mu_1 - \mu_2)$ is the birefringence of the untwisted structure. This expression is equivalent to equation (4.3.2) when $\lambda'^2 \ll 1$.

The high optical rotatory power observed in regions of the polypeptide liquid crystal in which no retardation lines were present indicates that in these regions the structure is being viewed along the Z direction. When measured values of $\mathscr{R}$ and P are used in equation (4.3.3), the birefringence of the untwisted structure can be calculated. The calculated $(\mu_1 - \mu_2)$ value agrees very well with an experimental value found in a nematic polypeptide liquid crystal. The nematic structure occurs in solutions of a racemic mixture of polypeptides, e.g., equal parts by weight of PBLG and PBDG [3].

In certain concentrated polypeptide solutions (PELG in ethyl acetate or PLGA in dimethylacetamide) S attains values comparable with the wavelength of visible light [6], [18]. In these liquid crystals the brilliant iridescence colours characteristic of thermotropic cholesteric liquid crystals are reflected from the solutions. When a sample (PELG, $C = 0{\cdot}3$, in ethyl acetate) is illuminated with a parallel beam of white light a continuous colour spectrum is reflected onto a white surface. Furthermore, experiments with concentrated PELG solutions in a thin cell (10^{-5}m) show that a considerable band of the reflected light is circularly polarized with the same sense as the incident light, while the transmitted light is polarized in the opposite sense [6]. Thus all of the optical properties of thermotropic liquid crystals with a cholesteric structure have been observed in the polypeptide liquid crystal. These observations show conclusively that the cholesteric structure is an accurate representation of the supramolecular arrangement in the polypeptide liquid crystal.

X-Ray Studies

X-ray diffraction patterns obtained from polypeptide liquid crystals give information on how the helical molecules are distributed in the cholesteric structure, on how the distribution changes with polymer concentration and, in certain cases, on details of the polypeptide molecular conformation. X-ray studies have been made on the PBLG-dioxane liquid crystal in the concentration range 0·1 to 0·5 volume fraction of polymer [4]. This liquid crystal exhibits a diffuse reflection in the low angle region. The sharpness of the reflection increased and the spacing decreased as the polymer concentration was increased. A quantitative interpretation was satisfactorily correlated with the cholesteric structure, if it was assumed that the packing of the helices is locally hexagonal on the microscopic scale. That is, a section taken through a microscopic region of the cholesteric structure at right angles to the helix axes would show that the cross sections of the helices define a hexagonal net in the plane of the section. This arrangement is similar to the arrangement of the micellar fibres in M_1 and M_2 mesophases of amphiphiles. It is shown below that the angular displacement of the helices which is required to produce a macroscopic cholesteric structure is very small and the distortion such small angular displacements cause in the hexagonal net at the microscopic level is negligible. Also, the high axis ratio of the polypeptide makes it reasonable to assume further that a reduction in the polymer concentration causes no dilution in a direction parallel to the helix axes, e.g., the net expands with the addition of solvent only because of increases in the lateral spacing between the helices. These assumptions made it possible to calculate how d, the distance between the $(10\bar{1}0)$ planes in the hexagonal net, should change as the volume fraction of helices is changed. The derived relation between d and the volume fraction of helices, C, is given by

$$d^2 = \frac{W\sqrt{3} \times 10^{24}}{2Nh\rho} \cdot \frac{1}{C} \tag{4.3.4}$$

where N is Avogadro's number, W is the molecular weight per monomer unit (219 for PBLG), ρ is the density of the polypeptide, and h is the length of the projection of one monomer unit along the helix axis (1·5 Å for the α-helix). The experimental values of d $v.$ polymer concentration (open

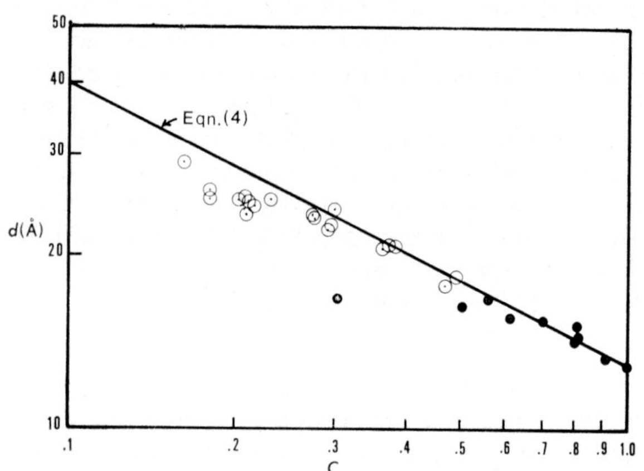

FIG. 4.3.5. The expansion d (Å) of the hexagonal lattice as the PBLG liquid crystal is diluted. C = PBLG vol. fraction.

⊙ Fluid liquid crystalline solution; PBLG–dioxane [4].
● Solid, plasticized PBLG film [27].
— Theoretical expansion of two dimensional hexagonal net—equation 4.3.4

circles) are compared with the theoretical curve in Fig. 4.3.5. The absolute values of d which were found confirm that the structure in the liquid crystal is composed of discrete single molecules and not aggregates of two or more helices. The general agreement with the calculated concentration dependence for d substantiates the assumption of local hexagonal packing.

From the dimensions of the hexagonal net it is possible to calculate the distance of closest approach between the exterior surfaces of two neighbouring helices. This distance, D, is obtained by subtracting the helix diameter (~25 Å for PBLG with extended side chains) from the spacing between helix centres, $2d/\sqrt{3}$. Values for D of +15 Å and −4 Å are found for the PBLG liquid crystal with $C = 0{\cdot}1$ and $C = 0{\cdot}5$, respectively. The negative sign for D at high polymer concentration indicates that the side chains on neighbouring helices could overlap in these solutions.

Also using the hexagonal model, it is possible to show that the angular rotations that the helices can execute about an axis perpendicular to the helix axis without making contact with neighbouring helices is rather small. Ideally, if the helix is fully extended (length $= nh$), the allowed angular displacement θ, is given by $\tan\theta = (2D/nh) \propto 1/C^{\frac{1}{2}}$. In the PBLG-dioxane liquid crystal at $C = 0{\cdot}1$, the helices are restricted to oscillations of $\theta \cong 1°$ of arc about the parallel orientation, with the restrictions more severe at higher polymer concentrations.

The angle of twist per molecule about the Z axis of the cholesteric structure which is required to give the observed pitch is much less than θ. The Z axis passes through S/d molecules over a distance equal to half the pitch. This number ranges from roughly 10^4 to 10^3 at PBLG concentrations $C = 0{\cdot}1$ and $C = 0{\cdot}5$, respectively. ϕ, the required angle of twist per molecule, is given by $d/S \times 180°$, and values of 0·018 and 0·18 degree of arc are calculated for $C = 0{\cdot}1$ and $C = 0{\cdot}5$, respectively. As noted earlier, $S \propto 1/C^2$ and $d \propto 1/C^{\frac{1}{2}}$, therefore, $\phi \propto C^{\frac{3}{2}}$. Not only is the angle of twist per molecule required to produce the observed value of P very small, but also it decreases as C is decreased.

The technique of low-angle X-ray diffraction has been employed to study the different ways in which polypeptide molecules are organized in the presence of solvents, especially in the concentration ranges where intermolecular interactions come into play. At least six structurally differing, concentration-dependent phases have been identified in two-component, polypeptide–solvent systems. Below, we have listed the phases and briefly described their structural characteristics.

A Dilute Isotropic Solutions—amorphous.

B Cholesteric (described above).

C Hexagonal: the solvent is uniformly distributed throughout a hexagonal array of parallel α-helices.

D Complex Hexagonal: bundles of multistranded, intertwined α-helices are packed on a hexagonal net. The net dimensions are independent of concentration, implying that a fixed amount of solvent is required for this phase.

E Tetragonal: single polypeptide molecules are packed on a square rather than a hexagonal net. The polypeptide chain is probably in a 4_{13}-helical conformation.

F Lamellar: this is composed of equidistant planar sheets, each containing extended polypeptide chains in the β-conformation. The distance between planes depends on the nature of the peptide residue and the amount of solvent.

Generally speaking, structures C–F occur at high polymer concentrations ($C > 0{\cdot}5$) and involve specific polymer–solvent and/or polymer–polymer interactions for their stabilization. These structures are, however, intriguing from the point of view of the analogies with structurally similar mesophases found in lyotropic amphiphilic systems (see Chapter 5). An introduction to the literature on polypeptide–solvent systems which exhibit these more complex structures may be found in Luzzati's review article, reference [19].

There has been considerable discussion on the exact dimensions of the molecular conformation of synthetic polypeptides in solution. Experimental evidence on PBLG has been employed to support the idea of two types of helices; the α-helix with the axial translation per residue, $h = 1{\cdot}5$ Å, and the 3_{10}-helix with $h = 2{\cdot}0$ Å. However, a fibre-like diffraction pattern obtained from a shear oriented PBLG-*m*-cresol liquid crystal shows the 1·5 Å spacing only [20]. The X-ray diffraction pattern obtained from an oriented PBLG–DMF liquid crystal also supports the α-helical conformation as the configuration of the polypeptide backbone chain in solution [21]. The usefulness of X-ray studies on polypeptide liquid crystals is further illustrated by the following recent results. Other features in the diffraction patterns obtained from the PBLG–DMF liquid crystal (attributed to a regular stacking of the benzene rings terminating the polypeptide side chains with a different symmetry from that of the backbone helix) are similar to features in the diffraction patterns of fibrous α-proteins [22]. For quite some time, these features in the α-protein diffraction patterns were thought to be indicative of a coiled-coil structure, e.g., two to three intertwined α-helices. However, in view of the recent observations on the polypeptide liquid crystal and the difficulties encountered in fitting the α-protein diffraction pattern to the coiled-coil model, the validity of the coiled-coil structure in fibrous α-proteins has been challenged [23].

Magnetic Field Effects

It has been known for some time that magnetic fields of the order of several hundred oersteds cause spontaneous orientation in nematic liquid crystals. As discussed above, on a microscopic level the local order in the cholesteric polypeptide liquid crystal is virtually nematic—an array of close-packed, parallel helices. Hence it is not surprising that a magnetic field produces dramatic changes in the structure of polypeptide liquid crystals. The application of a sufficiently strong magnetic field to these liquid crystals does in fact untwist the cholesteric structure and form a macroscopic nematic structure with the nematic optic axis oriented parallel to the field. The low degree of mobility of the polypeptide molecules in the viscous solutions accounts for the rather long times (up to several hours) required for completion of the magnetic field-induced cholesteric–nematic transition. The transition is accompanied by the disappearance of the optical characteristics of the cholesteric structure. Optically, the magnetically oriented fluid resembles a uniaxial single crystal having the direction of greatest refractive index parallel to the field. This observation, together with X-ray results from magnetically oriented polypeptide films, shows conclusively that the helices are aligned parallel to the field in the oriented nematic structure.

The cholesteric–nematic transition may be followed indirectly with nuclear magnetic resonance (NMR). The NMR spectrum of CH_2Cl_2 in PBLG-CH_2Cl_2 is split into a doublet because the direct dipole–dipole hyperfine interactions are not averaged to zero by the tumbling of the CH_2Cl_2 molecule in the liquid crystal. The tumbling is not isotropic since the solvent is partially oriented while translating between the close-packed, unidirectionally oriented helices.

The initial spectrum of CH_2Cl_2 in the PBLG liquid crystal is a "powder spectrum" because in a macroscopic sample with the cholesteric structure, the microscopic regions of parallel helices are randomly oriented with respect to the spectrometer field direction. The "powder spectrum" slowly changes to a discrete doublet as the cholesteric structure untwists in the spectrometer magnetic field and the oriented nematic structure is formed [24, 25]. The low mobility of the oriented nematic structure makes it possible to record the angular dependence of the hyperfine splittings as the axis of constraint (nematic axis) is turned to make different angles with the field. Recently, the angular dependence of hyperfine splittings was used to show that most of the helices ($\approx 85\%$) align with their long axes within $\pm 20°$ of the field direction in the oriented nematic structure [46].

The magnitude of the hyperfine splittings is a measure of the degree of orientation that the solvent experiences in the liquid crystal. The degree of orientation found in the lyotropic polypeptide liquid crystals is considerably smaller than that reported for thermotropic nematic phases; the components of the order matrix are of the order of 0·001. The low degree of order can be attributed to chemical exchange between isotropically and anistropically tumbling solvent molecules, the latter being more closely associated with the rather well oriented polypeptide molecules. Such an exchange process is reasonable, since, in the concentration range used for liquid crystal formation, there are from 5 to 10 solvent molecules per monomer of polypeptide and obviously all of the solvent molecules cannot be simultaneously associated with the polymer. Correspondingly, decreasing the solvent/monomer ratio does increase the degree of orientation of the solvent [26].

NMR experiments with polypeptide liquid crystals can provide information about solvent–polypeptide interactions as well as details about the solvent molecule itself. For example, the hyperfine dipole–dipole splittings in the proton NMR spectra of partially oriented DMF in the PLGA–DMF liquid crystal [27] are markedly different from those observed in the PBLG–DMF liquid crystal [28]. The induced nematic structure in both of these phases is composed of polymers with backbones identical in both composition and spatial conformation (α-helices). Therefore the differences in hyperfine interactions show that in contrast to observations with most thermotropic liquid crystals, specific interactions between the functional groups of DMF and those of the polypeptide side chain make significant

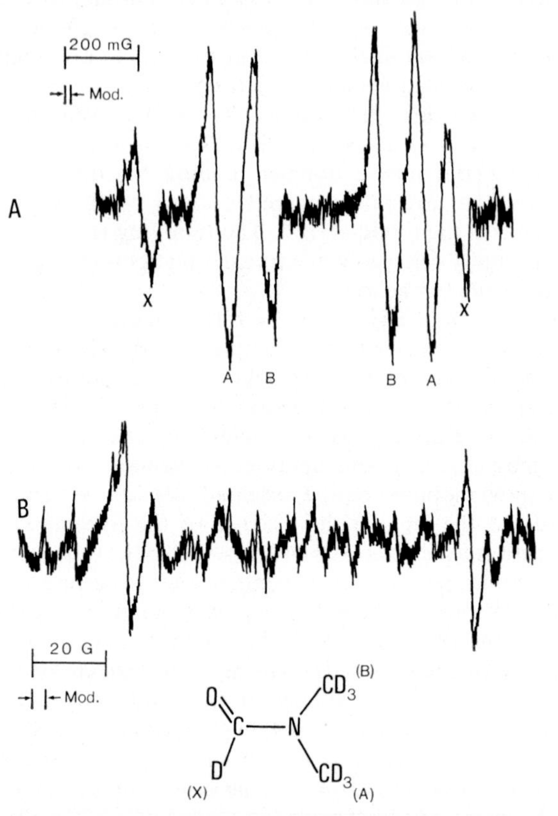

FIG. 4.3.6. Quadrupole splitting in the ^2H and ^{14}N magnetic resonance spectra of DMF (D_7) in the oriented PLGA nematic liquid crystal (molar ratios, PLGA with respect to the number of monomer units in the polypeptide: DMF:H_2O are 1:8:3). The deuterium spectrum, A, and the nitrogen spectrum, B, were recorded at 8 MHz and 4 MHz, respectively. The derivatives of the NMR absorption signals were obtained by field modulation with a frequency of 20 Hz. In the above figure, Mod. refers to modulation amplitude; G = gauss, mG = milligauss.

contributions to the mode of orientation of the solvent in these liquid crystals.

Hyperfine splitting in the NMR spectra of other nuclei (^{35}Cl and ^2H) of partially oriented solvents used for the polypeptides are also readily observed [26, 29]. The ^2H spectrum of fully deuteriated DMF in the PLGA–DMF (D_7) liquid crystal is shown in Fig. 4.3.6(A). The splittings arise because the nuclear quadrupole moment of ^2H interacts with the electric field gradient at the deuterium nuclei; the field gradient is not averaged to zero because of the anisotropic tumbling of the solvent in the liquid crystal. In addition, the relatively low degree of order of the solvent molecules in the sample enables the direct observation of the quadrupole splitting in the ^{14}N NMR spectrum of DMF (Fig. 4.3.6(B)). When the DMF order matrix can be obtained from the proton spectrum, the deuterium and nitrogen quadrupole coupling constants can be calculated from their respective splittings giving information about the local electric field gradients in the DMF molecule [30].

Origin of the Cholesteric Structure

The origin of the very regular, macroscopic supramolecular structure in concentrated polypeptide solutions is a point for much speculation. Robinson *et al.* suggested that long range, electric dipole forces play an important rôle in the formation of the cholesteric structure. The asymmetric array of permanent dipoles on the polypeptide molecule was thought to impose a unidirectional twist on the nearly parallel helices. It was also pointed out that very small oscillations of the helices would produce periodic contacts between nearest neighbours, indicating that it is not necessary to assume that long range forces are essential for the existence of the cholesteric structure [4–6]. We present evidence supporting the idea that short range forces are responsible for the cholesteric structure. Before examining this evidence, it is useful to present some additional observations on the polypeptide liquid crystal which must be consistent with any model proposed to explain the origin of the cholesteric structure.

The negative sign in equations (4.3.2) and (4.3.3) means that the sign of the form optical rotation $\mathscr{R}$ is opposite to that of the twist of the cholesteric structure. This sign can be determined experimentally, and it has been shown that the sense of the twist in the polypeptide–solvent liquid crystal is dependent on the solvent. In most solvents (dioxane, *m*-cresol, $CHCl_3$, etc.), $\mathscr{R}$ is positive for the PBLG liquid crystal. However, $\mathscr{R}$ is negative in the PBLG-CH_2Cl_2 liquid crystal. In an appropriate mixture of dioxane and CH_2Cl_2, the PBLG liquid crystal is nematic. The helix backbone is, however, a right-handed α-helix in both solvents [5]. Also, for a given solvent, the sign of $\mathscr{R}$ for the PBLG solution is opposite to that for the solution of PBDG a left-handed α-helix.

It is also of interest to point out that solutions of other rigid helical biological macromolecules display mesomorphism. Birefringent phases have been observed *in vitro* with concentrated aqueous solutions of DNA [5] and RNA [37a]. Furthermore, these solutions exhibit the equally spaced pattern of retardation lines. Although form optical rotation has not been reported, the striking similarity between the appearance of these solutions and the polypeptide liquid crystal makes it reasonable to accept that the cholesteric structure is present [6]. There is also evidence for cholesteric DNA *in vivo* in Dinoflagellate and bacterial chromosomal material [37b].

In view of the wide variety of polypeptides which form the cholesteric structure (Table 4.3.1) and the strong possibility that it occurs in the nucleic acid phases, we suggested that a model based on short range dispersion forces and dependent on the one feature all of these molecules have in common—an asymmetric helical molecular conformation—may provide an explanation for the cholesteric structure [38]. In effect, we said that the oscillations of the helices relative to their neighbours in the liquid crystal would be biased to one side of the parallel position because of the chirality of the van der Waals surface of the polypeptide molecule. This is analogous to the barrier to rotation on one side of the parallel position which can be demonstrated with two real screws of the same screw-sense, when their threads are interlocked. The biasing of oscillations would in turn lead to a unidirectional twist over large distances and the sense of this twist would be dependent on the sense of the chirality of the polypeptide outer surface.

This simple idea as realized by Robinson [3, 4] would account for the fact that the sign of $\mathscr{R}$ is opposite for cholesteric structures composed of polypeptides with different helix sense (D- or L-polypeptides) and that a mixture of equal parts of D- and L-polymers gives a nematic structure; a random mixture of screws with opposite screw-senses cancels out the unidirectional twist over large distances. In order to explain the change in sign of $\mathscr{R}$ with solvent, one has to assume that the chirality of the polypeptide exterior surface could change screw-sense, on change in solvation, without affecting the sense of the backbone helix. Molecular models show that the sense of the surface chirality of PBLG can be changed by changing side chain conformations only [38].

Further evidence supporting the dispersion force model comes from a theory developed to explain the source of the twist in thermotropic cholesteric liquid crystals [39]. The theory is based on the assumption of anharmonicity in the forces resisting the twist of neighbouring planes of molecules about the axis of torsion in the cholesteric structure. The macroscopic twist is treated as a rotational analogue of thermal expansion with the dominant anharmonic forces coming from nearest neighbours. The theory gives the following relation between the pitch (P) of the cholesteric structure and the temperature:

$$P = \frac{4\pi d I \omega}{AkT} \qquad (4.3.5)$$

k is the Boltzmann constant, d the spacing between neighbouring planes of molecules, ω_0 the frequency of the excited twist modes, I the moment of inertia of the molecule and A the constant in the cubic anharmonicity term of the anharmonic equation of motion describing the system. In the applications to thermotropic liquid crystals, A and ω_0 are assumed to be insensitive to temperature. In Fig. 4.3.7 our observed values for $P(=2S)$ in a PLGA liquid crystal (molar ratio of PLGA:DMF:H_2O, 1:6:5) and data reported for PELG in ethyl acetate [6] are plotted against $1/T$. It was recently reported that the $P \propto 1/T$ relation holds quite well for a number of thermotropic liquid crystal systems [40], and the temperature dependence of P in Fig. 4.3.7 shows that this relation is valid at temperatures removed from the cholesteric–isotropic transition point in the lyotropic polypeptide liquid crystal systems. Furthermore, in the thermotropic systems, a value for $\theta_0 = 20°$, the angle through which molecules in the liquid crystal instantaneously twist, was calculated from the theory, whereas the residual twist required to produce the cholesteric pitch is only 10 minutes of arc. Using the values of d and S found for the PBLG-dioxane liquid crystal at $C = 0{\cdot}1$, θ_0 is calculated to be 3·6°. This is close

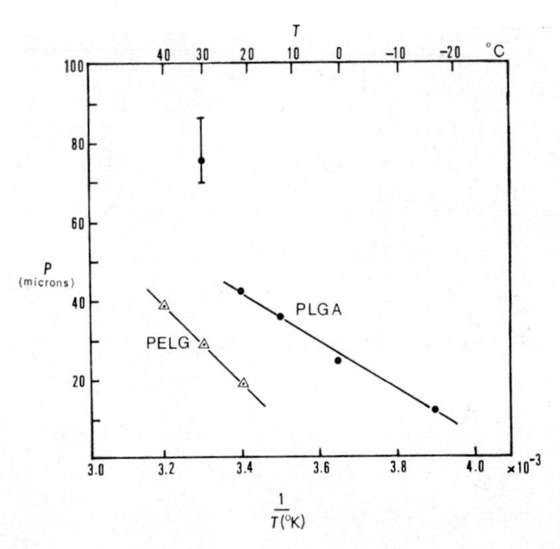

FIG. 4.3.7. The change in the pitch of the polypeptide cholesteric structure with temperature. Liquid crystals:

● Molar ratio of PLGA with respect to the number of monomer units in the polypeptide: DMF:H_2O (1:6:5); the amorphous isotropic solution–liquid crystal transition temperature ≃ 30°C.

△ PELG–ethylacetate, $C = 0{\cdot}3$ (plotted as S v. $1/T$, data from ref. [6]).

to the value calculated from packing considerations in the polypeptide liquid crystal (see X-Ray Studies). The residual twist per molecule required to produce the pitch observed in the polypeptide system is a small fraction of θ_0, also in agreement with the results for the thermotropic systems.

In view of the many similarities between the cholesteric lyotropic system and the thermotropic systems, perhaps it is not too surprising that a theory developed for one satisfactorily describes the behaviour of the other.

Solid Polypeptide Films

PLASTICIZED FILMS

Solid homogeneous films can be prepared by evaporating solutions of PBLG in volatile solvents such as $CHCl_3$ and CH_2Cl_2. Tobolsky *et al.* have investigated the supramolecular structure in the solid films and conclude from X-ray evidence [31] and NMR studies on swollen films [32] that the PBLG helices lie in the plane of the film and that the direction of the helix axes is randomly oriented in the film plane. The anisotropic swelling characteristics of the polypeptide films and the observation of iridescence colours reflected from the film surfaces prompted the suggestion that the solid films partially retain the cholesteric structure of the fluid polypeptide liquid crystal [33].

FIG. 4.3.8. Retardation lines in a solid, birefringent film of PBLG + plasticizer [25]. PBLG: 3,3′-dimethylbiphenyl, 3:7 (by vol.); $S = 2$ microns.

Plasticized PBLG films, films containing an inert, non volatile component added to the polypeptide solution before evaporation, have also been investigated. X-ray studies on the plasticized films show that d, the spacing between the planes of hexagonally packed helices, changes as the polypeptide/plasticizer ratio in the solid film is changed. Figure 4.3.5 shows that d has the same concentration dependence in the solid films (filled circles) as that calculated to describe the effect of dilution on the structure in the fluid liquid crystal. The relation between d and C given by equation (4.3.4) is valid for the plasticized films for polymer concentrations in the range $C = 0{\cdot}5$ to $C = 1{\cdot}0$. When $C < 0{\cdot}5$ phase separation occurs (the plasticizers are not completely miscible with PBLG) and the lattice does not continue to expand. Carefully prepared films do exhibit the pattern of optical retardation lines characteristic of the cholesteric structure in the polypeptide solutions. A photograph of the retardation lines in a thin section of a plasticized solid film is shown in Fig. 4.3.8, and can be compared with the appearance of the fluid liquid crystal (Fig. 4.3.2). This evidence is quite convincing that the unusual supramolecular structure of the polypeptide liquid crystal does exist in the solid state of certain polypeptide films.

In the 1930's it was shown by Vorländer that some thermotropic liquid crystals could be quick-frozen to a metastable state [34]. More recently,

Chistyakov and Kosterin have studied vitrified liquid crystals with smectic, nematic and cholesteric structures, prepared by supercooling thermotropic liquid crystals [35]. Specimens prepared by these techniques are stable for only a few days at room temperature before they gradually crystallize. The fact that the plasticized PBLG films are solid with regard to mechanical properties yet possess a thermodynamically stable liquid crystal structure probably results from the large size of the polypeptide molecules.

At this point it may be useful to call attention to the similarities between these results and those from studies of the solid states of certain biological systems. Neville and Caveney have confirmed an earlier suggestion by Robinson and shown that the *Scarabaeid* beetle exocuticle manifests the optical properties of cholesteric liquid crystals [36]. The structure in the exocuticle is cholesteric and analogous to the polypeptide system; the structural units are microfibrils which are unidirectionally oriented in layers. In some instances the pitch of the structure is of the order of the wavelength of light and consequently, iridescence colours are reflected from the surface of the beetles. Examples are shown in ref. [36]. These workers do not imply that the solid, optically active cuticles are liquid crystalline, but do suggest that the structure may have formed from cholesteric liquid crystals "stabilized in a thin zone of transition next to the secretory epidermal calls". It is apparent (see ref. [36]) that the *colour* as well as the form of the insect is symmetrical about a plane, one half being the mirror image of the other. However, examination with a suitable analyser shows that the sense of circular polarization of the coloured reflected light and hence of the helicoidal structure is the same over the whole surface. This lends support to the idea that the cholesteric structure responsible for the sense of polarization and localized on the exterior of the secretory epidermal cells has its origin in an extracellular self-assembly process.

Magnetically Oriented Films

When solutions of PBLG are slowly evaporated in a strong magnetic field, highly oriented polypeptide films are obtained. The orientation occurs while the evaporating solutions pass through the concentration range in which they are liquid crystalline and becomes permanently locked in when the mixture of solvent plus PBLG becomes solid. The solid, magnetically oriented nematic PBLG films produce X-ray diffraction patterns which show uniaxial orientation and are very similar to the fibre diffraction patterns obtained from mechanically oriented PBLG (fibres drawn from concentrated solutions). In the diffraction pattern exhibited by a nematic PBLG film cast from CH_2Cl_2, the layer line pattern consists of streaks with Bragg reflections occurring only on the equator. This implies that the lateral spacing between the helices is fairly regular, but there probably are random displacements of the molecules relative to one another parallel to the helix axis. The layer line distribution of the diffraction pattern can be indexed with helix parameters of 18 residues in 5 turns. This corresponds to the normal α-helix conformation for the polypeptide backbone. The 18/5 α-helix was also observed in nematic films of PBLG cast from *cis*-1,2-dichloroethylene.

Figure 4.3.9 shows the diffraction pattern of a nematic PBLG film cast from $CHCl_3$. The layer line pattern is markedly different to that exhibited by the polypeptide in the 18/5 α-helix. The strong layer line appearing at 10·54 Å and the absence of other layer lines between the equator and the "turn" layer line (5·29 Å) clearly demonstrates that the helix parameters are 7 residues in 2 turns [41]. When specimens containing the 7/2 α-helix are heated for several hours above 100°C, the layer line distribution changes to that of the normal 18/5 α-helix. No orientation was observed in films cast from dioxane, *trans*-1,2-dichloroethylene and DMF. However, NMR studies of the PBLG liquid crystal in these last three solvents do indicate that an oriented nematic phase is formed in the magnetic field. Apparently tnere are strong polymer–polymer interactions occurring in the more concentrated solutions of these latter solvents and such interactions destroy the magnetic orientation as the solutions are evaporated to dryness. There is evidence from low-angle X-ray studies that a more complex structure is formed in the PBLG–DMF solutions at high concentrations of polymer [19].

Since 1951, when Perutz used oriented fibres of PBLG in the first crystallographic proof of Pauling and Corey's model of an α-helix,

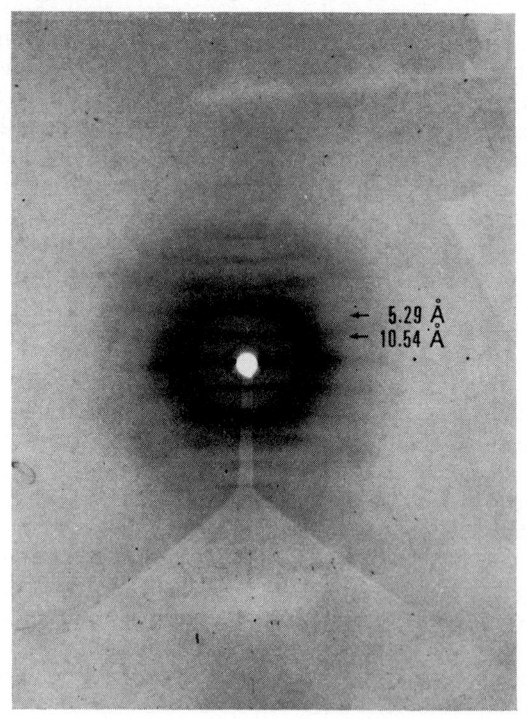

FIG. 4.3.9. X-Ray diffraction pattern from magnetically oriented PBLG evaporated from $CHCl_3$ solution. CuK_α irradiation; 200 micron collimation [27]

considerable detailed information concerning the possible molecular conformations adopted by polypeptide chains has been accumulated (see reference [42] for a review). During this time, only the normal 18/5 α-helix has been reported for the solid state of PBLG. On the other hand, conformations in which the polypeptide backbone is distorted from the 18/5 α-helix have been employed to describe the diffraction data from the solid state of racemic polybenzyl glutamate. In the mechanically oriented racemic fibres, it has been suggested that the distorted backbone conformation facilitates energetically favourable side chain–side chain interactions between neighbouring left- and right-handed helices in the crystal lattice. As previously mentioned, Squire and Elliott have reported X-ray diffraction studies on an oriented, liquid crystal phase of PBLG and the racemic polymer. The oriented liquid crystal was prepared by wetting mechanically oriented polypeptide fibres with DMF. Their observations support a polypeptide conformation in which there is intramolecular stacking of the benzene rings terminating the side chains in polybenzyl glutamate. The quasi-helical intramolecular stacks can in turn interact in a regular manner with the stacks of neighbouring helices. In this PBLG–DMF liquid crystal, the backbone of the polypeptide appears to remain in the normal α-helical conformation [22].

The 7/2 α-helix does appear to satisfy the stereochemical criteria for polypeptide conformations. Ramakrishnan and Ramachandran have calculated allowed polypeptide conformations for specified, minimum van der Waals contact distances between the atoms in adjacent residues. The 7/2 α-helix lies close to the region of allowed conformations when plotted on a diagram of "number of residues per turn" *v.* "axial translation per residue"—for the N—C_α—C′ angle, $\tau = 110°$ [43].

The above results on magnetically oriented films cast from $CHCl_3$ indicate that large deviations from the 18/5 α-helix do exist in the solid state containing only the L-isomer of polybenzyl glutamate. The fact that the 7/2 α-helix reverts to the normal 18/5 α-helix when it is heated suggests that the 7/2 α-helix may correspond to a local minimum in the PBLG conformational energy.

Concluding Remarks

The lyotropic polypeptide liquid crystal is clearly very closely related to thermotropic cholesteric liquid crystals and the cholesteric–nematic transition which occurs when the polypeptide liquid crystal is exposed to a magnetic field makes the system amenable to further investigation. Recently Toth and Tobolsky [44] have shown that the PBLG liquid crystal can be oriented by electric fields of about 100 kV m^{-1}. Again there appears to be a field-induced transition from a cholesteric to a nematic structure.

A considerable amount of information on the polypeptide molecular conformation itself can be obtained from the oriented solutions. Both NMR and X-ray diffraction are useful techniques for studying polypeptide–solvent interactions, and it is likely that the infrared dichroism and ultraviolet absorption spectra of these solutions could also give useful information. Further investigations of the oriented system offer the possibility of detecting and characterizing small, solvent-induced perturbations of the polypeptide molecular conformation. It seems that additional research on this liquid crystal system and the concentrated solutions of other rod-like biological macromolecules would be very profitable.

REFERENCES

[1] ELLIOTT, A. and AMBROSE, E. J. *Discuss. Faraday Soc.* **9,** 246 (1950).
[2] ROBINSON, C. *Trans. Faraday Soc.* **52,** 571 (1956).
[3] ROBINSON, C. and WARD, J. C. *Nature, Lond.* **180,** 1183 (1957).
[4] ROBINSON, C., WARD, J. C. and BEEVERS, R. B. *Discuss. Faraday Soc.* **25,** 29 (1958).
[5] ROBINSON, C. *Tetrahedron* **13,** 219 (1961).
[6] ROBINSON, C. *Molec. Crystals* **1,** 467 (1966).
[7] DOWNIE, A. R., ELLIOTT, A., HANBY, W. E. and MALCOLM, B. R. *Proc. R. Soc.* **A242,** 325 (1957).
[8] ONSAGER, L. *Ann. N.Y. Acad. Sci.* **56,** 627 (1949).
[9] FLORY, P. J. *Proc. R. Soc.* **A243**, 73 (1956).
[10] COTTER, M. A. and MARTIRE, D. E. *J. chem. Phys.* **52,** 1909 (1970).
[11] HERMANS, J. Jr. *J. Colloid Sci.* **17,** 638 (1962).
[12a] FLORY, P. J. and LEONARD, W. J. *J. Am. chem. Soc.* **87,** 2102 (1965).
[12b] FLORY, P. J. *J. Polym. Sci.* **49,** 105 (1961).
[13] AMBROSE, E. J. and ELLIOTT, A. *Proc. R. Soc.* **A205,** 47 (1951).
[14] KELLER, A. *J. Polym. Sci.* **39,** 151 (1959).
[15] STUMPF, F. *Phys. Z.* **11,** 780 (1910).
[16] DE VRIES, H. *Acta Crystallogr.* **4,** 219 (1951).
[17] CHANDRASEKHAR, S. and RAO, K. N. S. *Acta Crystallogr.* **A24,** 445 (1968).
[18] SAMULSKI, E. T. results now published for the first time.
[19] LUZZATI, V. and SALUDIJAN, P. in *Poly-α-amino Acids* (edited by G. Fasman), Marcel Dekker, Inc., New York (1968).
[20] PARRY, D. A. D. and ELLIOTT, A. *Nature, Lond.* **206,** 616 (1965).
[21] PARRY, D. A. D. and ELLIOTT, A. *J. molec. Biol.* **25,** 1 (1967).
[22] SQUIRE, J. M. and ELLIOTT, A. *Molec. Crystals Liqu. Crystals* **7,** 457 (1969).
[23] PARRY, D. A. D. *J. theor. Biol.* **26,** 429 (1970).
[24] SOBAJIMA, S. *J. phys. Soc. Japan,* **23,** 1070 (1967).
[25] SAMULSKI, E. T. and TOBOLSKY, A. V. *Molec. Crystals Liqu. Crystals* **7,** 433 (1969).
[26] FUNG, B. M., GERACE, M. G. and GERACE, L. S. *J. phys. chem.* **74,** 83 (1970).
[27] SAMULSKI, E. T. and TOBOLSKY, A. V. *Liquid Crystals and Ordered Fluids* (edited by J. F. Johnson and R. S. Porter), Plenum Press, New York (1970), p. 111.

[28] PANAR, M. and PHILLIPS, W. D. *J. Am. chem. Soc.* **90,** 3880 (1968).
[29] GILL, D., KLEIN, M. P. and KOTOWYCZ, G. *J. Am. chem. Soc.* **90,** 6870 (1968).
[30] SAMULSKI, E. T. and BERENDSEN, H. J. C. *J. chem. Phys.* **56,** 3920 (1972).
[31] MCKINNON, A. J. and TOBOLSKY, A. V. *J. phys. Chem.* **70,** 1453 (1966).
[32] SAMULSKI, E. T. and TOBOLSKY, A. V. *Macromols* **1,** 555 (1968).
[33] SAMULSKI, E. T. and TOBOLSKY, A. V. *Nature, Lond.* **216,** 997 (1967).
[34] VORLÄNDER, D. *Trans. Faraday Soc.* **29,** 907 (1933).
[35] CHISTYAKOV, I. G. and KOSTERIN, Y. A. *Rost Kristallov.* **4,** 68 (1964).
[36] NEVILLE, A. C. and CAVENEY, S. *Biol. Rev.* **44,** 531 (1969).
[37a] SPENCER, M., FULLER, W., WILKINS, M. H. F. and BROWN, G. L. *Nature, Lond.* **194,** 1014 (1962).
[37b] BOULIGAND, Y., SOYER, M. O. and PUISEUX-DAO, S. *Chromosoma* **24,** 251 (1968).
[38] SAMULSKI, E. T. Ph.D. Thesis, Dept. of Chemistry, Princeton University (1969).
[39] KEATING, P. N. *Molec. Crystals Liqu. Crystals* **8,** 315 (1969).
[40] BAESSLER, H. and LABES, M. M. *J. chem. Phys.* **52,** 631 (1970).
[41] SAMULSKI, E. T. and TOBOLSKY, A. V. *Biopolymers* **10,** 1013 (1971).
[42] ELLIOTT, A. in *Poly-α-amino Acids* (edited by G. Fasman), Marcel Dekker Inc., New York (1968).
[43] RAMAKRISHNAN, C. and RAMACHANDRAN, G. N. *Biophys. J.* **5,** 909 (1965).
[44] TOTH, W. J. and TOBOLSKY, A. V. *Polym. Lett.* **8,** 531 (1970).
[45] WEE, E. L. and MILLER, W. G. *J. phys. Chem.* **75,** 1446 (1971).
[46] ORWOLL, R. D. and VOLD, R. L. *J. Am. chem. Soc.* **93,** 5335 (1971).

Additional References

STRALEY, J. P. *Molec. Crystals Liqu. Crystals* (1973), in press.
MIYATA, N., TOHYAMA, K. and GO, Y. *J. phys. Soc. Japan* **33,** 1180 (1972).
IIZUKA, E. *J. phys. Soc. Japan* **34,** 1054 (1973).
WEE, E. L. and MILLER, W. G. *J. phys. Chem.* **77,** 182 (1973).
RAI, J. H. and MILLER, W. G. *J. phys. Chem.* **76,** 1081 (1972).
HINES, W. A. and SAMULSKI, E. T. *Macromol.* (1973), in press.
MILLER, W. G., 12th Prague Microsymposium on Macromolecules. *Pure appl. Chem.*, in press.

The following papers presented at the Symposium on Ordered Fluids and Liquid Crystals, 166th A.C.S. Meeting, Chicago, August, 1973; Proceedings (edited by J. F. Johnson and R. S. Porter), Plenum Press, New York, in press.

(*a*) Magnetohydrodynamics in Liquid Crystalline Poly-γ-benzyl-L-glutamate: Visual Observations by Dupré, D. B. and Hammersmith, J. R.

(*b*) Nuclear Magnetic Resonance in Polypeptide Liquid Crystals by Hines, W. A. and Samulski, E. T.

(*c*) Magnetic Relaxation of Poly-γ-benzyl-L-glutamate Solutions in Deuterochloroform by Fung, B. M. and Martin, T. H.

5

The Influence of Composition and Temperature on the Formation of Mesophases in Amphiphilic Systems. The *R*-Theory of Fused Micellar Phases

P. A. Winsor

GENERAL DISCUSSION

Introduction

SOLID CRYSTALLINE PHASES

A fundamental structural feature of all amphiphilic phases is the tendency for the mutual orientation of the amphiphile molecules with their hydrocarbon (= lipophilic) and polar (= hydrophilic) moities grouped like-to-like.

In solid crystalline phases this leads to the long-range arrangement of the molecules in parallel sheets with alternate orientation of the sheets. For example, in the one or more crystalline phases formed by the alkali metal soaps at room temperatures, the extended hydrocarbon chains are arranged in three-dimensional crystalline order either normal to or inclined at an angle to the sheets [1] (Fig. 5.1).

At higher temperatures a crystalline form is sometimes found in which the hydrocarbon chains undergo thermal rotation about their long axes. In this form the hydrocarbon chains lie normal to the sheets in two-dimensional hexagonal array.

MESOPHASES OF PARTLY CRYSTALLINE AND PARTLY FUSED CHARACTER

At still higher temperatures, and/or through the effect of solvents, the process of fusion results in the three-dimensional order of the hydrocarbon chains and of the polar groups within the crystal being broken down. In certain systems this may occur in one step giving either a mesomorphous

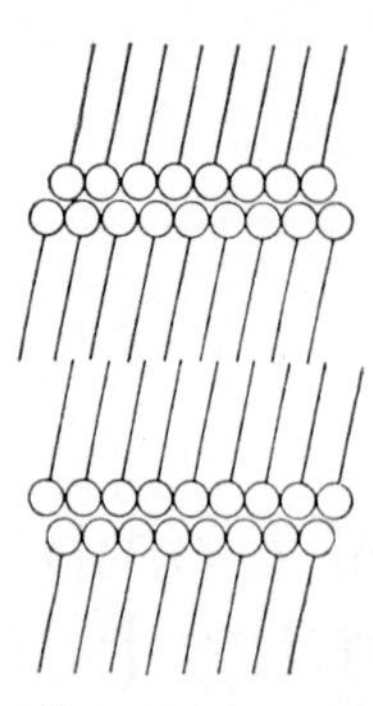

FIG. 5.1. Character of solid lamellar soap crystal. Both the hydrocarbon groups and the polar groups are arranged in regular three-dimensional solid crystalline order.

phase of the fused type or an amorphous liquid melt. It has been proposed that in other systems fusion occurs in stages involving a series of semi-crystalline mesophases. Thus, with certain amphiphiles in the presence of water, mesophases are produced for which it has been suggested that, while fusion of the polar groups has occurred, crystalline order of the hydrocarbon chains is retained (p. 234). In other semi-crystalline mesophases a measure of crystalline order is thought to be retained by the polar groups although the hydrocarbon chains are in the fused state (p. 225).

MESOPHASES OF FUSED CHARACTER

In this, the most widely studied type of mesophase, the three-dimensional intermolecular order of the crystalline solid is lost. In these mesophases the amphiphile molecules are clustered on the like-to-like principle in essentially liquid groupings, termed "micelles" or "aggregates", of various conformations. The mesomorphous character of the phases arises from the extended arrangement of these micelles with long-range order in one, two or three dimensions. On rise of temperature, and/or in the presence of solvents, this long-range intermicellar order may be broken down, the mesophase undergoing a first-order phase transition to an amorphous liquid phase. This transition is closely analogous to melting. It should not however be regarded as the last stage in the fusion of the three-dimensional crystal. The mesophases of the fused type are themselves already in the fused state and the micelles within them are of liquid, as distinct from solid, character.

Phase Sequences in Amphiphilic Systems under Conditions Which Give Rise to Mesophases of the Fused Type: the R-Theory

In amphiphilic systems under conditions in which solid crystalline phases or mesophases of partly crystalline and partly liquid character are not produced, several distinct fused mesophases, each with its own individual micellar character, are encountered. These mesophases arise in diverse systems of one, two or many components but their formation and, in systems where several mesophases arise, the order in which one phase succeeds another with progressive change in composition or temperature conform to an orderly pattern which may be interpreted on the basis of the *R*-theory [2, 3, 4, 5].

This theory is based on considerations of the effects of changes in composition and temperature on the short-range *intra*-micellar forces which regulate the shape and size adopted by the micelles. It considers further how, through the operation of long-range *inter*-micellar forces, at suitable temperatures and intermicellar distances, micelles of a particular shape and size can adopt long-range intermicellar order and constitute a particular mesophase.

The *R*-theory applies only to systems in thermodynamic equilibrium, whether individual micellar phases (amorphous liquid or mesomorphous) or mixtures of conjugate phases. Further, it applies only to phases in which the micelles are liquid in character and thus readily deformable, i.e. to "fused" phases. It does not apply to mesophases of "semi-crystalline" character in the crystalline parts of which some degree of the rigidity of a crystal is retained.

The Lamellar Mesophase, G

When considering the regularities shown by the fused types of amphiphilic phases it is convenient to commence with the lamellar mesophase G ("gel-like" phase, also termed the "smectic" or "neat"* phase) which, in many systems of one, two or more components, exists over a broad range of temperature and/or composition. This phase usually has a fairly thin gel-like consistency although in systems of high amphiphile content it may be quite thick. It is semi-transparent and almost always birefringent. Its general structure has been well established, mainly by studies of its optical properties [5] and by X-ray diffraction measurements [6].

In this phase, first exemplified by binary aqueous systems, the amphiphile molecules are arranged alternately in planar double-layers (Fig. 5.2(*a*)) with a constant repeat distance between the double-layers. This recalls the arrangement found in the crystal but, in contrast to the strict three-dimensional molecular arrangement within the crystal, except for being assembled in parallel layers the molecules within the G mesophase have no strict three-dimensional order and enjoy considerable mobility. The G phase is thus similar to the "smectic A" phase in non-amphiphilic systems (cf. Chap. 2.1). Except in so far as is limited by their attachment to the polar groups, the hydrocarbon chains constitute a disordered and fluid region. Likewise the polar groups, apart from being restricted to statistically planar sheets, are in a mobile non-crystalline condition (cf. however, p. 208). The water molecules within the aqueous zone, except in so far as

* The designation "neat", although currently frequently applied to the aqueous G-phase, was originally used for the anhydrous "neat soap" mesophase first formed on cooling the amorphous melt (p. 222). The G-phase is to be identified with the aqueous "soap boiler's neat phase", a phase which is apparently distinct from the anhydrous "neat soap" (p. 207).

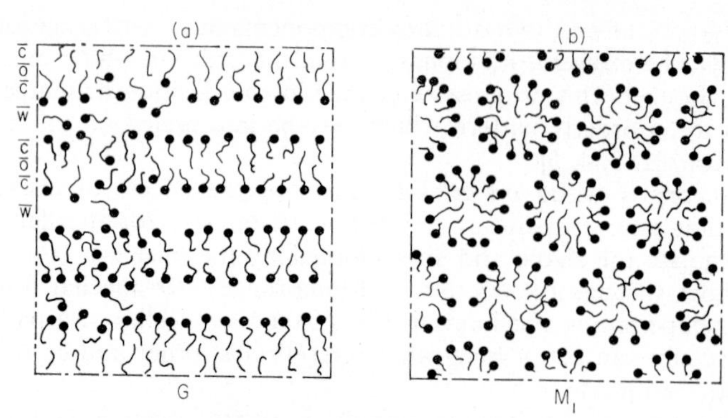

FIG. 5.2. Diagrammatic representation of the structure of (*a*) the lamellar mesophase, G, and (*b*) the middle phase, M, in binary amphiphile/water systems. In (*a*) the situation in principle is shown on the left hand side: the $\bar{C}$ layer contains most of the amphiphile but both amphiphile and water are distributed throughout the $\bar{C}$, $\bar{O}$ and $\bar{W}$ regions so as to maintain their respective activities statistically uniform throughout (p. 273). In (*a*) on the right hand side is shown the working approximation usually employed in X-ray studies: the $\bar{C}$ layer is regarded as containing solely and entirely all the amphiphile present. In (*b*) only the case of the working approximation is represented.

In the diagrams the amphiphile molecules nearest to the plane of the paper are represented as projected on the plane of the paper which in (*a*) is at right angles to the lamellae and in (*b*) at right angles to the fibrous micelles. An attempt is made to indicate (1) that the mean hydrocarbon density at all points within the micelles is approximately uniform (2) that the hydrocarbon chains wave about thermally in all directions with respect to the plane of the paper, i.e., above and below and within it.

they are ordered by orientation effects close to the amphiphilic sheets (p. 273) are again in the disordered liquid state. As discussed later (p. 238) the interfacial area per polar group on the lamellar face or, more shortly, "the area, S, per polar group" varies with the bulk concentration but is usually considerably greater than the actual "area" of the polar group itself. Water molecules must therefore interpenetrate between the polar groups so that the polar layer is an aqueous interfacial solution zone (p. 243). The micellar unit of the G phase i.e. the individual amphiphilic lamella, in accordance with these considerations, has a fluid, flexible structure. Within certain limits it is capable of dimensional changes, with variation in temperature and concentration, so that its thickness and the interfacial area per polar group (which in a lamellar micelle is equal to the effective cross-sectional area per amphiphile molecule) are dependently adjustable (p. 185 of reference [3]; cf. p. 204).

The experimental fact that in the G phase, except where subject to mechanical constraint, the amphiphilic lamellae retain their parallel and planar arrangement, indicates that each individual lamella, $\bar{C}$, although flexible, has no overall tendency preferentially to become convex towards either its lipophilic, $\bar{O}$, or polar, $\bar{W}$, environment. This has been expressed by saying that for each individual lamella throughout the G phase, the ratio R (where R = tendency of amphiphilic layer, $\bar{C}$, to become convex towards its lipophilic environment, $\bar{O}$/tendency of amphiphilic layer, $\bar{C}$, to become convex towards its polar environment, $\bar{W}$) must statistically be unity in all directions within the lamellae. On a molecular scale R will show local thermal fluctuations (Fig. 5.2) but only at the limits of stability of the G phase will these fluctuations become statistically significant. The barred symbol $\bar{C}$ has been used to indicate that, in multicomponent systems containing more than one amphiphilic species, the $\bar{C}$ layer will contain these various species jointly, e.g., in the sodium caprylate/decanol/water system (p. 279), the $\bar{C}$ layer will contain a mixture of sodium caprylate and decanol. Similarly, the barred symbols $\bar{O}$ and $\bar{W}$ have been used to indicate that the lipophilic region, $\bar{O}$ (consisting of the juxtaposed hydrocarbon chains of the oppositely oriented $\bar{C}$ layers with associated dissolved hydrocarbons, etc., if present) and the polar region, $\bar{W}$ (consisting of the juxtaposed polar groups of the oppositely oriented $\bar{C}$ layers together with water, inorganic salts, water-soluble organic compounds, etc., if present) again may be multicomponent (see also p. 274).

Nature of the Molecular Interactions Determining R

The nature of the electrostatic and electrokinetic molecular interactions operative in the systems under discussion is now briefly considered.

Electrostatic (hydrophilic) interactions

These arise from the charges on ions and from the uneven distribution of charge in certain electrically neutral molecules (dipoles). Interaction between ions and dipoles favours attractive orientations. These, however, tend to be destroyed by the thermal motion of the molecules and consequently decrease with rising temperature. The physical character of water and its solvent properties are largely determined by the electrostatic forces due to the $^{-}O\text{–}H^{+}$ dipole. Electrostatic interactions thus contribute mainly to hydrophilic character and have therefore been denoted by A_H.

Electrokinetic (lipophilic) interactions

Molecules which are electrically non-polar and thus show no electrostatic interaction still attract one another. The attractive forces are due to the movement of the electrons within the molecules. In neighbouring molecules these movements tend to become in phase and thus produce attraction. This electrokinetic interaction is almost wholly responsible for the interaction between paraffin molecules and conditions their lipophilic character. Electrokinetic interactions are therefore denoted by A_L.

In a binary solution of two species, X and Y, the molecular interactions may be represented by

$A_{XX} = A_{HXX} + A_{LXX}$ tending to promote clustering of molecules of X and ultimately phase separation

$A_{YY} = A_{HYY} + A_{LYY}$ tending to promote clustering of molecules of Y and ultimately phase separation

$A_{XY} = A_{HXY} + A_{LXY}$ tending to promote mixing of molecules of X and Y.

All these interactions will depend on temperature, i.e. molecular motion, and on concentration.

When these considerations are applied to the concept of the ratio R it can be seen that the tendency of the $\bar{C}$ layer preferentially to spread out into the $\bar{O}$ region, i.e. to become convex towards $\bar{O}$, will be assisted by $A_{\bar{C}\bar{O}}$ and resisted by $A_{\bar{O}\bar{O}}$. (In this locally segregated system A_{XY} will represent the interfacial interaction energy between the X and Y species per unit area of surface.) Similarly, the tendency of the $\bar{C}$ layer to become convex towards $\bar{W}$ will be assisted by $A_{\bar{C}\bar{W}}$ and resisted by $A_{\bar{W}\bar{W}}$. Thus R will increase with increase in the ratio

$$(A_{\bar{C}\bar{O}} - A_{\bar{O}\bar{O}})/(A_{\bar{C}\bar{W}} - A_{\bar{W}\bar{W}})$$

where $A_{\bar{C}\bar{O}}$, etc., are energies of interaction per unit area of interface at the $\bar{C}/\bar{O}$ and $\bar{C}/\bar{W}$ interfaces respectively.

Such a relationship we denote by

$$R \rightarrow [(A_{\bar{C}\bar{O}} - A_{\bar{O}\bar{O}})/(A_{\bar{C}\bar{W}} - A_{\bar{W}\bar{W}})]$$

Further, for a given $\bar{O}$ and $\bar{W}$, R will increase with increase in the ratio

$$A_{\bar{C}\bar{O}}/A_{\bar{C}\bar{W}}.$$

The intermolecular attractions responsible both for $A_{\bar{C}\bar{O}}$ and for $A_{\bar{C}\bar{W}}$ will be partly electrokinetic (forces of lipophilic character, A_L) and partly electrostatic (forces of hydrophilic character, A_H).

The effects of changes in composition and temperature on the ratio R, and consequently on micellar shape, may therefore be considered from the point of view of their probable effects on

$$[A_{\bar{C}\bar{O}}/A_{\bar{C}\bar{W}}]$$

or more explicitly, on

$$[(A_{L_{\bar{C}\bar{O}}} + A_{H_{\bar{C}\bar{O}}})/(A_{L_{\bar{C}\bar{W}}} + A_{H_{\bar{C}\bar{W}}})].$$

For example, if on a stable lamellar solution, G, such as that formed by water and Aerosol MA* or Aerosol OT† (Fig. 5.3) at a given temperature, one imposed a gradual change (e.g. dilution with water) which would tend relatively to increase $A_{\bar{C}\bar{W}}$ and thus to diminish R, the stability of the lamellar phase could be maintained only by some internal rearrangement (increase in interfacial area per polar group, i.e. reduction in the number of polar groups per unit area, with consequent decrease in $A_{\bar{C}\bar{W}}$, accompanied by the dependent contraction in thickness of the $\bar{O}$ region) that restored the value of $R = 1$ (p. 185 of reference [3]). If this degree of rearrangement were not energetically possible within the limits of the imposed change, then the lamellar phase, G, would break down giving a new phase (at first conjugate with remaining G) in which the $\bar{C}$ region was, on balance, convex towards $\bar{W}$.

* Aerosol MA
$CH_2{\cdot}CO{\cdot}O{\cdot}CH(CH_3){\cdot}CH_2{\cdot}CH(CH_3)_2$
|
$CH(SO_3Na)CO{\cdot}O{\cdot}CH(CH_3){\cdot}CH_2{\cdot}CH(CH_3)_2$

† Aerosol OT
$CH_2{\cdot}CO{\cdot}O{\cdot}CH_2{\cdot}CH(C_2H_5){\cdot}CH_2{\cdot}CH_2{\cdot}CH_2{\cdot}CH_3$
|
$CH(SO_3Na){\cdot}CO{\cdot}O{\cdot}CH_2{\cdot}CH(C_2H_5)CH_2CH_2CH_2CH_3$

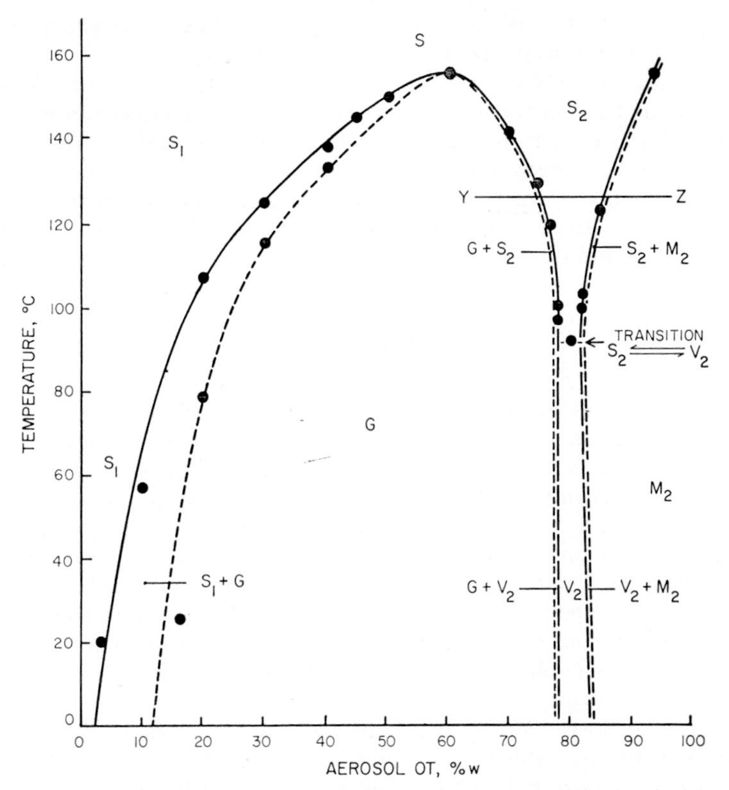

FIG. 5.3. Phase diagram for the Aerosol OT/water system [7]. The boundaries indicated by broken lines are tentative.
$S(S_1, S_2)$ = Mobile isotropic phase; G = Neat phase;
M_2 = Inverse middle phase; V_2 = Inverse viscous isotropic phase

Similarly, if on a stable lamellar phase one imposed a change (e.g. removal of water or addition of an inorganic salt to $\overline{W}$) which would tend relatively to decrease $A_{\overline{CW}}$ and thus to increase R, the stability of the lamellar phase could be maintained only by some internal rearrangement (decrease in interfacial area per polar group with dependent expansion in thickness of the $\overline{O}$ region) that restored the value $R = 1$. Beyond the limits where such rearrangement was energetically possible, the lamellar phase would break down giving a new phase in which the $\overline{C}$ region was on balance concave towards $\overline{W}$.

The possibility of the restoration of the value $R = 1$ by rearrangement (e.g. by increase in leaflet area, diminution in leaflet thickness) implies that $A_{\overline{CO}}$ and $A_{\overline{CW}}$ do not vary as precisely similar functions of area and thickness. From experimental evidence (p. 204), it appears that, at least in systems where the $\overline{O}$ region is wholly paraffinic in composition, $A_{\overline{CO}}$ is comparatively little affected by the thickness of the $\overline{O}$ region. On the other hand, an expansion in the interfacial area per polar group would diminish $A_{\overline{CW}}$ since the number of polar groups per unit area available for interaction with the $\overline{W}$ region would be diminished. The effect of any change in composition tending to increase $A_{\overline{CW}}$ (cf. Table 5.16, p. 268) could therefore, within limits, be offset and the stability of the G phase maintained, by a lateral expansion and transverse contraction of the $\overline{C}$ region and, necessarily, of the accompanying $\overline{O}$ and $\overline{W}$ regions. The reasons why dimensional changes occur in this way to maintain the value of $R = 1$ and the stability of the lamellar phase over a considerable range of compositions at a given temperature, are not wholly clear. Other energetic factors, in addition to the ratio R, must favour persistence of the lamellar structure (8).

Breakdown of the G Phase on Dilution to give the Amorphous Solution Phase, S_1

With certain binary aqueous G phases, e.g. that of Aerosol OT over a wide range of temperatures (Fig. 5.3) or that of sodium laurate over an intermediate range (Fig. 5.4), the effect of dilution beyond a certain limit leads to a discontinuous phase transition producing an amorphous solution phase, S_1. Initially, the G and S_1 phases exist conjugately, each, at a given temperature, having a definite composition in accordance with the phase rule. The S_1 phase is always more dilute than the conjugate G phase, and in some cases very markedly so—see Fig. 5.3 for lower temperatures. With further overall dilution the amount of the S_1 phase increases while that of the G phase diminishes until finally the transition is complete.

The breakdown of the G phase is associated with the tendency of dilution to decrease R. Up to a limit this may be offset by dimensional changes within the lamellae (see above), but with increasing dilution and increasing tendency of R to fall below unity, thermal fluctuations will tend more and more to produce local areas on the $\overline{C}$ face convex towards $\overline{W}$. This process, in the examples now considered, finally leads to separation of the S_1 phase. If the temperature is raised the thermal fluctuations are intensified and separation of the S_1 phase takes place at a higher concentration (Figs. 5.3 and 5.4; cf., however, p. 272).

Except where the difference in concentrations of the conjugate G and S_1 phases is large, there is evidence that the lamellar arrangement, indefinitely extended within the G phase, persists locally over short distances within the amorphous phase (cf. pp. 227, 258, 259). However, on further progressive dilution with consequent further decrease in R the average micellar form approaches spherical (Hartley* micelles, S_1 micelles).

It is emphasized that all the effects of dilution considered above are reversible; thus, with the systems now considered, commencing with a dilute S_1 solution and progressively increasing the concentration, the mean value of R, which throughout will be subject to local thermal fluctuation, gradually increases and the mean micellar form tends to become of diminished curvature. As the transition concentration is approached, lamellar loci arise within the amorphous solution. These increase in extent and concentration and finally give separation of the lamellar phase, G.

* Cf. Hartley, G. S. Paraffin-Chain Salts, Hermann et cie, Paris (1936).

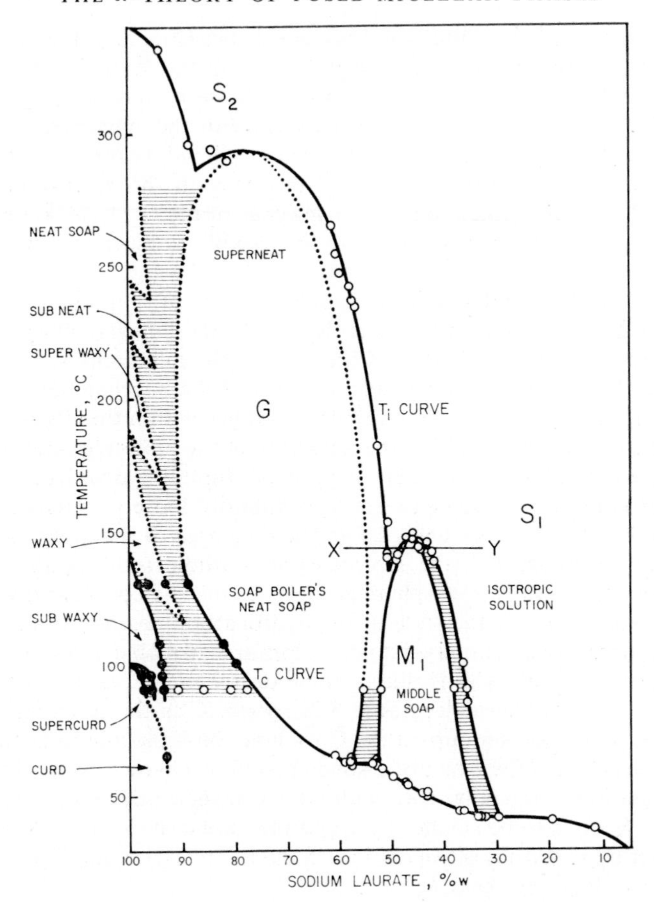

FIG. 5.4. Phase diagram for sodium laurate/water system [9].

BREAKDOWN OF THE G PHASE ON DILUTION TO GIVE THE MESOMORPHOUS FIBROUS HEXAGONAL SOLUTION PHASE, M_1: FURTHER IMPLICATIONS CONCERNING THE STRUCTURE OF THE G PHASE

Although in the examples just examined the G phase on dilution undergoes a discontinuous transition to the amorphous S_1 phase, in certain other systems (see p. 216, Table 5.2), and in the sodium laurate/water system itself at room temperature, a discontinuous transition to a second mesomorphous phase intervenes.

This "middle" phase, on further dilution, itself ultimately undergoes a discontinuous phase transition to the amorphous phase, S_1. The middle phase, M_1, like the G phase, is translucent and birefringent but, in spite of its great dilution, of a thicker consistency. When fractured it often tends to be drawn out into "strings". Its general structure has been established, mainly by X-ray diffraction measurements [6] and studies of its optical properties [5].

In the M_1 phase the amphiphile molecules are associated in parallel, indefinitely extended, cylindrical, fibrous micelles (Fig. 5.2(*b*)) arranged in a regular two-dimensional hexagonal array with the $\overline{W}$ region forming a continuum between the cylinders. The M_1 phase is thus nematic in type (cf., p. 9). The polar groups lie on the surface of the cylinders, where, as in other micellar phases, they should be considered as forming an interfacial-solution region (p. 243) in equilibrium with the $\overline{W}$ region. These fibrous M_1 micelles, like the lamellar micelles of the G phase, are essentially fluid and flexible in character and readily subject to mechanical deformation.

In such a thermodynamically stable system of parallel cylindrical micelles, the unit value of R, characteristic of the lamellar phase G, has been retained along the length of the fibres although circumferentially R has fallen below unity. The cylindrical form of the micellar fibres implies that the force fields of the amphiphile molecules within the fibres are not statistically symmetrical about the long axes of the molecules and that the molecules, in addition to their lengthwise liquid hydrocarbon-tail-to-hydrocarbon-tail arrangement, are also laterally loosely ordered. If this were not so and the force fields about the long axes of the molecules were statistically symmetrical (the molecules as a whole rotating about their long axes) only lamellar or spherical forms of micelle could be envisaged. Since it is known that the state of the hydrocarbon chains within the M_1 micellar fibres is essentially liquid it appears probable that the lateral ordering of the molecules is due to some degree of mutual orientation of the polar ends of the molecules (—CX, where C is the terminal carbon atom and X the polar group) at the $\overline{C}/\overline{W}$ junction, this ordering being of a liquid crystalline rather than of a solid crystalline character. In relation to the orientation of the polar ends within the $\overline{C}$ layer we denote the direction in which R tends to be smaller by x and the direction in which R tends to be greater by y. With this convention, in the M_1 phase, around the micelles $R_x < 1$, while along them $R_y = 1$.

It is unreasonable to expect that the mutual orientation of the polar ends within an M_1 phase would be lost in the corresponding G phase of higher concentration, in which the polar ends are closer together (p. 240). Such orientation would not be incompatible with the stability of the lamellar structure ($R = 1$ in all directions with respect to the molecular axes) since, by suitable rearrangement of the spacings between the polar ends on the face of the lamellae (with dependent adjustment in the lamellar thickness), both R_x and R_y could attain unit value. Ordering of the polar ends within the lamellae would be expected to confer optical properties on an individual lamella differing in the x and y directions. The fact that the G phase is observed to be optically uniaxial could then be taken to indicate that successive lamellae are randomly stacked and show no statistically preferred inter-relation with respect to their individual x and y axes. A further consequence of mutual orientation of the polar ends within both the lamellae of the G phase and the fibrous micelles of the M_1 phase would be the possible existence of more than one form of either of these phases, differing in the details of the orientations of the polar ends (cf. p. 272).

Breakdown of the Mesomorphous Solution Phase, M_1, on Dilution to give the Amorphous Solution Phase, S_1

As mentioned, dilution of the M_1 phase beyond its limit of stability gives the mobile amorphous solution phase S_1. This breakdown may be regarded as caused by the general tendency of dilution to diminish R. With increasing dilution R_y, in the direction along the length of the fibrous micelles, will tend to fall below unity. Thermal fluctuations (not indicated in Fig. 5.2(*b*) but analogous to those indicated in Fig. 5.2(*a*)) will therefore tend more and more to produce local convexities along the length of the fibres of magnitude sufficient to lead to fragmentation. Up to a certain limit of dilution, this tendency may be offset by dimensional changes (increased interfacial area per polar group, diminished radius of fibre), as already discussed for the G phase. Beyond this limit, breakdown of the indefinitely long fibres occurs with formation of the amorphous liquid phase, S_1, which at first exists in equilibrium with remaining M_1 phase (Figs. 5.5 and 5.6). There is evidence [10] that within this initial S_1 phase the fibrous hexagonal micellar arrangement of the M_1 phase still exists locally but without extended long-range order. On further dilution, with consequent increase in R, the local fibrous regions are progressively replaced by spherical micelles.

Breakdown of the G and M_1 Mesophases on Rise of Temperature

It is seen from Figs. 5.3 and 5.4 that rise of temperature, with both the G and M_1 phases, leads to the breakdown of the liquid crystalline solution phases to give the amorphous solution phase, S. This breakdown is due to increasing thermal fluctuations of the type discussed above within the micellar units of the mesophases. With both the G and M_1 mesophases there is a composition at which thermal stability is at a maximum. Between the maxima for the individual mesophases there is a downward projecting salient of the S_1 phase. Consequently, in proceeding through a range of compositions along a line such as XY in Fig. 5.4 the phase sequence with increasing dilution is:

$$G \rightarrow (G + S_1) \rightarrow S_1 \rightarrow (S_1 + M_1) \rightarrow M_1 \rightarrow (M_1 + S_1) \rightarrow S_1$$

This may be interpreted as follows. At the temperature of the observations, at the transition $G \rightarrow (G + S_1)$, owing to a diminishing value of R, the G phase breaks down giving an S_1 phase in which lamellar groupings are still present but of insufficient extension and concentration to constitute the G phase (cf. p. 218). With further dilution and reduction of R these lamellar groupings are progressively replaced by fibrous hexagonal groupings, the two existing in the amorphous solution in equilibrium. If at some stage in the dilution process the fibrous hexagonal groupings attain a sufficiently preponderant concentration and extension, then, on account of their mutually orientating effects, separation of the M_1 phase occurs. This will happen the more readily the lower the temperature and the less the general disintegrating effect of the thermal motion. Likewise, close to the G phase boundary, the lower the temperature the more readily will the lamellar groupings produce separation of the G phase. With falling temperature, the salient of S_1 phase thus progressively narrows to a point. At the temperature of the eutectic the G, S_1 and M_1 phases can exist in equilibrium. Below it the G and M_1 phases are in equilibrium.

In certain circumstances, the lamellar and fibrous groupings present within the S_1 phase in the zone intermediate on the phase diagram between the G and M_1 regions on fall of temperature, instead of selectively associating to form separate conjugate G and M_1 phases, probably associate jointly (either individually or as some type of hybrid unit) to form an intermediate phase of a hybrid constitution (cf. p. 212 and Table 5.1).

Breakdown of the G Phase on Increase in Concentration to Give the Mobile Amorphous Solution Phase, S_2

In Figs. 5.3 and 5.4 it is seen that over the appropriate ranges of temperature we encounter with increasing concentration the further phase sequence:

$$G \rightarrow (G + S_2) \rightarrow S_2.$$

These changes are complementary to the changes $G \rightarrow (G + S_1) \rightarrow S_1$ which occur with dilution, and show analogous features. However, whereas the latter sequence is related to an increasing tendency of the $\bar{C}$ layer to become convex towards $\bar{W}$ ($R < 1$), the former sequence presents the reverse case with the $\bar{C}$ layer increasingly tending to become convex towards $\bar{O}$ ($R > 1$). In the mobile amorphous solutions, S_2, we therefore encounter the tendency to develop aqueous regions, $\bar{W}$, dispersed in an amphiphilic continuum, while in the mobile amorphous solutions, S_1, we typically have micelles of amphiphile dispersed in an aqueous continuum. It is emphasized that, as illustrated in Figs. 5.3 and 5.4, the amorphous solutions which have been termed S_1 and S_2 form part of a continuous amorphous solution region, S. The division of the S region, into S_1 on the water-rich side of the G phase and S_2 on the amphiphile-rich side, is largely for convenience in discussion. Within the S region, by following a suitable succession of temperatures and compositions, one may pass continuously from solutions in which the predominant form of aggregation is S_1 to those in which it is S_2, with G, M_1 and M_2 groupings occurring locally under favourable intermediate conditions (cf. p. 215).

Formation of the Inverse Middle Phase, M_2

In Fig. 5.3 for the Aerosol OT/water system along YZ is represented the phase sequence with increasing concentration

$$G \rightarrow (G + S_2) \rightarrow S_2 \rightarrow (S_2 + M_2) \rightarrow M_2$$

which is complementary to the sequence

$$G \rightarrow (G + S_1) \rightarrow S_1 \rightarrow (S_1 + M_1) \rightarrow M_1$$

represented with increasing dilution along XY (Fig. 5.4)—see above.

The phase, M_2, or "inverse middle phase", is translucent, birefringent and of a very stiff consistency. Its optical properties and its X-ray diffraction pattern indicate that its structure is complementary to that of the M_1 phase, i.e., the polar groups with their associated aqueous region, $\bar{W}$, form the fibre cores while the hydrocarbon groups form an essentially liquid continuum between them. It is to be noted that in the straight-chain soap/water systems such as that represented in Fig. 5.4, a further fused mesomorphous phase, "neat soap", occurs at high temperatures on the high concentration side of the aqueous G phase. This phase, however, as demonstrated by its X-ray diffraction characteristics and by its optical behaviour (p. 222, Fig. 5.10) is not the inverse middle phase but a further lamellar phase. It seems probable that this lamellar phase differs from the more dilute lamellar phase, G ("soap boilers' neat phase"), on account of a differing mutual orientation of the polar ends (cf. p. 208). In both phases the polar groups are in the fused or labile state (cf. pp. 227, 238).

The reason why the straight chain alkali soaps do not form an M_2 phase at high concentrations is probably that volume filling by n-alkyl groups diverging round a cylindrical core of $-CO_2{\cdot}M$ groups would involve geometrical difficulties. In ternary systems such as sodium caprylate/decanol-l/water (p. 279) inclusion of part of the alcohol between the n-alkyl chains of the soap apparently eliminates these difficulties and such systems frequently give M_2 phases.

Formation of the Inverse Viscous Optically Isotropic Cubic Mesophase, V_2

On gradual increase in concentration in the Aerosol OT/water system (Fig. 5.3) at temperatures below 90°C one encounters the phase sequence

$$S_1 \rightarrow (S_1 + G) \rightarrow G \rightarrow (G + V_2) \rightarrow V_2 \rightarrow (V_2 + M_2) \rightarrow M_2$$

The phase V_2 thus replaces at lower temperatures the phase S_2 as it occurs intermediate between G and M_2 at higher temperatures. Accordingly, the mobile amorphous S_2 solutions of appropriate composition as indicated on Fig. 5.3, pass at temperatures below 90°C into the viscous optically isotropic phase, V_2.

The phase V_2 is optically clear, exceedingly viscous and non-birefringent, although when a sample observed between crossed polars is pressed with a glass rod transient strain birefringence is seen. The V_2 phase, although optically isotropic, is not amorphous but of cubic symmetry and under certain conditions it may separate in well-defined polyhedral forms (p. 224). Its constitution has not yet been definitively established (p. 251) but it seems very probable that it possesses some form of cubic lattice constructed from globular micellar units which arise as intermediate stages between the indefinitely extended G and M_2 micelles of the anisotropic neat and inverse middle phases (cf. p. 217, Fig. 5.6). Whatever the details of this structure, it is probably, on the micellar level, of a comparatively mobile character, since the V_2 phase shows the same high resolution NMR spectrum as is shown at higher temperatures by the amorphous mobile S_2 solution of the same composition but which is not shown by the G or M_2 mesophases. This suggests that in the V_2 phase the globular micellar units possess a rotational mobility analogous to that of the globular molecules in the non-amphiphilic cubic plastic crystal mesophases (Chap. 2.2).

FORMATION OF THE VISCOUS OPTICALLY ISOTROPIC CUBIC MESOPHASE, V_1

In certain systems (cf. Tables 5.1 and 5.2) a viscous optically isotropic phase, V_1, complementary to the inverse viscous optically isotropic phase, V_2, is found intermediate between the G and M_1 phases. This phase, like the V_2 phase, is optically clear, exceedingly viscous and non-birefringent, shows transient strain birefringence, possesses cubic symmetry and, in certain circumstances, may separate in polyhedral forms (p. 223). Its structure is not yet established (p. 251) but it seems highly probable that it is constituted from globular micellar units which arise as intermediate stages between the indefinitely extended G and M_1 micelles of the anisotropic neat and middle phases (cf. p. 217, Fig. 5.6). Whatever the detailed structure of the V_1 phase, it is probably, on a micellar level, of a comparatively mobile character, since the V_1 phase, like the V_2 and S phases but unlike the M_1, G and M_2 phases, gives a high resolution NMR spectrum. This suggests that in the V_1 and V_2 phases the micellar units possess rotational mobility.

As well as the authenticated cubic phase V_1, a number of other mesophases have been characterized, with greater or smaller definitiveness, as existing in soap/water systems intermediate between the M_1 and G phases [6, 11]. These birefringent phases were first diagnosed from X-ray diffraction measurements and although information is thus available concerning their lattice dimensions, their detailed inner structures are not established [11]. The successive phases have been named hexagonal (M_1), deformed hexagonal, rectangular, complex hexagonal, cubic (V_1), lamellar (G). Not all the phases arise in any particular system but those which do appear always follow one another with increasing concentration in the order given (Table 5.1).

It seems possible that all these phases ,like the cubic phases V_1 and V_2, possess structures, possibly hybrid, constituted from micellar units intermediate between the G and M_1 types, the individual phases differing in the mode of assembly and degree of anisometry of the intermediate units. Two further mesophases, "C" and "K", have been described by Ekwall *et al.* [12] as occurring in certain three-component systems, intermediate between the M_1 and G phases and between the G and M_2 phases respectively. Cf. p. 279.

FORMATION OF A VISCOUS OPTICALLY ISOTROPIC CUBIC PHASE, S_{1c}, INTERMEDIATE BETWEEN THE S_1 AND M_1 PHASES

An additional mesophase, not previously mentioned, arises in certain systems intermediate between the amorphous phase, S_1, and the mesophase, M_1. This mesophase apparently contains S_1 micelles, ordered in some form of cubic lattice [12(*a*), 12(*b*)]. Examples of its occurrence are given on pp. 255, 282, and Figs. 5.31 and 5.46. Like the V_1 and V_2 phases, it shows a high resolution NMR spectrum and marked strain birefringence. It is apparently not structurally analogous to the V_1 phase [12(*b*)]. In this article the connotation S_{1c} will be used to indicate that this phase is here tentatively assigned a rotational structure constituted from

TABLE 5.1. *The occurrence of phases intermediate between M_1 and G in some amphiphile/water systems* [6(*b*)]. *The concentrations of the phase boundaries are given. The phases existing in each system are shown by full lines*

Amphiphile	Temperature, °C	Lamellar G		Cubic V_1		Complex hexagonal		Rectangular		Deformed hexagonal		Isotropic liquid S_1		Hexagonal M_1	
Sodium laurate	100	—	0·59	...		...		...		...		...	0·59	—	
Sodium myristate	100	—	0·59	...	0·59	—	0·55	...	0·55	—	0·54	...	0·54	—	
Sodium palmitate	100	—	0·56	...	0·56	—	0·52	...	0·52	—	0·51	...	0·51	—	
Sodium stearate	100	—	0·54	...	0·54	—	0·51	...	0·51	—	0·50	...	0·50	—	
Potassium laurate	100	—	0·69	—	0·61	...		...		...		...	0·61	—	
Potassium myristate	100	—	0·66	—	0·59	...		...		...		...	0·59	—	
Potassium palmitate	100	—	0·65	—	0·59	—	0·55	...	0·55	—	0·54	...	0·54	—	
Potassium stearate	100	—	0·65	...	0·65	—	0·59	...	0·59	—	0·58	...	0·58	—	
Sodium oleate	65	—	0·69	...	0·69	—	0·59	—	0·52	...		...	0·52	—	0·28
Potassium oleate	20	—	0·72	...	0·72	—	0·68	—	0·60	...		...	0·60	—	0·21
Sodium lauryl sulphate	75	—	0·69	...	0·69	—	0·62	...		...		...	0·62	—	0·38
Aerosol MA (p. 204)	20	—		...		...		...		...		...		...	
Cetyl trimethyl-ammonium bromide	70	—	0·84	—	0·78	...		...		...		...	0·78	—	0·38
Arkopal 9*	20	—	0·61	...		...		...		...	0·61	—	0·48	—	0·45
Arkopal 13	20	...		...		...		...		...		...	0·63	—	0·43

* Condensate of *p*-nonylphenol and polyethylene glycol.

S_1 micelles ordered in some form of cubic array. The complementary inverse structure would be denoted as S_{2c} but a phase to which this structure can be ascribed with assurance does not appear to have been identified as yet. Formal structures of the S_{1c} and S_{2c} types have on various occasions been proposed for the V_1 and V_2 phases but such structures are not conformable with the sequential positions of these phases as indicated in Figs. 5.5 and 5.6.

A clathrate structure for the S_{1c} phase, consisting of spherical micelles caged within a network of rods, has been proposed by Tardieu & Luzzati [12(*c*)] on the basis of X-ray measurements with the S_{1c} phase formed in the sodium caprylate/xylene/water system (p. 281). However, it appears that in their calculations no account was taken of the inclusion of part of the sodium caprylate within the aqueous zone of the mesophase. A reasonable allowance for this (cf. p. 274) would reduce the ratio of (the volume of the $(\bar{O} + \bar{C})$ zone)/(the volume of the $\overline{W}$ zone) from 0·46/0·54 to about 0·36/0·64. Such a change would apparently permit a revision of Luzzati's structure in which the spatial arrangement of rods was replaced by a related arrangement of globular micelles giving a structure of the S_{1c} type suggested above.

Idealized Comprehensive Phase Diagram including the Amorphous Solution Phase, S (S_1, S_2), and the Mesomorphous Solution Phases, S_{1c}, M_1, V_1, G, V_2 and M_2, Encountered in Amphiphile/Water Systems

The principal types of phase change encountered in amphiphile/water systems of the fused type may be represented on an idealized phase

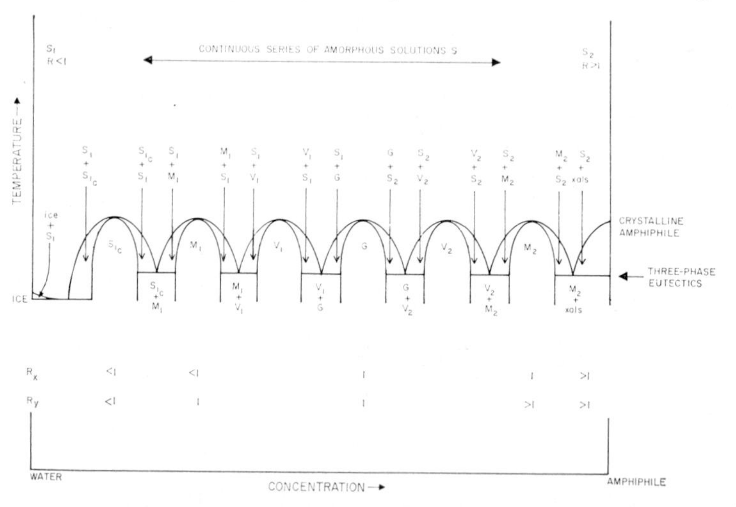

FIG. 5.5. Idealized phase diagram including the amorphous solution phase S(S_1, S_2) and the succession of fused mesomorphous solution phases realizable in binary amphiphile/water systems. In individual real systems not all the phases arise, but those that do succeed one another with increasing concentration in the order indicated; further, the heights of the different peaks and the depths of the eutectics vary greatly.

diagram (Fig. 5.5). In this diagram only the V_1 and V_2 phases are indicated intermediate between the M_1 and G, and the G and M_2 phases, respectively. Other "intermediate" phases (p. 213) are omitted pending further corroboration.

In Fig. 5.5 the peaks representing the maximum upper temperature limits of stability of the different liquid crystalline phases are all drawn of equal height and the eutectics of equal depth. In real systems, however, the ranges of existence of the individual phases, the heights of the various peaks and the depths of the eutectics differ widely, and certain of the phases may not arise. However, it seems to be a general rule that those phases which do appear always follow the sequence shown with increasing concentration. Further, the more hydrophilic the amphiphile the higher the minimum concentration required for the appearance of a particular phase (cf. p. 268, Table 5.16, Method Ic).

In Fig. 5.5 the effect of rising temperature is shown as tending everywhere to break down the liquid crystalline phases. Although, broadly speaking, this tendency is general, in certain specific ranges of composition and temperature in particular systems the reverse effect may occasionally be encountered (cf. p. 271).

A list showing those phases indicated in Fig. 5.5 which arise in some particular amphiphile/water systems at particular temperatures is given in Table 5.2. In certain other systems it appears that additional phases may also arise, as noted above (p. 212 and Table 5.1), intermediate between the G and M_1 phases.

It seems reasonable to associate the regular order of phase succession illustrated in Fig. 5.5 and Tables 5.1 and 5.2 with a progressive increase with concentration in the ratio *R*. This viewpoint has been embodied in Fig. 5.6 which illustrates the relationships between the micellar characteristics of the different phases in so far as these have been established with reasonable assurance. The details of the structures suggested for the V_1 and V_2 phases need further elucidation while proposals for the structures of the other "intermediate phases" mentioned on p. 212, must, for the present, remain in abeyance.

The Amorphous Solution Phase, S(S_1, S_2); Some Further Considerations

It is possible in binary amphiphile/water systems to pass continuously (if necessary first raising and then lowering the temperature) between any two points which represent the S phase on the phase diagram, without encountering any change of phase or, apart from the changes at the critical concentration for micelle formation (cmc), any abrupt discontinuity in properties. None the less, it is clear that when so traversing a wide range of compositions, large changes in micellar structure must occur. It is therefore necessary to envisage a thermal equilibrium between different micellar forms within the S phase, that is, the occurrence within this phase of local thermal fluctuations in concentration and in the ratio *R*. At concentrations below the cmc the amphiphile is molecularly dispersed and, from a formal point of view, *R* may be regarded as zero. At concentrations not too greatly

TABLE 5.2. *The occurrence of the crystalline phase, C, the amorphous phase and the principal fused mesomorphous phases in some amphiphile/water systems. The phases are listed in the order in which they arise with increasing water content (cf. Fig. 5.5)*

Amphiphile	Reference	Temperature, °C	C	S_2	M_2	S_2	V_2	S_2	G	S_1	V_1	S_1	M_1	S_1	S_{1c}	S_1
Sodium caprylate	Fig. 5.44	20	×										×			×
Potassium caprylate	Fig. 5.45		×								×		×			×
Sodium laurate	Fig. 5.4	20	×						×				×			×
		140	×						×			×	×			×
		200	×						×							×
Potassium laurate	Table 5.1	100	×						×		×		×			×
Aerosol OT	Fig. 5.3	20			×		×		×							×
		100			×	×			×							×
		160			×	×	(= S_2, S_1)									
Sodium di(2-ethylhexyl)acetate	p. 224	20			×		×	×	×							×
Potassium di(2-ethylhexyl)acetate	p. 222	20			×	×			×							×
Sodium n-octyl sulphate		20	×										×			×
Sodium 2-ethylhexyl sulphate	[18]	20							×		×					×
Sodium undecane-3-sulphate	Table 5.15								×		×					×
Sodium hept-6-enyl-sulphate	[45]								×							×
Cetyltrimethyl-		25		×									×			×
ammonium bromide	Table 5.1	70		×					×		×		×			×
Dodecyltrimethylammonium chloride	Fig. 5.31	40		×					×		×		×		×	×
Decaethylene glycol monolauryl ether		20		×									×		×	×
Triton X 100*	[62]			×									×			×

* A C_8 phenol condensed with 8–10 ethylene oxide units

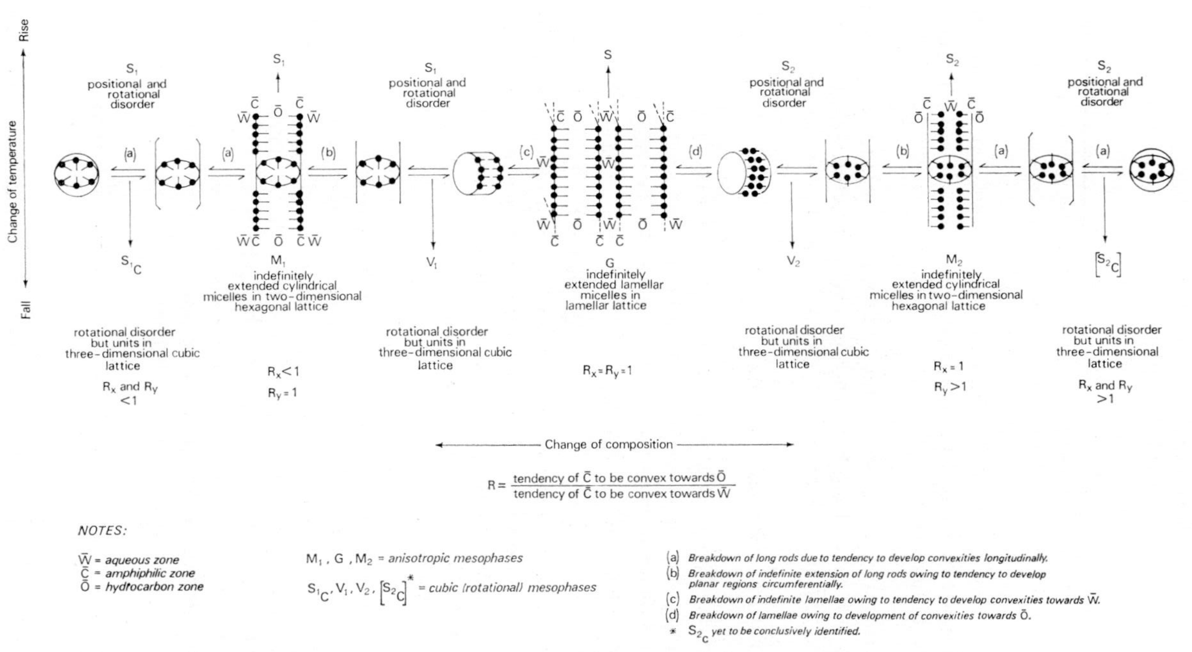

FIG. 5.6. The nature of the succession of amphiphilic mesophases of the fused type in relation to the underlying micellar equilibria.

above the cmc $R \ll 1$ and the amphiphile will be aggregated predominantly as spherical, Hartley or S_1 micelles. At high concentrations, where $R \gg 1$, micellar forms that are concave towards the $\overline{W}$ region will predominate. At intermediate concentrations a thermal equilibrium of fluctuating conformations will be present. If conditions (composition and/or temperature) are adjusted so as sufficiently to favour the concentration and extension of an individual form of micelle, mutual interaction of these micelles through intermicellar forces, may occur, producing long-range order and leading to the separation of a liquid crystalline phase (e.g., S_{1c}, M_1, G and M_2).

The formation of particular individual micellar forms within a binary aqueous S phase by a particular amphiphile at a particular bulk concentration will be influenced by:

(1) the size and character of the hydrocarbon group (e.g., degree of branching, presence of double bonds or cyclic structures etc.);
(2) the point of attachment of the polar group to the hydrocarbon group;
(3) the nature of the polar group.

For homologous amphiphiles of given constitution at a given weight concentration, the intermicellar equilibrium is displaced to the right by increase of molecular weight, i.e., a given stage will be reached at progressively lower concentrations the higher the molecular weight. If the change

$$\text{micellar form X} \rightarrow \text{micellar form Y}$$

proceeds over a concentration range below a certain limit, the concentration of the X micellar form necessary to produce the mutual ordering of the micelles by intermicellar forces that would be required for the formation of the corresponding mesophase at the temperature prevailing, may not be reached. This may explain why the formation of the S_{1c} mesophase, for example, is somewhat critical in relation to the chain length and nature of the polar group. Hydrocarbon chain branching and/or medial attachment of the polar group would be expected to diminish the number of polar groups per unit area of the micellar face under given conditions of concentration. This will diminish $A_{H\overline{CW}}$, increase R and displace the intermicellar equilibrium to the right in Figs. 5.5 and 5.6 [2]. Further, for a given type of polar group, the increased separation should diminish the degree of dissymmetry of polar interaction which, it is proposed (p. 208), gives rise to fibrous micellar forms. In accordance with this, highly branched medial amphiphiles such as Aerosol OT and Aerosol MA (p. 204) do not form M_1 phases, but change $S_1 \rightarrow G$ directly. Conversely, with amphiphiles with relatively large polar group ($—O(C_2H_4)nOH$, $n > 9$), formation of the M_1 phase is favoured.

It is conceivable that when two individual micellar forms exist in equilibrium, both at a sufficient concentration, mutual arrangement of the two forms jointly could form a "hybrid" mesophase. Such a process is probably involved in the formation of V_1, V_2 and certain other "intermediate" phases (p. 212).

The Shapes and Structures Adopted by Particles of the M_1, V_1, G, V_2 and M_2 Mesophases when Formed by Separation from a Second Phase

In the general phase progressions discussed above, many types of pairs of conjugate phases are possible (Fig. 5.5). When an amorphous liquid separates as droplets from a second amorphous liquid, on account of the isotropy of the mechanical properties (including the surface tension) of amorphous liquids the separating particles adopt the form of spheres. In the case of the uniaxial mesophases, M_1, G and M_2, and the cubic mesophases, S_{1c}, V_1 and V_2, the mechanical properties vary with direction with respect to the crystallographic axes. This can lead to non-spherical forms and, under controlled conditions, to particles of characteristic shapes and structures [13, 14]. Such particles may, in some instances, be formed by the agitation together of two conjugate phases but better developed examples may often be obtained by the slow precipitation of one of the conjugate phases from the other as the result of a gradual change in temperature or composition.

A convenient method of effecting a gradual change in composition under conditions where the resulting phase separation may be clearly observed, is to subject a thin film of a particular phase between microscope slide and coverslip to conditions which bring about a change in peripheral composition that, through diffusion [13], gradually extends towards the centre of the slide. If a thin layer of the aqueous S_1 phase containing Aerosol MA* (35% w) is subjected to slow peripheral dehydration by exposure to the atmosphere, bands of the successive phases G, V_2 and M_2 appear at the periphery and gradually follow each other inwards towards the centre of the slide. Alternatively, if one exposes a thin layer of the M_2 phase of Aerosol MA to air of suitably chosen humidity, the reverse sequence of phases, V_2, G, S_1, gradually moves towards the centre of the slide. In suitable circumstances the phase transition phenomena and particle growth may be followed visually over many hours.

Bâtonnets and Platelets of the M_1 Phase

If a sodium caprylate solution (30% w; S_1 phase) is flooded into the cross between four 1 cm square coverslips (0·2 mm thick) arranged on a slide and covered with a 2 cm square coverslip and the sample is allowed to undergo peripheral evaporation at room temperature, a band of the M_1 phase appears at the edge of the slide and extends slowly inwards along the arms of the cross. After about two days the residual S_1 phase at the centre of the cross becomes saturated with respect to the M_1 phase and individual particles of the M_1 phase commence to grow within it both as bâtonnets and as platelets (Figs. 5.7, 5.8 and 5.9; see also Fig. 5.8(*a*).

In their most perfect form, M_1 bâtonnets are shaped as figures of revolution which usually have their greatest elongation along their geometric axis and frequently lie approximately parallel to slide and coverslip in "longitudinal aspect". More occasionally, a bâtonnet is initiated at a

* Cf. footnote, p. 204.

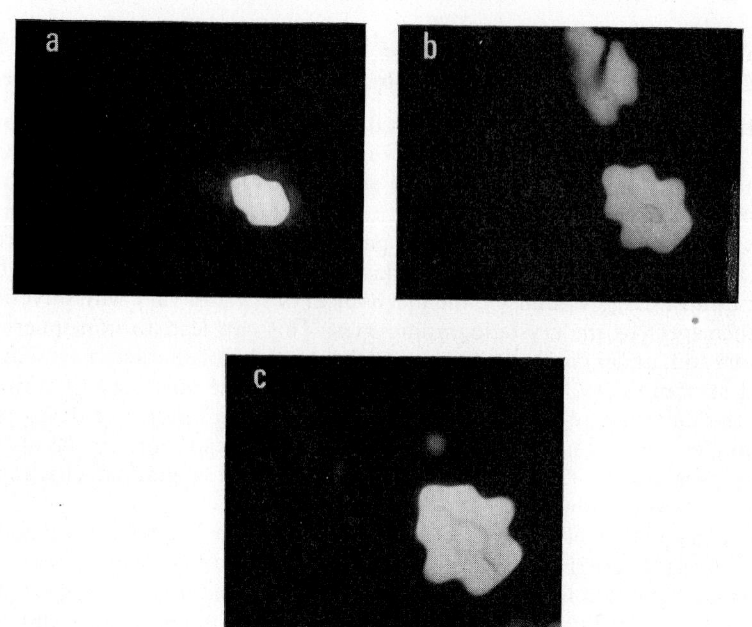

FIG. 5.7. Developing M_1 bâtonnets and M_1 platelets produced on peripheral evaporation of sodium caprylate solution between slide and coverslip [13] (approx. $\times 66$ crossed polars).
Exposures b and c made at 60 and 210 minutes after exposure a.

centre lying either on slide or coverslip and develops with its axis approximately normal to slide and coverslip in "transverse aspect". The optical properties [70] of an M_1 bâtonnet (see Fig. 5.8(a)) indicate that within it the flexible parallel M_1 micellar fibres (with intervening water) are constrained so as to lie similarly to fibres wound on a spindle (or possibly as successive rings centred on the spindle as axis). Necessarily there must be some

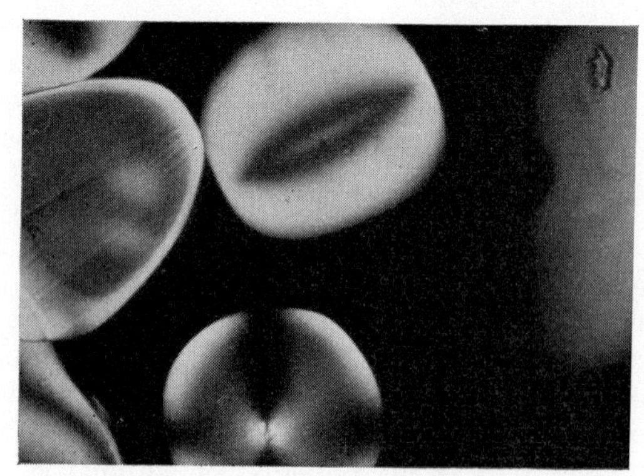

FIG. 5.8. M_1 bâtonnets in longitudinal and transverse aspects [13] (approx. $\times 100$ crossed polars).

FIG. 5.8(a). M, bâtonnet in 45° position between crossed polars showing bright areas (interference colours) and pseudo-isotrophy (homeotrophy).

modification of structure along the spindle itself and this is usually microscopically discernible (Fig. 5.9).

The optical properties of the M_1 platelets indicate that in the platelets the M_1 micellar fibres lie normal to the plane of the platelet. Those platelets which lie accurately parallel to slide and coverslip are therefore viewed along their optic axis and appear optically isotropic (dark) between crossed polars.

BÂTONNETS AND PLATELETS OF THE M_2 PHASE

M_2 bâtonnets, analogous in structure and appearance to M_1 bâtonnets, may be obtained by the peripheral evaporation technique from a solution

FIG. 5.9. M_1 bâtonnet in longitudinal aspect [13] (approx. ×170 non-polarized light).

(S_2) of Aerosol OT in toluene or from an aqueous solution (S_2) of potassium di(2-ethylhexyl)acetate (70% w). With this salt the phase succession observed at room temperature with increasing concentration is:

$$\begin{array}{ccccccccccccc} S_1 & \rightarrow & (S_1 + G) & \rightarrow & G & \rightarrow & (G + S_2) & \rightarrow & S_2 & \rightarrow & (S_2 + M_2) & \rightleftarrows & M_2 \\ < 6\%\,\mathrm{w} & & 6\text{–}18 & & 18\text{–}47 & & 47\text{–}56 & & 56\text{–}80 & & 80\text{–}82{\cdot}5 & & > 82{\cdot}5 \end{array}$$

The M_2 bâtonnets from potassium di(2-ethylhexyl)acetate may be up to 1 cm long and may be observed between crossed polars without magnification. M_2 platelets, analogous to M_1 platelets, are concurrently deposited from the potassium di(2-ethylhexyl)acetate solution.

Bâtonnets of the G Phase

The smectic or G phases of amphiphilic systems, like the smectics of non amphiphilic systems, frequently separate as bâtonnets at the amorphous-

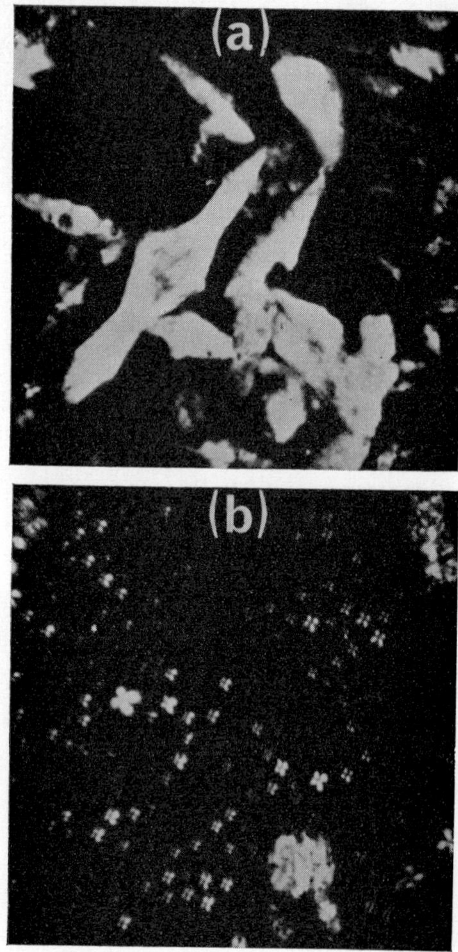

Fig. 5.10. Formation of (a) bâtonnets and (b) spherical droplets of the anhydrous neat phase on cooling the amorphous melt of sodium palmitate [17].

to-smectic phase transition. These bâtonnets were described by Friedel [15, 16]. Bâtonnets of the anhydrous "neat soap" phase are formed on cooling the amorphous melt of sodium palmitate (Fig. 5.10). The aqueous S_2 phase of potassium di(2-ethylhexyl)acetate, which yields bâtonnets of the M_2 phase on increase of concentration as discussed above, yields bâtonnets of the G phase on dilution.

The G bâtonnets form figures of revolution but differ diagnostically from the M bâtonnets in appearance and in optical properties [13]. In smectic bâtonnets, as detailed by Friedel, the lamellae lie normal to the axis of the bâtonnet. In accordance with this structure G bâtonnets show no "spindle" and do not develop striations on ageing.

Rounded Droplets of the G Phase

At phase boundaries where the phase transition S $\longrightarrow$ G occurs, the G phase often separates as rounded droplets. Such rounded droplets (Fig. 5.10(*b*)) arise in the binary aqueous systems containing Aerosol MA, Aerosol OT or the higher 1-monoglycerides. In these droplets the lamellae are constrained to lie as concentric spherical shells like the layers of an onion. When viewed between crossed polars with the polarizing microscope the droplets exhibit an extinction cross (Fig. 5.11).

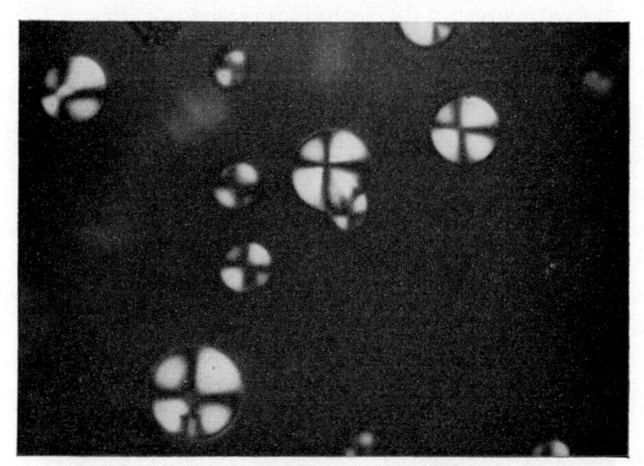

Fig. 5.11. Rounded droplets of the G phase separating from the S_1 phase in the Aerosol MA/water system [13] (approx. ×430 crossed polars).

Polyhedral Particles of the V_2 and V_1 Phases

When the aqueous G phase from either Aerosol MA or Aerosol OT (p. 204) undergoes peripheral evaporation between slide and coverslip, bands of the successive phases V_2 and M_2 diffuse towards the centre. After a time the residual G phase becomes saturated with respect to the V_2 phase and optically isotropic polyhedral particles of this phase commence to grow. The residual G phase in the region of the separating V_2 phase frequently adopts the homeotropic texture and may consequently readily be distinguished from the optically isotropic V_2 polyhedra using crossed

FIG. 5.12. Polyhedra of the V_2 phase separating from residual homeotropic G phase on peripheral evaporation of the aqueous G phase from Aerosol OT [13] (approx. $\times 270$ crossed polars convergent light).

polars and convergent light (Fig. 5.12). In certain circumstances not yet precisely defined, but probably connected with the presence of traces of inorganic salt, the peripheral evaporation method leads to separation of well-grown V_2 polyhedra from the S_2 phase. This was observed both with Aerosol OT and with sodium di(2-ethylhexyl)acetate (Fig. 5.13). The phase sequence involved is probably:

$$G \rightarrow (G + S_2) \rightarrow S_2 \rightarrow (S_2 + V_2) \rightarrow V_2 \text{ (cf. Fig. 5.5.)}$$

Separation of the V_1 phase as distinct individual polyhedra has not yet been observed. However, when a homeotropic area of the aqueous G phase from N,N,N-trimethylaminododecanoimide between slide and coverslip is exposed to air at 88% humidity at 20°C additional water is absorbed and the phase changes $G \rightarrow G + V_1 \rightarrow V_1$ occur. A band of the V_1 phase gradually extends from the periphery centrewards. Well-formed crystal faces of the V_1 phase develop at the V_1 G boundary.

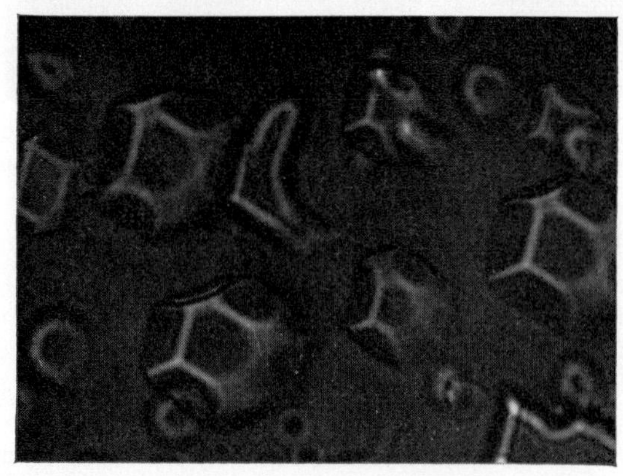

FIG. 5.13. Polyhedra of the V_2 phase separating from the S_2 phase in the sodium di(2-ethylhexyl)acetate/water system [13] (approx. $\times 470$).

Probable pseudomorphs of the precursive V_1 phase have been observed in electron micrographs of the anhydrous neat phase formed at room temperature by drying down an aqueous solution of sodium 2-ethylhexyl sulphate [18] (p. 253, Table 5.15).

MESOPHASE FORMATION IN SOME SELECTED SYSTEMS

To illustrate the various types of mesophase, both semi-crystalline and fused, noted in the above general discussion, the behaviour observed in some selected systems will be considered in more detail.

SYSTEMS OF ONE AND TWO COMPONENTS

Alkali Metal Soap/Water Systems

High-temperature Anhydrous Mesophases

The sodium soaps of the fatty acids

At room temperature the sodium soaps form lamellar crystalline solids or solid hydrates. When heated, the anhydrous solids undergo a process of stepwise melting proceeding, through a series of individual apparently semi-crystalline mesophases, to the wholly fused "neat-soap" lamellar mesophase and finally to the amorphous melt. These sequences were extensively studied by McBain and Vold and their associates and by others [9, 11, 19]. Techniques including X-ray diffraction and polarization microscopy, and measurements of heats of transition, electrical conductivity, vapour pressure and volume changes, were employed for detecting the phase changes. The type of phase behaviour observed is illustrated on the left of Fig. 5.4. Similar diagrams were obtained for sodium palmitate, myristate, stearate and oleate [9]. The temperature limits of existence of the

TABLE 5.3. *The range of existence of the high-temperature mesophases formed by sodium laurate, myristate, palmitate and stearate* [19]
(The range of existence as determined by X-rays [11] is given in brackets)

Phase	$C_{12}Na$	$C_{14}Na$	$C_{16}Na$	$C_{18}Na$
Subwaxy	100–141°C	107–141°C	117–135°C	117–132°C
	(–142)	(–142)	(–140)	(–133)
Waxy	141–182	141–176	135–172	132–167
	(142–183)	(142–182)	(140–176)	(133–175)
Superwaxy	182–220	176–217	172–208	167–198
	(183–200)	(182–210)	(176–211)	(175–210)
New phase	(200–215)	—	—	—
Subneat	220–255	217–245	208–253	198–257
	(215–252)	(210–248)	(211–254)	(210–256)
Neat	255–336	245–310	253–295	257–288
	(252–325)	(248–305)	(254–300)	(256–290)

individual anhydrous mesophases formed by the saturated soaps are given in Table 5.3 [19]. It is seen from Fig 5.4 that all the high-temperature mesophases can incorporate a limited amount of water. This leads to the production of two phase zones between the individual mesophases.

The structures of the anhydrous high-temperature soap phases have been studied by Luzzati *et al.* by X-ray diffraction methods (for review see [11]) and by Grant and Dunnell [20] and Lawson and Flautt [21] using nuclear magnetic resonance techniques. These studies supplement and largely corroborate the earlier work. The interpretation of their X-ray diffraction measurements given by Skoulios and Luzzati [22] and their collaborators will now be discussed.

At lower temperatures ($< 100°C$) the sodium soaps form three-dimensional lamellar crystals [1]. The paraffin zones of the lamellae commence to melt at about 120°C and, according to Luzzati, the resulting

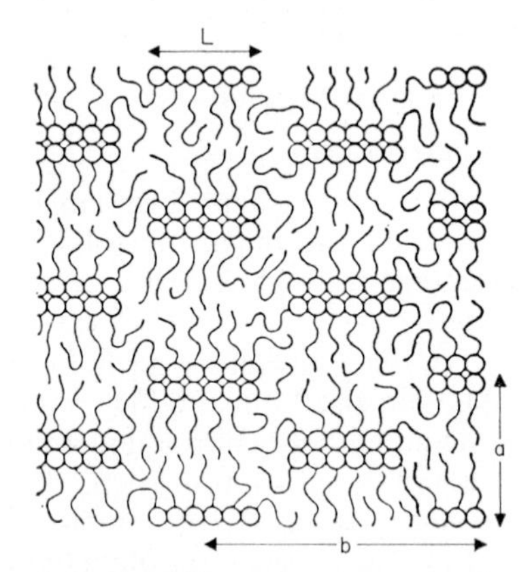

FIG. 5.14. Structure of the two-dimensional rectangular-centred lattice. Cross section of ribbons [22].

tensions break up the lamellae into ribbons. These ribbons contain the polar groups in a quasi-crystalline type of organization dispersed in a liquid matrix comprised of the hydrocarbon chains. The ribbons are of indefinite length and lie parallel to one another in a two-dimensional rectangular centred lattice (Fig. 5.14). The breadth of the ribbons is determined by the equilibrium between the thermal agitation of the paraffin chains and the cohesion of the polar groups. These polar groups largely retain their crystalline order and the interfacial area per polar group on the face of the ribbons remains approximately the same in all the semi-crystalline mesophases and equal to the area found in the crystal.

With increasing temperature the breadth of the ribbons decreases discontinuously in steps and leads to the series of sharp, reversible phase transitions, sub-waxy $\rightleftharpoons$ waxy $\rightleftharpoons$ super-waxy $\rightleftharpoons$ sub-neat. The dimensional changes accompanying the last three transitions are recorded in Table 5.4. In the sub-waxy phase, equilibrium is difficult to attain. The transition super-waxy $\rightleftharpoons$ sub-neat appears to be more complex than the previous transitions and to be accompanied by some rearrangement of the polar groups. In the sodium laurate system an additional phase, not noted in Fig. 5.4 was detected between the super-waxy and sub-neat phases by the X-ray technique (Table 5.4). Its suggested structure is shown in Fig. 5.15 in which parallel ribbons have been replaced by parallel disks. In NMR studies [23] the laurate again showed a behaviour somewhat different from that of the higher soaps. The transition sub-neat → neat, is accompanied by fusion of the polar groups. The organization becomes "lamellar labile", LL, the interfacial area per polar group increases markedly and becomes dependent on temperature. The fusion of the polar groups is further supported by the observation that, alone among the high-temperature mesophases, the neat phase shows temperature/electrical conductivity relationships similar to those of a fused salt [24, 70]. According to Skoulios and Luzzati the organization in the high-temperature, anhydrous neat phase is similar to that in the aqueous neat phase G,

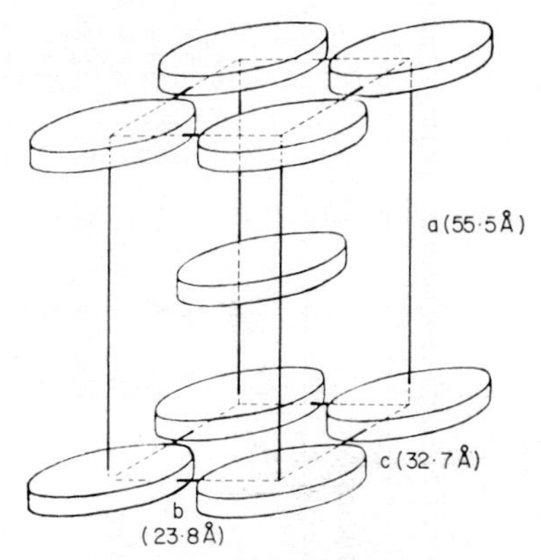

FIG. 5.15. Proposed structure of the three-dimensional orthorhombic body-centred phase in the sodium laurate system (Table 5.4). The disks are the loci of the polar groups and are separated by the liquid hydrocarbon chains [22].

and they consider that the polar groups form "a disordered structure analogous to that which they adopt in the presence of water". A rather different view has been suggested on p. 211.

At higher temperatures the anhydrous neat phase undergoes a sharp transition to the amorphous liquid. It is interesting that X-ray diffraction diagrams of this amorphous liquid melt indicate that, at temperatures near to the transition point, lamellar regions, of long spacing similar to that of the anhydrous neat phase but without extended long-range order, still persist. When the amorphous melt is cooled, the lamellar regions undergo mutual orientation. Long-range order is regained and under favourable conditions the neat soap phase separates as characteristic smectic bâtonnets (p. 222, Fig. 5.10).

Soaps of the other alkali metals: general relationships

These soaps show a behaviour broadly similar to that of the sodium soaps. The soaps of Li, K, Rb and Cs have been studied by Gallot and

TABLE 5.4. *Some dimensional characteristics of the high-temperature phases of the anhydrous sodium soaps* [11, 22]

	Name of the phase	Lattice symmetry			C_{12}	C_{14}	C_{16}	C_{18}
Semi-crystalline	Subwaxy (cf. [22])							
			t,	°C	142	142	140	133
	Waxy	Rectangular, centred	a,	Å	75·2	80·0	86·6	80·0
			b,	Å	30·9	34·5	38·3	40·3
			ϕ		45	47	48	53
			N_2	mole/Å	2·83	2·87	3·14	2·88
			L,	Å	34	33	36	36
			S,	Å^2	24	23	23	25
			t		183	182	176	175
	Superwaxy	Rectangular, centred	a		68·5	69·0	74·2	72·5
			b		30·3	33·6	36·0	37·9
			ϕ		48	52	52	55
			N_2		2·53	2·50	2·53	2·46
			L			30		
			S			24		

			t	200	210	211	210
	New Phase	Orthorhombic, body-centred	a	55·5			
			b	28·3			
			c	32·7			
			t	215	210	211	210
	Subneat	Rectangular, centred	a	49·8	53·8	56·2	62·4
			b	27·0	29·1	30·6	34·4
			ϕ	57	57	57	58
			N_2	1·64	1·58	1·67	1·82
			L		19	20	20
			S		24	24	22
			t	252	248	254	256
Fused	Neat	Lamellar	t_1	290	271	278	285
			d	25·3	27·5	29·1	30·6
			S	36	38	40	42
			t	325	305	300	290
	Liquid melt	Amorphous					

t transition temperature as determined from X-ray diffraction
a, b, c, L dimensions as indicated on Figs. 5.14 and 5.15
ϕ angle between the diagonals of the rectangular centred lattice
S interfacial area per polar group on the plane of the ribbon or on the lamellae of the neat phase
N_2 number of molecules per unit length of ribbon
t_1 temperature of measurements for the neat phase; for other phases the dimensions are independent of temperature
d long spacing in the neat phase

Skoulios [25]. In all the high-temperature high-concentration phases the hydrocarbon chains are essentially fused while (except in the neat and liquid amorphous phases where the polar groups likewise are fused) the polar groups are apparently essentially crystalline. Ribbon and disk structures are formed as with the sodium soaps but the number and types of such semi-crystalline mesophases formed by a particular soap depend on the hydrocarbon-chain length and on the cation. Table 5.5 presents a general review of the structures which appear with the various soaps. Fig. 5.16 summarizes the correlation between the appearance of ribbon or disk structures with the nature of the cation and the chain length of the

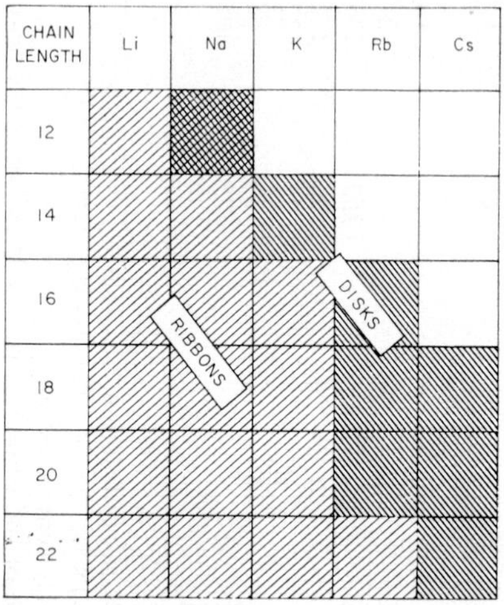

FIG. 5.16. Correlation between the appearance of ribbon or disk structures, the nature of the cation and the chain length of the alkali metal soap [25].

individual soaps. For a given cation the appearance of ribbon or disk structures requires a chain length above some minimum value which is higher for ribbons than for disks (Fig. 5.16). Above this chain length, ribbons arise with the cations of lower atomic weight and both ribbons and disks with those of higher atomic weight. The ribbons are arranged according to a two-dimensional rectangular centred lattice (Fig. 5.14) for the soaps of lithium and sodium and in a two-dimensional oblique lattice for the potassium soaps (Fig. 5.17, Table 5.6). This difference in the

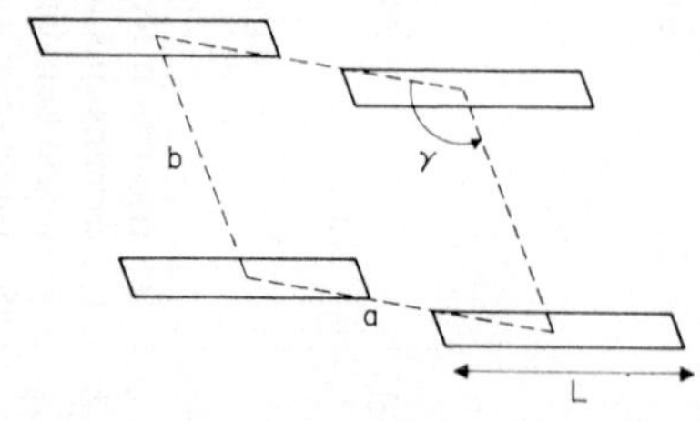

FIG. 5.17. Diagrammatic representation of the disposition of the polar ribbons in a two-dimensional oblique lattice [26].

TABLE 5.5. *Phases appearing in the course of the fusion of the alkali metal soaps* [25]

Polar groups	← Crystalline			→	← Fused	→
Paraffin chains	Crystalline ←			Fused		→
Structural units	Lamellar	Lamellar	Disk	Ribbon	Lamellar	—
$C_{8-10}Li$	LC					Liq.
$C_{12-14-16-18-20-22}Li$	LC			R		Liq.
$C_{14-16-18}Na$	LC			R	LL	Liq.
$C_{8-10-12}K$; $C_{12-14-16}Cs$	LC				LL	Liq.
$C_{14}K$	LC		D		LL	Liq.
$C_{16-18-20-22}K$	LC			O	LL	Liq.
$C_{12-14}Rb$	LC	LSC			LL	Liq.
$C_{16-18-20}Rb$	LC	LSC	D		LL	Liq.
$C_{22}Rb$	LC			O, R	LL	Liq.
$C_{18-20-22}Cs$	LC		I		LL	Liq.

For $C_{12}Na$: LC → R → D → R → LL → Liq.

R ribbons in two-dimensional rectangular lattice
O ribbons in two-dimensional oblique lattice
D disks in three-dimensional lattice
I disks in form of islets in equidistant parallel planes but with only loose lateral correlations

TABLE 5.6. *Dimensions of the high-temperature two-dimensional oblique phases of potassium stearate* [11]

t (°C)	170	185	210	225	238	272
a (Å)	80·0	69·0	50·6	48·3	45·5	
b (Å)	43·9	42·2	41·2	40·8	40·0	Lamellar
γ (°)	113	114	116	118	119	fused
N (mole/Å)	5·45	4·40	3·12	2·83	2·50	

t temperature of phase transitions
N molecules per unit length, Å, of ribbon
a, b, γ as in Fig. 5.17.

symmetry of the arrangement of the ribbons reflects a difference in the symmetry of the ribbons themselves. The paraffin chains although, broadly speaking, disordered are necessarily in some degree ordered near to the polar ribbon structure. Here they retain the inclination imposed on them by the "crystalline" arrangement of the polar groups i.e., on average, normal to the ribbons for lithium and sodium, and inclined for potassium. With rubidium behenate (C_{22}) four oblique phases arise at lower temperatures and two rectangular phases at higher temperatures. The change is attributed to a polymorphic change in the arrangement of the polar groups. In the disk phases of the soaps of potassium and rubidium, the disks are arranged in a face B-centred orthorhombic lattice (Fig. 5.18) [26]. In the

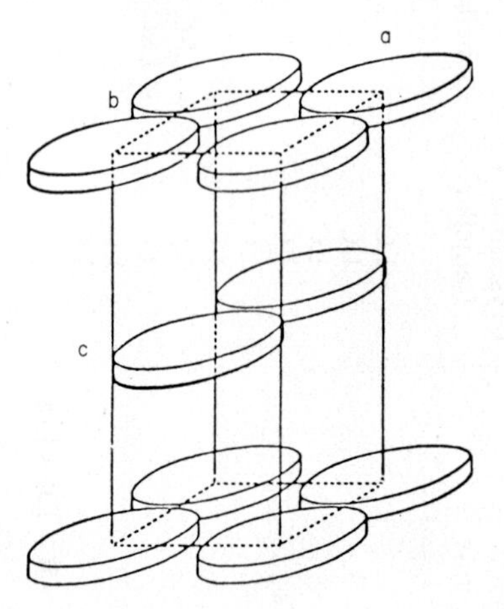

FIG. 5.18. Arrangement of the disks in the face B-centred orthorhombic lattice [26].

disk phases, I, of the C_{18}, C_{20} and C_{22} soaps of caesium (Table 5.5) the disks (islets) are located in parallel and equidistant planes but their lateral correlations appear to be loose.

With either the ribbon or disk phases formed by a given soap, the breadth of the ribbons or the size of the disks decreases discontinuously with rise of temperature, each decrease corresponding to a sharp phase transition. The ribbons necessarily become more symmetrical with decrease in breadth and as a consequence the lattices of the successive ribbon phases, whether rectangular or oblique, tend towards two-dimensional hexagonal in the higher temperature phases (cf. Table 5.4, ϕ approaches 60°; Table 5.6, γ approaches 120°).

The fused lamellar ("lamellar labile, LL") or anhydrous neat soap phase (see pp. 211, 229) is formed on rise of temperature by all the soaps (except those of lithium) as the phase immediately preceding the amorphous liquid.

In addition to the fused lamellar phase formed at high temperatures and the polymorphic crystalline lamellar phases formed at lower temperatures, a third type of lamellar phase, lamellar semi-crystalline, LSC, immediately succeeds the crystalline phases on rise of temperature with the C_{12}–C_{22} rubidium soaps [27]. In this phase the polar groups are crystalline while the

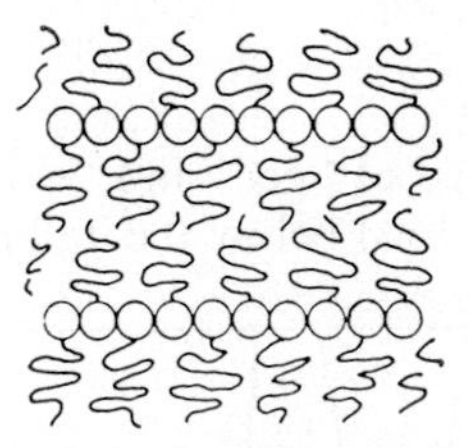

FIG. 5.19. Probable structure of the lamellar semi-crystalline phase formed by the rubidium soaps [27].

paraffin chains are liquid. Its suggested structure is shown in Fig. 5.19. In the planes of the lamellae the polar groups occupy a regular two-dimensional hexagonal array. The interfacial area per polar group is ca. 20 Å^2.

C_{16} and C_{22} ω-disoaps of the alkali metals

In addition to the monosoaps discussed above, Gallot and Skoulios [28] have made a comparable X-ray study of the C_{16} and C_{22} ω-disoaps of Li, Na, K, Rb and Cs. With the potassium and rubidium disoaps, the crystalline lower temperature lamellar phases, LC_1 and LC_2, analogous to the corresponding crystalline phases of the monosoaps, pass, on heating, directly into the fused lamellar or neat phase and then into the amorphous liquid. With the lithium soaps the LC_2 form melts directly to the amorphous liquid. With the C_{22} caesium disoap the LC_2 phase decomposes before melting. With the sodium disoaps the form LC_1 passes into a more complex crystalline lamellar phase and then into the fused lamellar phase (LL). The non-appearance of semi-crystalline mesophases with the disoaps accords with the structures assigned to these mesophases in the case of the monosoaps. It is clear that the disposition of the liquid hydrocarbon chains among the polar ribbons or disks postulated for the monosoap mesophases would not be possible with the disoaps.

It may not be inappropriate to add here the cautionary note that the structural models discussed above for the "semi-crystalline" high-temperature high-concentration mesophases formed by the alkali-metal soaps, should be regarded as proposed rather than proven. Alternative interpretations more along the lines given for the non-amphiphilic smectic polymorphic phases (Chap. 2.1) may yet prove more satisfactory.

LOW-TEMPERATURE SEMI-CRYSTALLINE MESOPHASES BELOW THE T_c LINE: THE "GEL" PHASES AND THE "COAGEL"

Complementarily to the anhydrous high-temperature mesophases considered above in which the polar groups are believed to retain crystalline order while the hydrocarbon chains are fused, low-temperature semi-crystalline mesophases are found in soap/water systems in which it is proposed that while the hydrocarbon chains are crystallized the polar groups have been brought into the ionized fused state by the solvent action of water. These "gel" phases have been studied by X-ray diffraction and polarization microscopy by Vincent and Skoulios who have also reviewed earlier work [29, 30].

Occurrence

The region of occurrence of "gel" and "coagel" is that represented in Fig. 5.20 (or similar diagrams) below the T_c line.

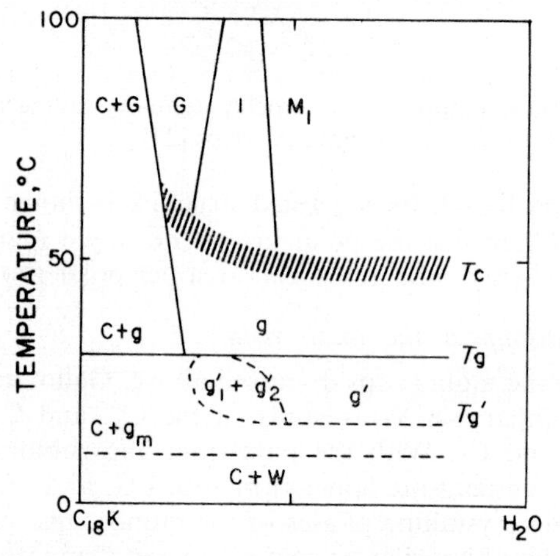

FIG. 5.20. Phase diagram for the system potassium stearate/water. M_1, middle phase; G, neat phase; I, intermediate phases; C, crystalline soap; W, dilute aqueous solution; g, thermodynamically stable "gel"; g′, "gel" in zone of metastability; $g'_1 + g'_2$, zone of separation of conjugate metastable "gels"; $C + g_m$, soap crystals + metastable "gel" [29].

A mixture of soap and water, when strongly heated and then cooled to a temperature below the T_c line, can show, according to circumstance, two different appearances. In one case it appears as a translucent jelly, in the other as an opaque mass. For example, a mixture of potassium stearate and water cooled from 100°C to room temperature appears (if it contains less than 70% w potassium stearate) as a homogeneous and clear jelly; if it contains more than 70% w soap, it yields an opaque product which is microscopically heterogeneous. According to current terminology, in the first case we have the "gel" and in the second the "coagel".

The X-ray diffraction diagram of the "coagel" contains lines characteristic of solid soap, which are independent of the water content of the system; the "coagel" is thus characterized as a dispersion of hydrated solid soap.

The diffraction diagram of the "gel" is, however, quite different both from that of the "coagel" and from those of the liquid crystalline phases above the T_c line. With the "gel", the spacings indicated by the central lines, in contrast to the comparable spacings for the "coagel", depend on water content. At high angles the "gel" gives a characteristic narrow band in contrast to the broad high-angle band of low intensity, indicative of the disordered configuration of the hydrocarbon chains, found for all phases indicated above the T_c line.

An approximate phase diagram for the potassium stearate/water system by Vincent and Skoulios is given in Fig. 5.20. Above the T_g line, the "gel" is thermodynamically stable. Below the T_g line it passes eventually to "coagel" but is able to remain supercooled for long periods above the curve $T_{g'}$. Between T_g and $T_{g'}$ over a limited range of concentration two conjugate "gels", g_1 and g_2, exist apparently in equilibrium although supercooled.

The formation of "gel" phases in soap/water systems depends on the nature of both cation and fatty acid. Lithium and sodium soaps do not yield "gels" and pass directly from liquid crystalline phases to "coagels".

The potassium C_{14}, C_{16}, C_{18} and C_{22} soaps form similar "gels", the T_c line being displaced to lower temperatures the lower the molecular weight. With equimolar mixtures of pairs of these potassium soaps differing in

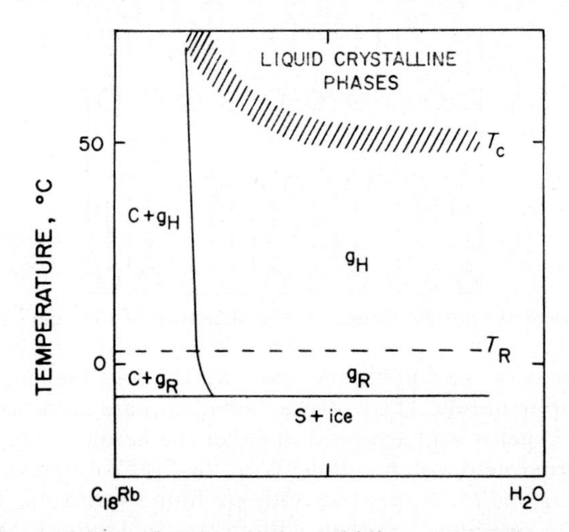

Fig. 5.21. Phase diagram for the system rubidium stearate/water; g_R and g_H, "gel" (indices R and H refer to a lateral arrangement of soap molecules according to a rectangular or hexagonal lattice); C, crystalline soap; S, soap ("gel" or crystals). The hachured zone corresponds to conjugate "gel" and liquid crystalline phases [30].

chain length by C_2 or C_4, X-ray diffraction diagrams indicate the formation of a mixed "gel" phase containing both soaps; with a difference in chain length of C_6, however, two conjugate "gels" form simultaneously. How the soaps are distributed individually between the separate "gel" phases has not been determined.

Rubidium and caesium soaps form thermodynamically stable "gel" phases with water over a wide range of composition and temperature. The approximate phase diagram for rubidium stearate is given in Fig. 5.21.

As an extension of studies on alkali metal soaps, Vincent and Skoulios examined the system cetyltrimethylammonium bromide/water. The phase diagram for this cationic soap is analogous to that for rubidium stearate; the "gel" exists, in the thermodynamically stable form only, for water contents greater than about 20% w and for temperatures between −5°C and +20°C. At lower temperature ice crystallizes from residual "gel". Within the range of thermodynamically stable "gel", two conjugate "gel" phases are found at 15°C at water contents (expressed as $(1 - c)/c$) between 0·96 and 0·30. These are analogous to the conjugate "gels" formed in the potassium stearate system in the supercooled "gel" region.

Structure of "gel" phase

The model proposed by Vincent and Skoulios for the "gel" phase in all the above instances is illustrated by Fig. 5.22.

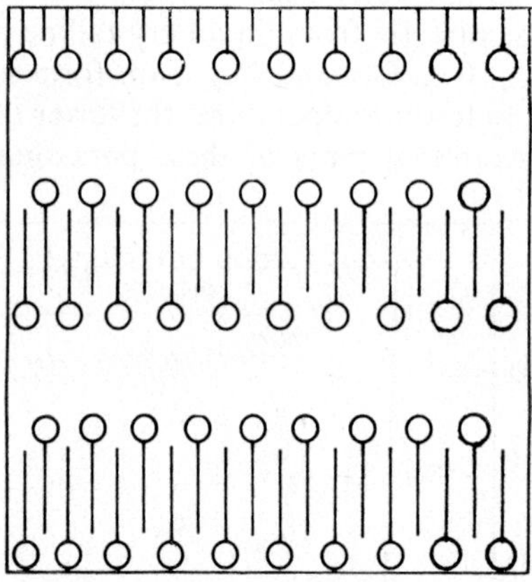

FIG. 5.22. Schematic representation of the structure of the "gel" phase [29].

The thickness of the amphiphile layer, d_a, is about the length of a fully extended soap molecule. The hydrocarbon chains are perpendicular to the plane of the lamellae and arranged in either the hexagonal (g_H) or rectangular (g_R) form found with paraffins. With the "gel" of potassium stearate, only the hexagonal form appears; both are found with rubidium stearate. The change of one into the other with change in composition appears to be a continuous process.

The interfacial area per polar group, calculated from the X-ray long spacing on the basis of the above model, is, within the accuracy of experiment, independent of concentration and of the order of 39·8 Å². The cross-sectional area per paraffin chain, ε (derived from the line at high angles), is 19·9 Å². Since, on the model in Fig. 5.22, the polar groups are alternately on one side or the other of the palisade of hydrocarbon chains, ε should equal half the interfacial area per polar group. This is so (Table 5.7).

TABLE 5.7. *Dimensions of the "gel" phase of several soaps* [30]

Soap*	Temperature, °C	d_a, Å	S_a	S, Å²	2_ε, Å²	d_w, Å
$C_{14}K$	0	20·0	1·11	39·7	39·0	12·0
$C_{16}K$	25	22·7	1·09	39·6	39·4	10·7
$C_{18}K$	25	25·2	1·09	38·8	38·8	10·8
$C_{18}K$	45	25·2	1·06	39·8	39·8	9·3
$C_{22}K$	25	30·3	1·06	39·9	39·6	10·8
$C_{16}Rb$	25	23·6	1·21	39·6	39·6	7·4
$C_{18}Rb$	25	26·2	1·17	39·9	39·2	8·0
$C_{18}Cs$	25	26·8	1·33	38·9	39·2	6·8
$C_{16}H_{33}N(CH_3)_3Br$	15	27·6	1·13	38·8	39·2	10·4

* The lengths of the fully extended alkali metal soap molecules are approximately 18·5, 21·0, 23·5 and 28·5 Å for the C_{14}, C_{16}, C_{18} and C_{22} soaps, respectively. d_a = thickness of the amphiphile layer, S_a = specific gravity of amphiphile, S and 2_ε = interfacial area per polar group calculated from the low- and high-angle lines, respectively, and d_w = minimum thickness of water layer.

For potassium stearate at 45°C, when the water content is below $(1 - c)/c = 0{\cdot}34$, the system consists of crystalline soap and "gel" in equilibrium. The thickness, d_w, of the water layer between the leaflets is 9.3 Å, corresponding to about six water molecules per molecule of soap. For water contents $0{\cdot}34 < (1 - c)/c < 3{\cdot}00$, only gel is present. The thickness of the water layer increases continuously from 9·3 to about. 80 Å, while the area per polar group remains sensibly constant at 39·8 Å². For higher water contents, the system ceases to be transparent and homogeneous. The measured long spacing, $d = (d_w + d_a)$, no longer varies linearly with $(1 - c)/c$ but tends to a maximum value. These observations show that below a certain concentration the system ceases to form a single phase.

At 25°C, for water contents in the range $0{\cdot}44 < (1 - c)/c < 0{\cdot}92$, the potassium stearate/water system is no longer composed of a single "gel" phase as at 45°C, but contains two conjugate metastable "gel" phases in which the thicknesses of the water layers between the leaflets are approximately 12 and 25 Å, respectively. The thickness of the soap leaflets for both "gels" are, as previously, sensibly equal to that of a fully extended soap molecule. Such a separation of two "gels" also occurs with the cetyltrimethylammonium bromide system, in this case not in a region of supercooling, as with the potassium stearate system, but in a region where the "gel" phase is thermodynamically stable. On thermodynamic grounds, the amphiphilic lamellae in these conjugate "gels" of greatly different water content cannot be entirely similar.

With these "gel" systems, in which the hydrocarbon region of the amphiphile leaflets is apparently crystalline, the R-theory, which relates only to systems entirely in the liquid state, cannot be applied. The following interpretation of the formation of the "gel" phase has been proposed by Vincent and Skoulios.

The minimum thicknesses of the water layers separating the leaflets of soap are identical for soaps of the same cation but decrease on passing from soaps of potassium to those of rubidium and caesium (Table 5.7). This agrees with the known decrease in hydration of alkali cations with increase in atomic number. This phenomenon, coupled with the lowering of the T_c line with increasing atomic number, shows the importance of ionization of the soaps to the stability of both liquid crystalline and "gel" phases. It seems that formation of the "gel" constitutes a compromise answer to the conflict between the tendency of the dissolved polar groups to ionize and spread apart from one another and the tendency of the hydrocarbon chains to crystallize and draw together. The case of potassium stearate may be considered. At low temperatures the two tendencies are of similar intensity, the polar groups are ionized and the chains crystallized, and the system adopts the "gel" structure. On this basis the soaps of sodium and lithium are incapable of yielding a "gel" phase because their tendency to ionize is insufficient to give a separation of polar groups compatible with formation of the "gel" structure.*

It is to be noted that the term "gel" is used with several different meanings in the literature on amphiphile solutions. It would seem preferable to reserve the term "gel" (with quotation marks) solely for phases arising and constituted as discussed in the foregoing section.

Fused Mesophases

The general types of sequence of fused mesophases encountered with aqueous solutions of the fatty acid soaps of the alkali metals have been considered in the general discussion. The anhydrous neat soap phase (lamellar labile), included in the preceding sections as a matter of convenience, is a fused mesophase. In the following paragraphs some additional points concerning the aqueous mesophases of the fused type, as formed by the alkali metal soaps, are discussed [31].

Variation in interfacial area per polar group with changes in composition and temperature

On general grounds, and particularly from the point of view of the R theory, information concerning the dimensional changes with composition and temperature that occur at the polar interface between the aqueous and hydrocarbon regions in amphiphilic phases is of great interest.

* In comparison with the fused mesophases of various types, the "gel" phase shows a very low electrical conductivity (François, J. and Skoulios, A. *Kolloid Z.u. Z. für Polymere* **219**, 144 (1967)). This seems hard to reconcile with the proposed ionized structure for the "gel" phase.

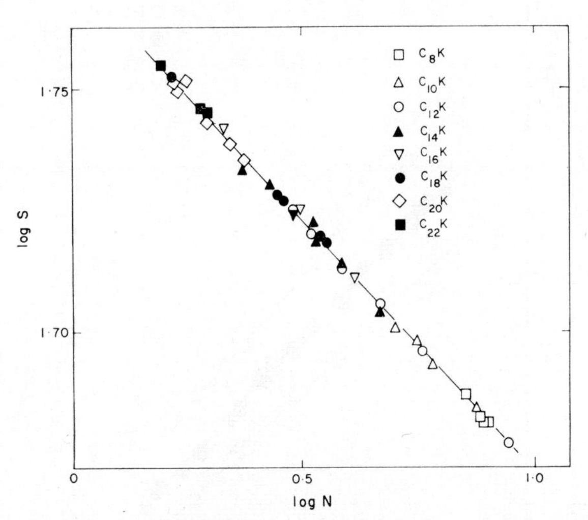

FIG. 5.23. Variation in the interfacial area S per polar group with change in concentration N (moles of soap/litre of water) for M_1 solutions of potassium soaps at 86°C [32].

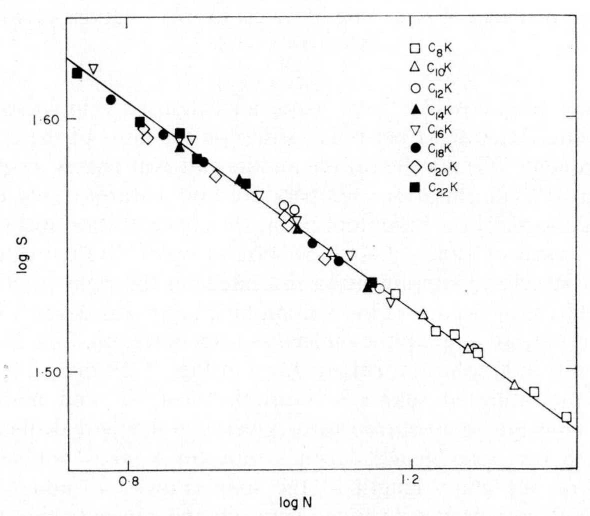

FIG. 5.24. Variation in the interfacial area S per polar group with change in concentration N (moles of soap/litre of water) for G solutions of potassium soaps at 86°C [32].

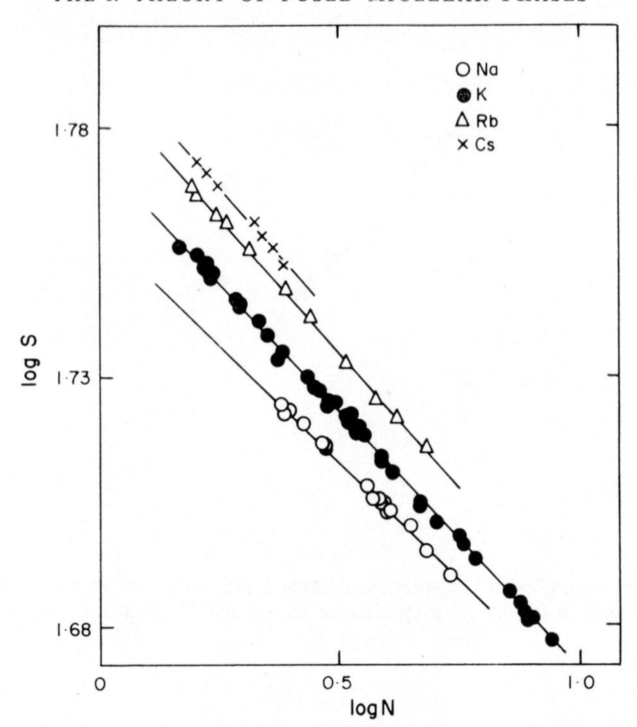

FIG. 5.25. Variation in the interfacial area S per polar group with change in concentration N (moles of soap/litre of water) for M_1 solutions of alkali metal soaps at 86°C [32].

Gallot and Skoulios [32] have made an extensive comparison of the calculated interfacial areas per polar group on the faces of the cylindrical (M_1) and lamellar (G) micelles of the middle and neat phases, respectively, as functions of concentration, temperature and nature of the dissolved alkali metal soaps.* For this comparison, the concentration was expressed as N, the number of moles of soap per litre of water. In their calculations the type of structural simplification indicated on the right of Fig. 5.2(a) was adopted throughout, i.e. the amphiphilic layer was taken to contain only and entirely all amphiphile molecules present (cf. pp. 273, 274). Some of their results in graphical form are given in Figs. 5.23 and 5.24; with the straight-chain saturated soaps, in both the neat, G, and middle, M_1, phases, at constant temperature for a given cation, the calculated interfacial area S per polar group depends only on N. It is not perceptibly dependent on the chain length of the soap (Figs. 5.23 and 5.24). For different alkali cations at a given temperature and concentration the interfacial area per polar group increases with atomic number (Fig. 5.25).

* More recently they (*Molec. Cryst. Liqu. Cryst.* **13**, 323 (1971)) have investigated the liquid crystalline phases formed by aqueous systems containing the α–ω disoaps of potassium and subidium. Only the lamellar mesophase, G, is encountered.

For both the M_1 and G phases, the variation of S with N is given by relationships:

$$S = S_0 N^{-p}$$

or

$$\log S = \log S_0 - p \log N$$

where S_0 and p are constants for a particular soap at a given temperature.

The constants for the lamellar (G) and cylindrical (M_1) phases at 86°C

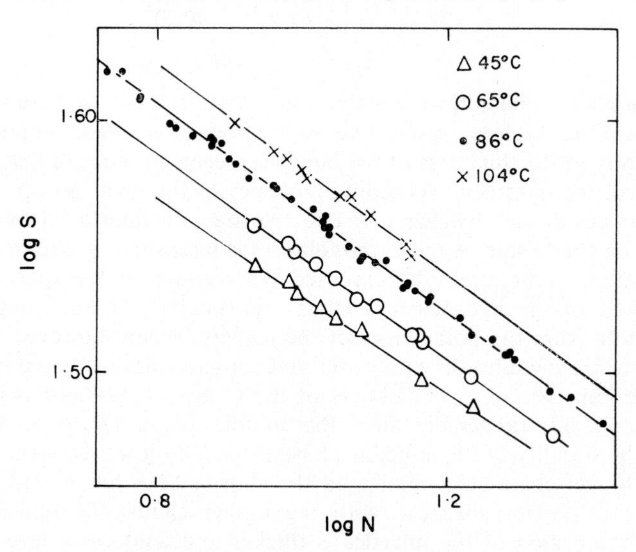

FIG. 5.26. Variation in the interfacial area S per polar group with change in concentration N (moles of soap/litre of water) for G solutions of potassium soaps at 45, 65, 86 and 104°C [32].

are in Table 5.8. The constant p is practically independent of temperature; the constant S_0 increases with temperature (Table 5.9 and Fig. 5.26).

According to the R-theory, the interfacial area S per polar group for the G phase at a given concentration represents that interfacial area per polar group at which the mean interactions of the $\bar{C}$ region with the $\bar{O}$ and $\bar{W}$ regions are balanced to give a planar monolayer. The observation

TABLE 5.8. *Values of S and* p *for* M_1 *and G solutions of fatty acid soaps at* 86° [32]

	M_1 phase		G phase	
Cation	S_0, Å²	p	S_0, Å²	p
Na	57·6	$0{\cdot}09_5$	63·1	0·24
K	58·9	$0{\cdot}10_5$	58·9	0·20
Rb	61·7	0·11	56·3	0·19
Cs	63·1	$0{\cdot}11_5$	55	0·18

TABLE 5.9. *Effect of temperature on values of S_0 and* p *for lamellar solutions of potassium soaps* [32]

Temperature, °C	S_0, Å^2	p
45	52·5	0·19
65	56·3	0·20
86	58·9	0·20
104	61·7	0·20

that, within the G phase, S for a given concentration N is, if not completely independent of, at least insensitive to hydrocarbon chain length (and consequently to the thickness of the lamellae) seems to indicate that while, as expected, the interfacial spreading tendency of the polar groups on the $\bar{C}$ face may be adjusted by change in effective area, the interfacial spreading tendency on the $\bar{O}$ side of the $\bar{C}$ monolayers is insensitive to the thickness of the leaflets. This might be expected on account of the quasi-liquid arrangement of the hydrocarbon chains, particularly of their segments more remote from the polar faces of the leaflets. When, however, on account of increasing dilution with water and consequently increased spreading of the ionic group, the thickness of the $\bar{O}$ layer is reduced beyond a certain limit, rearrangement into the middle phase ($R_x < 1$, $R = 1$) occurs. The stability of the lamellar phase extends to lower concentrations with the longer-chain salts than with the shorter ones (cf. p. 213, Table 5.1). This is possibly because with the longer chains the quasi-liquid hydrocarbon region of the micelles is thicker and contains a larger proportion of hydrocarbon chain remote from the polar group and consequently of more random character. Fig. 5.27, shows that, at a given value of N, the increase in the calculated interfacial area per polar group corresponding to the change in structure from G to M_1 is relatively large (ca. 20–25%).

Although only the interaction of water with the ionic groups has been mentioned, interaction between the water and the interfacial hydrocarbon region exposed between the polar groups will also occur. Although this will be similar for neighbouring homologues in a series of saturated soaps at a given concentration, N, it would be expected to be increased by introduction into the hydrocarbon chain of a relatively polarizable double bond. Such an increased interaction may account for the experimental observation that at a given concentration the interfacial area per polar group for the oleates is greater than that for the saturated soaps [6, 8]. This has been noted with both G and M_1 phases.

The increase in interfacial area per polar group with increasing dilution is sometimes regarded as a steric effect due to increase in size of the polar groups with their increasing hydration. It is preferable to regard the increase in interfacial area per polar group with this dilution as due to that adjustment in the concentration of the interfacial solution of the polar group which necessarily accompanies changes in bulk concentration

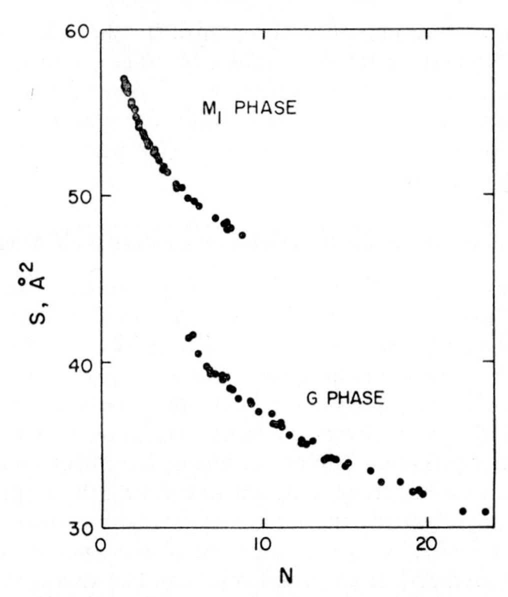

FIG. 5.27. Variation in the interfacial area S per polar group with change in concentration N (moles of soap/litre of water) in M_1 and G solutions of potassium soaps at 86°C [32].

in order to maintain uniform activity of the water throughout all regions of the system (p. 273). In the S_1 phase, for example, with increase in concentration above the cmc a decrease in interfacial area per polar group at the surface of the micelles (i.e., an increase in concentration of the interfacial solution of the polar groups) can permit an equilibrium between micelles and intermicellar liquid such that the concentration of the intermicellar liquid can increase with increasing bulk concentration (cf. p. 274).

The increase in interfacial area per polar group on dilution at constant temperature can be regarded as due to the tendency of dilution to increase $A_{\overline{CW}}$, leaving $A_{\overline{CO}}$ unaffected. This tendency is to some extent offset by the dimensional changes produced which, by decreasing the number of polar groups per unit area of interface, tend to diminish $A_{\overline{CW}}$ (p. 204). If the increase in interfacial area per polar group arising from dilution at constant temperature is accompanied by a change of phase, this is always in a direction from right to left in Figs. 5.5 and 5.6, (decrease in R).

An increase in the calculated area per polar group, as well as being produced by dilution at constant temperature, is also produced by rise of temperature at constant composition. This has been attributed to an increase in the disorder of the hydrocarbon chains with rise of temperature [6, 11] but other factors could contribute, e.g., reduction in $A_{H\overline{WW}}$. It seems anomalous, in view of the large increase in the calculated interfacial area per polar group that, with aqueous systems of the alkali soaps, accompanies the change in structure from G to M, that in certain systems the increase in the apparent area per polar group with rise of temperature at constant composition may be accompanied by phase changes from left to right in Fig. 5.5 and 5.6 (increase in *R*). In other systems, however, the phase changes may be in the opposite sense. It may be remarked that on change of temperature conditions are modified on both sides of the $\bar{C}$ layer while on dilution with water at constant temperature, only the polar side is primarily affected.

Soaps of the Metals of Group II: High-temperature Anhydrous Phases

The anhydrous soaps, $(RCO{\cdot}O)_2M$, of Group II metals, like those of the alkali metals, show stepwise melting with a series of sharp transitions between distinct phases. These phases and transitions, as for the sodium soaps (p. 225), have been characterized by a variety of techniques, mainly by Vold *et al.* [33, 34]. They have been extensively examined from a structural point of view by Spegt and Skoulios [35] using X-ray diffraction methods. The structures of a number of phases have been established with reasonable assurance but those assigned to certain others appear more in doubt and, although recently the subject of extensive revision still can not be regarded as definite. A summary of the phases encountered with the stearates of various metals is given in Table 5.10. The interpretation of their

TABLE 5.10. *Phase sequences in the anhydrous stearates of the metals of Group II* [11]

Mg	C_1 or C_2 (109°) H_1 (195°) H_2 (210°) melt
Ca	C_1 (100°) C_2 (123°) D(T) (152°) . . . (179°) H (>350°)
Zn	C_1 (130°) melt
Sr	C_1 (130°) . . . (170°) $\triangle$(R) (197°) Q (246°) H (>400°)
Cd	C_1 (99°) H_1 (211°) H_2 (~230°) melt
Ba	C_1 (150°) . . . (220°) Q (>400°)

The transition temperatures, °C, are given in brackets. The dotted lines indicate ranges where the phases have not been fully characterized. Phase nomenclature (C_1, C_2, H_1, H_2, etc.) as in text.

results originally given by Spegt and Skoulios [35] first will be summarized. Revised interpretations proposed by Luzzati and Spegt [36] and by Luzzati, Gulik–Krzywicki and Tardieu [37, 38], are discussed subsequently.

ORIGINAL INTERPRETATION OF THE PHASE SEQUENCES AS EXAMINED BY X-RAY DIFFRACTION

At temperatures below 100°C all the soaps (Table 5.10) form three-dimensional lamellar crystals (C_1) in which the completely extended hydrocarbon chains are either normal (e.g. Ca, Cd) or inclined to (Sr, Zn) the planes of the lamellae [39]. At about 100°C the hydrocarbon chains acquire rotational mobility about their long axes and, on that account, a modified lateral arrangement (C_2). At higher temperatures mesophases are produced the number of which tends to be greater the bulkier the metallic ion.

The soaps of zinc pass directly at about 130°C into the amorphous melt.

The soaps of magnesium and cadmium each give either a single fibrous two-dimensional hexagonal mesophase H (probably with the M_2 type of structure, p. 210), or two such phases (H_1 and H_2) which apparently differ in possessing modified arrangements of the polar groups which form the micellar cores, the higher temperature arrangement being slightly less compact [40].

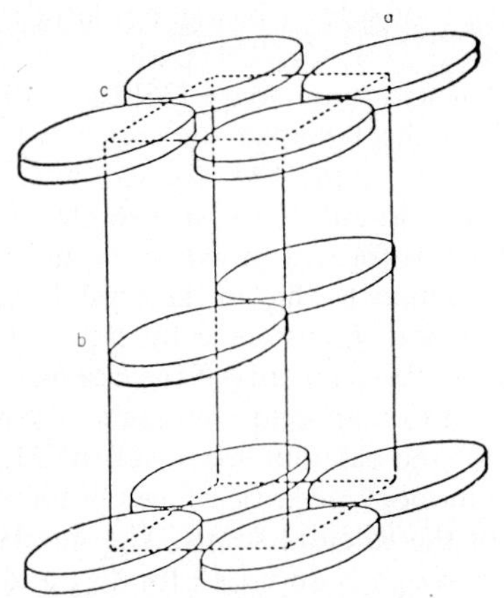

FIG. 5.28. Disk model of structure of phase D(=T) of the calcium soaps. Face C-centred orthorhombic lattice [41].

The calcium soaps give a number of disk structures of which the best characterized has an orthorhombic lattice (Phase D later revised to T) (Table 5.10, Fig. 5.28 [41]). Some structural parameters for the D phase of calcium myristate on this model are recorded in Table 5.11. Within this phase, the number of molecules, *M*, per disk appears to depend only on the molecular weight of the soap, not on temperature, being approximately for C_{14}, 21; C_{16}, 23; C_{18}, 25; C_{20}, 27.

TABLE 5.11. *Structural parameters for phase D (= T) of calcium myristate interpreted on the disk model* [41] (cf. *Table* 5.13)

t, °C	*a*, Å	*b*, Å	*c*, Å	δ, g cm^{-3}	*m*	*ac*/*m*, Å^2
133	30·5	57·4	21·6	0·91	21·0	31·4
140	30·7	57·1	21·6	0·91	20·9	31·6
145	31·0	56·5	21·7	0·90	20·9	32·0
152	31·0	56·1	21·9	0·90	20·8	32·3
160	31·2	55·7	22·1	0·89	20·9	32·8
165*	31·7	55·0	22·1	0·89	20·9	33·5

* At this temperature the structure has become pseudo-hexagonal.

a, *b*, *c* as in Fig. 5.28
δ density used in calculating *m*
m number of polar groups per disk.

At about 180°C the disk structures break down and undergo a sharp transition to the fibrous hexagonal phase, H. This exists over a wide temperature range (Table 5.10). The number of ions per unit length calculated for the fibrous micelles is virtually independent of temperature and molecular weight of the calcium soap, being 0·53 ion per Å. The radius of the polar core is about 3·6 Å and the distances between the centres of the polar cores (= length of side of hexagonal lattice) are for C_{12}, 23 Å; C_{14}, 24·5; C_{16}, 26·1; C_{18}, 27·5; C_{20}, 29·0; all at 240°C.

In the C_1 phase of the strontium soaps [35] the hydrocarbon chains are inclined at about 72° to the lamellae (in contrast to the perpendicular arrangement with calcium soaps). At about 130°C the chains acquire rotational mobility and at about 165°C a disk structure (phase △ (later revised to R), Table 5.10) is formed in which the polar groups apparently retain crystalline order similar to that in the crystal. Although, as in other disk phases, the hydrocarbon chains are in the liquid state the arrangement of the polar groups causes the segments of the chains immediately adjacent to the disks to be inclined to them and imposes a dissymmetry on the disks not found with those of the calcium soaps (cf. p. 232). Accordingly, the disks are arranged in a monoclinic lattice, face C-centred, in contrast to the orthorhombic lattice of the calcium soaps. The number of molecules per disk with the strontium soaps is about 12 for C_{12} and 14 for C_{14}.

At a temperature somewhat above 200°C, phase △ of the strontium soaps undergoes a transition to a body-centred cubic mesophase, Q, which at about 250°C undergoes a further transition to a hexagonal phase, H, which is analogous to the hexagonal phases of the Mg, Cd and Ca soaps. The characteristic dimensions of the cubic lattices of the strontium soaps are virtually independent of temperature and are for C_{12}, 58·4; C_{14}, 62·6; C_{16}, 66·2; C_{18}, 69·7; C_{20}, 73·3 and C_{22}, 76·0 Å.

The temperature range of the hexagonal phase of the strontium soaps extends from about 250°C to 400°C, the upper limit of study. As with the calcium soaps the calculated number of ions per unit length of micelle is virtually independent of temperature and chain length and is 0·496 ion per Å. The distances between the centres of the polar cores are: for C_{12}, 23·2; C_{14}, 24·9; C_{16}, 26·1; C_{18}, 27·5; C_{20}, 28·9; C_{22}, 30·0 Å; all at 265°C.

The barium soaps form a body-centred cubic mesophase, Q, which persists from about 200°C to the temperature of thermal decomposition. Calcium *ω*-*p*-ethylphenyl-undecanoate, an amphiphilic salt of the non-crystalline type (p. 252), exists as the cubic phase, Q, at room temperature. In their original publication Spegt and Skoulios were not able to suggest a satisfactory structural model for the cubic phase.

The state of the hydrocarbon chains and of the polar groups in the high-temperature mesophases of the soaps of the metals of Group II

In all the high-temperature mesophases of the soaps of the Group II metals X-ray measurements indicate that the hydrocarbon chains are in the liquid state. Whether, however, the polar groups are "crystalline" or "fused" (= "labile") is less clear. According to Spegt and Skoulios [35, 42] the state in all the phases, including the hexagonal phase, is compact and quasi-crystalline. To the writer, however, it seems equally possible that, while in phases D(T) and △(R) the polar groups may perhaps be "crystalline", as is thought to be the case in the disk phases of the alkali metal soaps, in the cubic Q, and hexagonal phases they are fused. The facts advanced by Spegt and Skoulios as evidence for the quasi-crystalline structure of the polar cores in the hexagonal phase are as follows:

(1) The calculated number of cations, *n*, per unit length (Å) of cylinder (for Mg, 0·68; Ca, 0·53; Sr, 0·49; Cd, 0·78) is independent of temperature and of chain length of the soap.

(2) Certain soaps of cadmium and magnesium each yield two hexagonal phases. It is argued that, since the hydrocarbon chains in both the hexagonal phases are in the liquid state, the two hexagonal phases must differ in the arrangement of the polar groups, which must therefore be crystalline. It is possible, however, that the two arrangements of the polar groups could be of liquid crystalline character (pp. 208, 211).

(3) Mixtures of soaps of the same cation give a single hexagonal phase of dimensions corresponding to the mean molecular weight of the soaps.

(4) At temperatures common to the ranges for the hexagonal phase for each of the individual soaps, (*a*) a mixture of a calcium soap and a strontium soap gives an X-ray diffraction diagram corresponding to a homogeneous hexagonal phase containing the two metals jointly (270°C); (*b*) mixtures of the stearates of cadmium and magnesium (150°C) or of the stearates of calcium and magnesium (180°C) give X-ray diffraction diagrams corresponding to heterogeneous mixtures of the pure hexagonal phases of the individual soaps. It is argued from these facts that, where the values of *n* for the two cations involved are closely similar, syncrystallization of the polar groups is possible (Ca and Sr), while where the values of *n* differ considerably (Cd and Mg, or Ca and Mg) syncrystallization is not possible. It must be remembered, however, that even in the fused state, while certain salts are miscible, others are not.

(5) When either a paraffin (octadecane or dotriacontane) or a fatty acid (octadecanoic or docosanoic) is added to a hexagonal phase at the appropriate temperature, the lattice dilates in direct proportion to the amount of solvent added up to a certain limit above which the solvent separates as a second phase, very poor in soap. Dilation of the lattice must occur without change in *n*.

To the writer it seems that while all the above facts are consistent with a crystalline arrangement of the polar groups, they are not inconsistent with a fused one. However, the very profound transition on rise of temperature from phases D(T) or △(R) (Table 5.10), to the hexagonal phase, H, either directly or through the intermediate cubic phase, Q, seems to indicate a rearrangement of polar groups that, by analogy with the transition of the disk of ribbon phases to the neat phase LL (p. 227), is likely to involve their fusion. Further, the Q and H phases appear to be closely analogous to the typical fused mesophases, V_2 and M_2 respectively (cf. Fig. 5.3 and 5.5.

Structural Model for the Cubic Phase, Q, and Revised Structural Models for the Phases, D(T) and △(R)

Structure of the cubic phase, Q

A structural model for the cubic phase, Q, was suggested by Luzzati and Spegt [36] based on X-ray measurements of Spegt and Skoulios [35]. In this model (Fig. 5.29) the strontium ions are located on short rods which

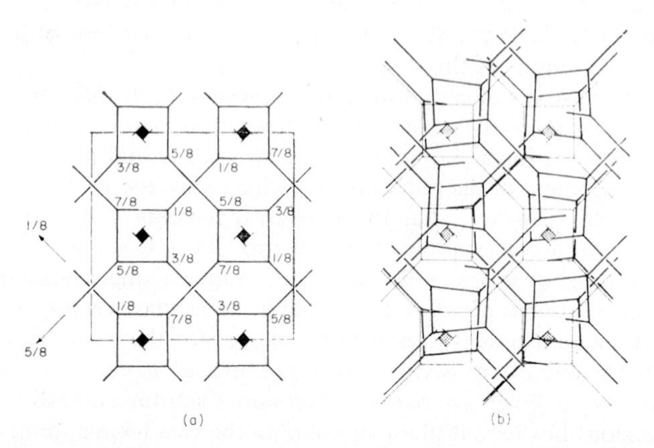

FIG. 5.29. Representation of the proposed rodlet structure of the cubic phase, Q. The loci of strontium ions, represented by heavy lines, are rods of finite length that lie along two-fold axes. The ends of the rods are joined three-by-three. The gaps at the ends of the rods are omitted.

(a) Projection on the plane *ab*, with the position of some of the symmetry elements. The thin lines show the limits of one unit cell. The fractional figures are the *z* co-ordinates of the intersections of three rods.

(b) Perspective view. The dotted lines are the projection of the heavy lines on the plane *ab*.

Note the three-dimensional networks mutually interwoven, otherwise unconnected [36].

are crystallographically equivalent, each rod containing 9–12 strontium ions (Table 5.12). The calculated number of strontium ions per unit length (Å) of rod is 0·51, very similar to that calculated for the hexagonal phase (0·50). Each end of a rod is grouped with those of two other rods with which the first rod is co-planar. The three rods are thus related by a three-fold axis perpendicular to the plane containing the rods. There is a gap (1·56 Å) devoid of strontium ions between this axis and the end of each rod. The length of each rod is ($0{\cdot}304a - 2 \times 1{\cdot}56$) where a is the cell parameter (Table 5.12).

If the short gaps are neglected the rods define two infinite three-dimensional networks, mutually interwoven but otherwise unconnected. Twenty-four rods are contained in one unit cell. The space between the rods is occupied by the paraffin chains in a disordered state; these form a continuous matrix throughout the crystal. It was originally considered that in this phase, as in the hexagonal phase, the polar groups are highly ordered

TABLE 5.12. *Structural parameters for the cubic phase, Q of the strontium soaps at 235°C* [36]

Soap	a, Å	δ, g cm^{-3}	m	l, Å	n, Å^{-1}
Laurate	58·4	$0{\cdot}87_3$	8·9	17·5	0·51
Myristate	62·6	$0{\cdot}85_8$	9·7	18·9	0·51
Palmitate	66·2	$0{\cdot}84_9$	10·3	20·3	0·51
Stearate	69·7	$0{\cdot}83_8$	10·9	21·5	0·51
Arachidate	73·3	$0{\cdot}83_2$	11·6	22·8	0·51
Behenate	76·0	$0{\cdot}82_2$	11·9	23·8	0·50

a = parameter of body-centred cubic cell
δ = density of the soap
m = number of molecules per rod
l = length of rod = $(0{\cdot}25\ \sqrt{2}a - 2 \times 1{\cdot}56)$ Å
n = number of strontium ions per unit length of rod

in contrast to the chaotic conformations of the paraffin chains. The finite length of the rods could then be explained by the interplay of the ordered and disordered regions (cf. p. 226). Assuming that the neighbouring chains tend to keep closer than is allowed by the separation of the crystalline polar groups, the structure becomes increasingly strained as the length of the rods increases. If the length of the rods is limited and the ends grouped three by three, this releases the "strain" and preserves efficient space filling.

Revised structures, T *and* R, *for phases* D *and* △

It was proposed [37], by analogy with the above structural models for the Q and H phases, that phases D and △ are also composed of rod-like elements. Inspection of data for these phases of the calcium and strontium soaps showed that the experimental results, originally taken to indicate a face-centred orthorhombic lattice for phase D [41], were in equally good agreement with a body-centred tetragonal lattice, T. Similarly the data for phase △, originally interpreted as indicating a face-centred monoclinic lattice [35], were in equally good agreement with a rhombohedral lattice, R.

The structure of the D(T) phase of the calcium soaps, now assigned a tetragonal lattice, was then re-interpreted as formed by a succession of parallel layers (separation distance, $c/2$) consisting of a square network of coplanar rods (side, a) of calcium ions, with a gap 3 Å long between the ends of each rod and the corner of the squares. The dimensions involved are illustrated by Table 5.13. The networks, stacked (according to a body-centred tetragonal lattice) but otherwise unconnected, were considered to be embedded in a matrix consisting of the hydrocarbon chains in chaotic disorder.

The structure of phase △(R) of the strontium soaps, now assigned a rhombohedral lattice, was re-interpreted as derived in a similar manner from successive layers (separation distance, $c/3$) of a regular planar

TABLE 5.13. *Structural parameters for phase D(= T) of the calcium soaps as interpreted on the tetragonal model* [37] (cf. *Table* 5.11)

Soap	t, °C	δ, g cm^{-3}	a, Å	c, Å	m	l, Å	n, Å^{-1}
Myristate	165	$0{\cdot}89_0$	31·7	55·0	15·1	25·7	0·59
Palmitate	159	$0{\cdot}88_5$	34·1	59·1	16·6	28·1	0·59
Stearate	152	$0{\cdot}88_0$	35·9	62·2	17·6	29·9	0·59
Arachidate	152	$0{\cdot}87_5$	38·3	66·4	19·4	32·3	0·60

a, c cell parameters
m number of molecules per rod
l length of rod = (a — 2 × 3) Å
n number of calcium ions per unit length of rod

hexagonal network of rods containing strontium ions with a gap of 1·56 Å between the end of each rod and the corners of the hexagon (cf. Table 5.14).

Phases T and R can be visualized as layered structures with stronger cohesion in the plane of the layers than between the layers and may be expected to show order–disorder phenomena if the inter-layer correlations become sufficiently loose.

TABLE 5.14. *Structural parameters for the phase* $\triangle$*(= R) of the strontium soaps on the rhombohedral model* [37]

Soap	t, °C	δ, g cm^{-3}	a, Å	c, Å	m	l, Å	n, Å^{-1}
Laurate	180	$0{\cdot}91_0$	29·0	78·0	7·1	13·6	0·52
Myristate	180	$0{\cdot}89_3$	31·0	84·8	7·8	14·8	0·53
Stearate	181	$0{\cdot}87_3$	36·0	93·3	9·3	17·7	0·53
Arachidate	200	$0{\cdot}85_0$	37·9	97·5	9·8	18·2	0·52

a, c parameters of hexagonal cell
δ density of soap
m number of molecules per rod
l length of rod = $((a/\sqrt{3}) - 2 \times 1{\cdot}56)$ Å
n number of strontium ions per unit length of rod

Extension of rod models to other systems

It has been suggested that rod-like models of the types proposed for phases H, Q, R and T above may also apply to other systems, e.g., to the hexagonal, cubic and rhombohedral high-concentration phases of the lecithin/water system [38]. If this is justifiable it would appear that it could also be extended to the V_2(Q) and M_2(H) phases of the Aerosol OT/water system and of other similar systems (p. 253). It was suggested that "models of opposite polarity, with the paraffin chains inside the threads of the fabric, could well exist in more highly hydrated preparations, e.g., the wide-spread 'middle' phase of soap/water systems is formed by paraffin-filled liquid rods embedded in water". It is emphasized that such hydrated mesophases as M_1, V_1, V_2 and M_2 are of the "fused" type and not of the semi-crystalline class so that stability considerations based on crystallinity of the polar cores would not be applicable in these cases. Indeed, as mentioned (p. 247), it appears possible that the high-temperature

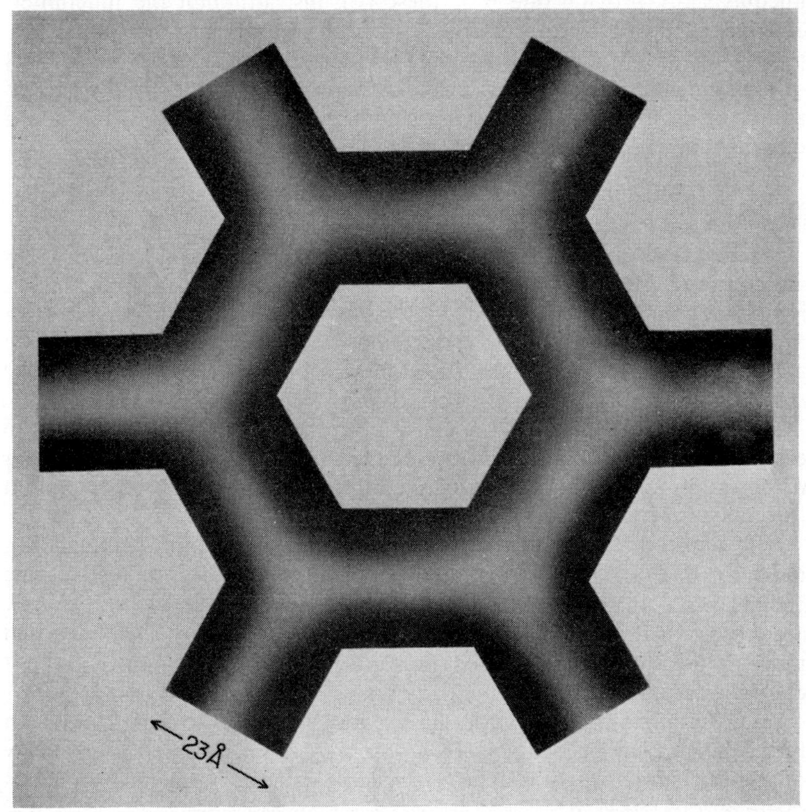

FIG. 5.30. Phase R: A tentative representation of one of the two-dimensional hexagonal networks in phase R of the lecithin/water system with the dimensions of the polar rods to scale [38].

Q and H mesophases of the alkaline earth soaps are themselves of the fused type.

A tentative representation of one segment of the two-dimensional hexagonal network proposed for phase R of the lecithin/water system is given in Fig. 5.30. In this model no gaps exist between the ends of the polar rods.

Comments concerning the validity of the proposed rodlet structures for phases T, R and Q and of analogous models for phases of complementary character

The fact that alternative disk or rod structures have been suggested foı the T and R phases, and also that earlier suggestions were made for alternative structures both for the cubic phase (V_1, Q_I) and the middle phase M_1 [cf. 43, 44, 5], emphasize that a given series of X-ray measurements may appear consistent with a number of structural models. Further, in the systems under consideration, although X-ray measurements may be used definitively to reject a proposed structural model, they cannot usually unequivocally provide one. For these reasons, although the indefinitely extended rod (or fibre) structures for the two-dimensional hexagonal phases, M_1 and M_2, may be accepted as being in accord with a broad body of experiments, it seems worthwhile to assess more critically the acceptability of the rodlet structures on general grounds. The following points may be remarked:

(*a*) It is difficult to see on the basis of what physical forces the rodlets should group in the various ways suggested.

(*b*) For fused amphiphilic mesophases there is evidence that both the aqueous and paraffinic structural elements are devoid of internal rigidity [43]. It is therefore difficult to see on the basis of what equilibrium of forces the network structures of the types proposed could be maintained in fused mesophases even if they might persist in mesophases of the semicrystalline type.

(*c*) The rodlet structures, unless it were possible to modify them in essential particulars in a manner to reconcile them with the suggestions made on pp. 210, 218, are not in accord with the R-theory as at present formulated for fused mesophases.

(*d*) Within the micellar units of the principal fused amphiphilic phases, indicated in Fig. 5.6, all the individual amphiphile molecules of the same chemical type apparently play statistically equivalent roles and statistically are geometrically and energetically equivalent (p. 273). The individual molecules within the proposed network structures apparently do not enjoy this equivalence.

(*e*) Alternative structures for the V_1 and V_2 phases in agreement with the R-theory have been suggested on pp. 217, 218. The detailed geometry of these structures might well be reconcilable with the X-ray measurements made by Luzzati *et al.*

Systems from Amphiphilic Anion Salts which are not Crystalline at Room Temperature

Amphiphilic anion salts in which the lipophilic groups are of a non-compact type frequently do not crystallize and the pure salts exist as mesophases, apparently of the fused type, at room temperature. The hydrocarbon groups, on account of their branched character, do not themselves crystallize; further, on account of their relatively large cross-section they prevent sufficiently close lateral approach of the polar groups for crystallization of these to occur.

Amphiphilic salts which are mesomorphous at room temperature, when heated, frequently undergo decomposition without previous transition to an amorphous liquid. However, the mesophases may often be converted

TABLE 5.15. *Mesophases formed by certain anhydrous amphiphilic salts at room temperature and the phase changes undergone by these mesophases on addition of solvent*

Amphiphilic salt	Mesophase stable at room temperature	Phase transitions on addition of solvent at room temperature
Sodium hept-6-enyl sulphate [45]	G	$G \rightarrow (G + S_1) \rightarrow S_1$ (water)
Sodium 2-ethylhexyl sulphate [18]	G	$G \rightarrow (G + V_1) \rightarrow V_1 \rightarrow (V_1 + S_1) \rightarrow S_1$ (water)
Undecane-3-sodium sulphate	G	$G \rightarrow (G + V_1) \rightarrow V_1 \rightarrow (V_1 + S_1) \rightarrow S_1$ (water)
Calcium *p*-ethylphenyl-ω-undecanoate [35]	V_2	not published
Aerosol OT (Fig. 5.3, p. 205)	M_2	$M_2 \rightarrow (M_2 + V_2) \rightarrow V_2 \rightarrow (V_2 + G) \rightarrow G \rightarrow (G + S_1) \rightarrow S_1$ (water)
Aerosol MA (p. 204)	M_2	$M_2 \rightarrow (M_2 + S_2) \rightarrow S_2$ (hydrocarbon, p. 221)
Sodium di(2-ethylhexyl)acetate [13]	M_2	$M_2 \rightarrow (M_2 + V_2) \rightarrow V_2 \rightarrow (V_2 + S_2) \rightarrow S_2 \rightarrow (S_2 + G) \rightarrow G \rightarrow (G + S_1) \rightarrow S_1$ (water)
Potassium di(2-ethylhexyl)acetate [13]	M_2	$M_2 \rightarrow (M_2 + S_2) \rightarrow S_2 \rightarrow (S_2 + G) \rightarrow G \rightarrow (G + S_1) \rightarrow S_1$ (water)
Tetradecane -3, -4, -5, -6 and -7-sodium sulphates. (The -1 and -2-sulphates are crystalline [2, 3])	Not determined	Mesomorphous solutions (not characterized) $\rightarrow S_1$ (water)

to the amorphous liquid, frequently with intermediate formation of other mesophases, on the addition of water or organic liquids at room temperature. Before any phase change occurs the initial mesophase absorbs considerable amounts of the added liquid (cf. Fig. 5.3). Some examples of mesophases formed by pure amphiphilic salts at room temperature and of the phase changes undergone on the addition of a liquid (usually water) are in Table 5.15. Mesomorphous amphiphilic salts like those listed in Table 5.15 are usually readily soluble in polar organic liquids, such as acetone or ethanol in the presence of small amounts of water, giving mobile amorphous solutions. The "Aerosol" series (cf. p. 204) are soluble in hydrocarbons even under anhydrous conditions.

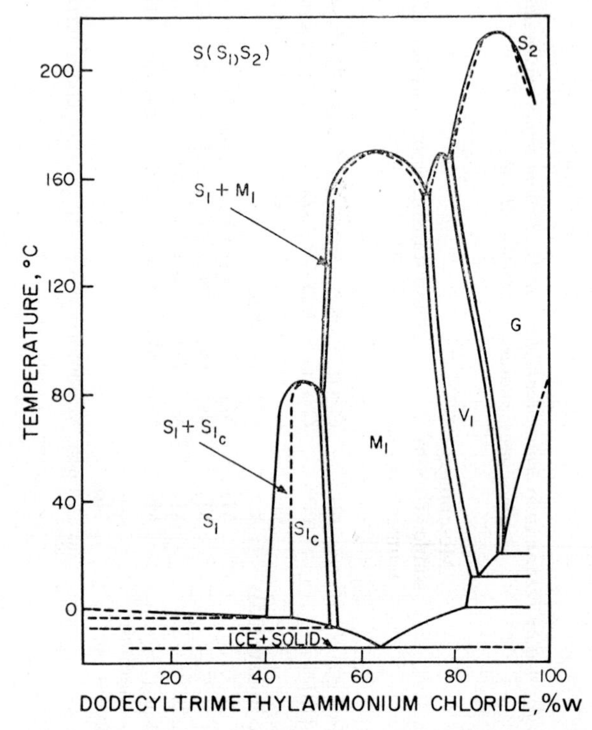

FIG. 5.31. Phase diagram for the dodecyltrimethylammonium chloride/water system [12(*b*)].

——, Experimental boundary; – – –, interpolated boundary.

Aqueous Systems from Amphiphilic Cation Salts

Mesophases analogous to those arising with aqueous solutions of amphiphilic anion salts are also encountered with cationic amphiphiles. An interesting system exemplifying the formation of the phase S_{1c} (p. 212), was described by Balmbra, Clunie and Goodman [12(*b*)] (Fig. 5.31).

Although the phase S_{1c} does not appear to have been identified in a binary aqueous anionic amphiphile system it has been found in ternary aqueous anionic amphiphile systems containing hydrocarbon (p. 281) and in certain aqueous non-ionic amphiphile systems (12).

The phase diagram (Fig. 5.31) for the dodecyltrimethylammonium chloride/water system was determined by optical microscopy, differential thermal analysis, nuclear magnetic resonance spectroscopy and X-ray diffraction.

Two cubic phase, S_{1c} and V_1, arise at room temperature. They are optically isotropic but show strain birefringence. They are extremely viscous, particularly the V_1 phase, but give high resolution NMR spectra. There are conspicuous qualitative differences between the low angle X-ray diffraction patterns given by the S_{1c} and V_1 phases of Fig. 5.31. The patterns for the V_1 phase appear very similar to those given by the V_1 phase of the dimethyldodecylamine oxide/water system (p. 256) and are consistent with a face-centred cubic lattice (lattice parameter $a = 56$ Å at 85% amphiphile), although the absence of higher order reflexions makes the allocation of lattice type uncertain. The patterns for the S_{1c} phase indicate the presence of large uniformly oriented regions, and specimens giving predominantly "single crystal" type diffraction patterns of a primitive cubic lattice (lattice parameter $a = 86$ Å at 43% amphiphile) were obtained by slow cooling.

The formation of the S_{1c} phase depends on the length of the alkyl chain and the nature of the counter ion. For example, this phase is formed by the decyl-, dodecyl- and tetradecyl-trimethylammonium chlorides but not by the hexadecyl or octadecyl homologues nor by any of the corresponding bromides. No model structure was assigned [12*b*] to the S_{1c} phase but it was remarked that any proposed theory must take account of the effects of composition just noted.

According to Fig. 5.6 the formation in a particular system of any individual homomicellar type of mesophase* (S_{1c}, M_1, G, M_2) depends on:

(1) the formation of the particular micellar form characteristic of the mesophase and

(2) a concentration and mutual proximity of the micelles of this form which are sufficient, at the temperature prevailing, to result in the mutual ordering of the micelles by intermicellar forces, into that array characteristic of the particular mesophase.

As regards (1), it is found generally that with aqueous solutions of any homologous series of amphiphiles corresponding states with respect to Figs. 5.5 and 5.6 are reached at progressively lower concentrations, the higher the molecular weight of the amphiphile. A likely reason why the S_{1c} phase is formed by C_{10}-, C_{12}- and C_{14}-trimethylammonium chlorides, but not by the C_{16} and C_{18} homologues, is that with the higher homologues the transition from S_1 to fibrous (M_1) micellar forms occurs at a concentration which is too low to permit that concentration of S_1

* The "intermediate" mesophases V_1, V_2, etc., are possibly heteromicellar, or hybrid in character (p. 218).

micelles which is necessary for their mutual ordering to give the S_{1c} mesophase. The effect of replacement of the Cl^- ion by the less hydrophilic Br^- ion in inhibiting the formation of the S_{1c} phase in this series, may again be due to the displacement of the intermicellar equilibrium (Fig. 5.6) to the right (*R* increased) by this change, again resulting in the transformation of spherical to fibrous forms of micelle at concentrations which are too low to permit mutual ordering of the micelles to give the S_{1c} phase. In support of this view it is noted that according to X-ray diffraction studies [10] the transition between spherical and fibrous micelles at 27°C occurs at a concentration close to 40% w for the C_{16}-trimethylammonium chloride but at a concentration of only 5% w for the C_{16}-trimethylammonium bromide.

Non-ionic Amphiphile/Water Systems

Mesophases analogous to those which arise with ionic amphiphiles are also produced in a variety of non-ionic amphiphile/water systems.

Polyethylene Glycol Derivatives

Mesophase formation in aqueous systems containing polyethylene glycol derivatives has been extensively studied [6(*a*), 6(*b*), 6(*c*), 11, 12(*a*)]; cf. also Chap. 4 in reference [70]. The S_{1c}, M_1 and G mesophases have been encountered but apparently not the M_2 phase.

The N,N,N-trimethylaminododecanoimide/Water and the Dimethyldodecylamine Oxide/Water Systems

The phase diagrams for these two systems are in Fig. 5.32(*a*) and (*b*). It is seen that their phase behaviour makes both systems amenable to experimental study over a convenient temperature range, including room temperature and these systems have been carefully investigated by X-ray diffraction [44, 48], NMR spectroscopy [21, 46, 49], electrical conductivity [46] and polarization microscopy [13, 14]. These studies, not detailed here, when taken collectively support the fibrous hexagonal structure for the M_1 phase and the lamellar structure for the G phase but leave the structure of the V_1 phase not fully decided.

Glyceride/Water Systems

Studies have been made of the mesophases in glyceride/water systems, those by Lutton [50] and Larsson [51], who provide key references, being particularly noted.

Fused mesophases

While neither pure triglycerides nor diglycerides yield mesophases, either in presence or absence of water, the lower 2-monoglycerides on cooling from the amorphous melt yield lamellar liquid crystals. These mesophases are formally analogous to the anhydrous high-temperature neat phase of the alkali metal soaps (p. 227). Further the amorphous glyceride

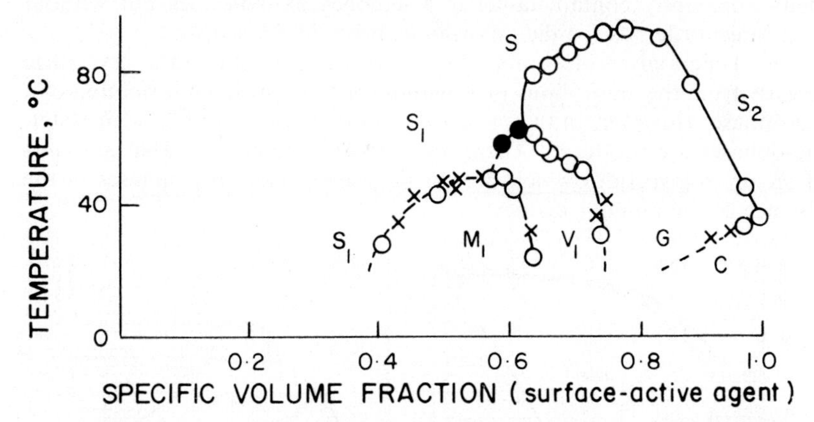

FIG. 5.32(a). Phase diagram for the N,N,N-trimethylaminododecanoimide/water system. Phase boundaries were determined from optical O, density × and X-ray diffraction ● measurements. S(S_1, S_2), mobile isotropic solution (amorphous solution). M_1, middle phase; V_1, viscous isotropic phase; G, neat phase; C, crystals [44].

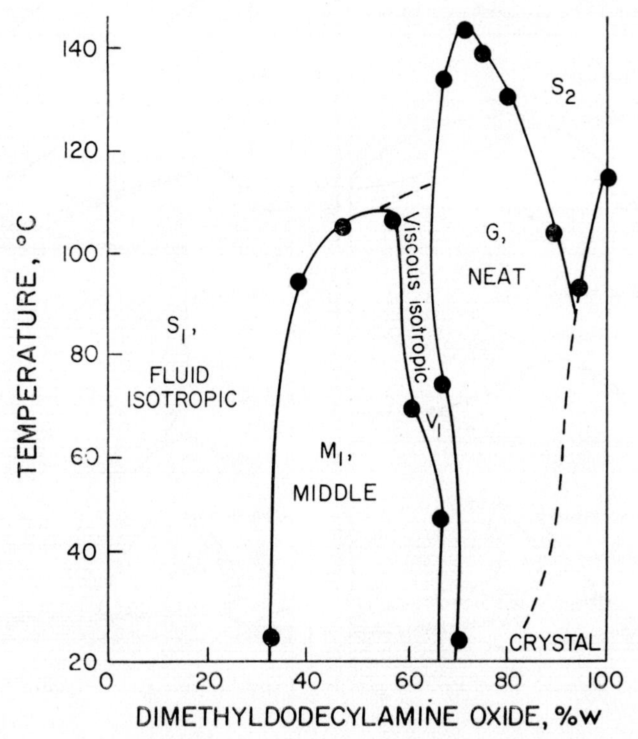

FIG. 5.32(b). The dimethyldodecylamine oxide/water system [47].

melts apparently contain lamellar assemblies of molecules but without long-range order, as for the amorphous melts of the soaps.

The 1-monoglycerides and the higher 2-monoglycerides crystallize directly from the amorphous melt without the intermediate formation of a mesophase. However, in the presence of more than about 5% w of water, mesophases are produced. These may incorporate considerable amounts of *di*- or *tri*-glycerides which do not themselves form mesophases in the absence of the monoglycerides.

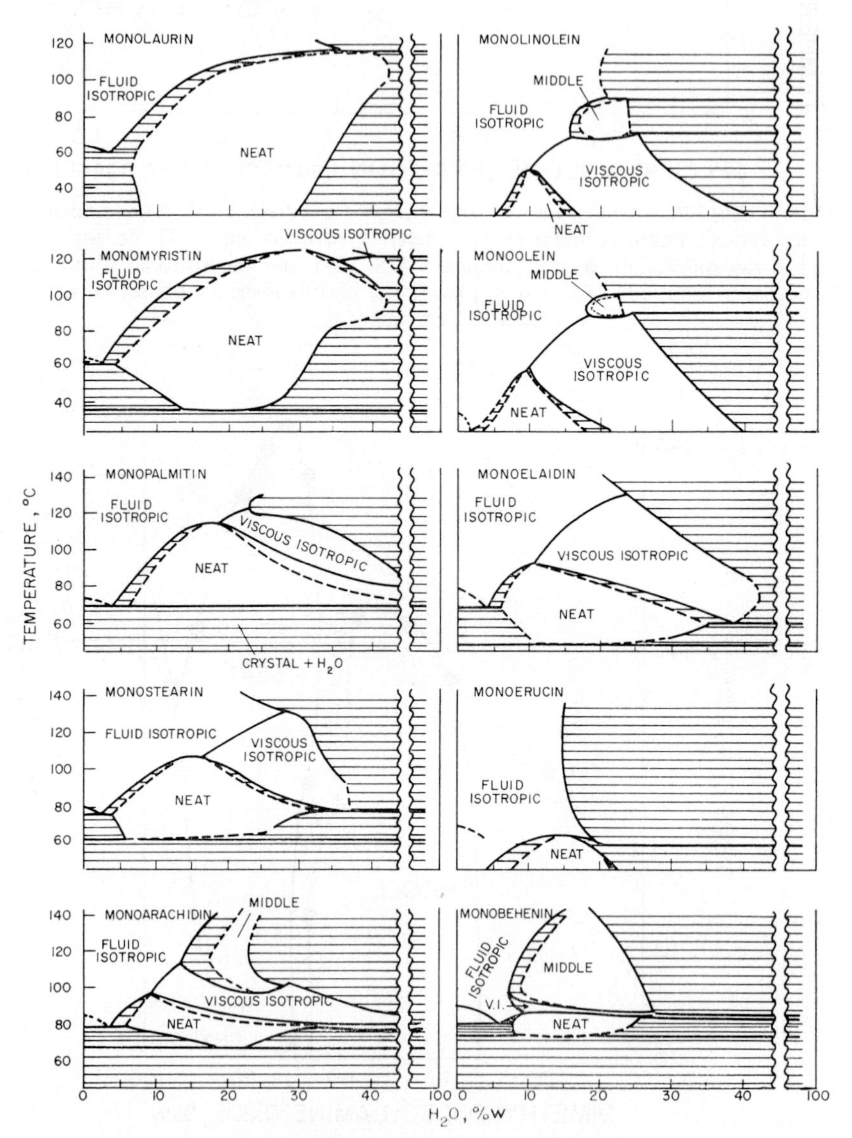

FIG. 5.33. Aqueous monoglyceride systems [50].

Some phase diagrams obtained by Lutton for 1-monoglyceride/water systems are reproduced in Fig. 5.33. The middle phase is formed only by the longer chain monoglycerides. However, although monostearin alone does not give a middle phase with water, a middle phase is formed on addition of 10% w of tristearin to a 4:1 monostearin-water mixture at 100°C. Similarly, Larsson observed a middle phase with mixtures of monopalmitin and dipalmitin.

Phase diagrams (Fig. 5.34(*a*), (*b*), (*c*)) for the C_6, C_8 and C_{10} 1-mono-

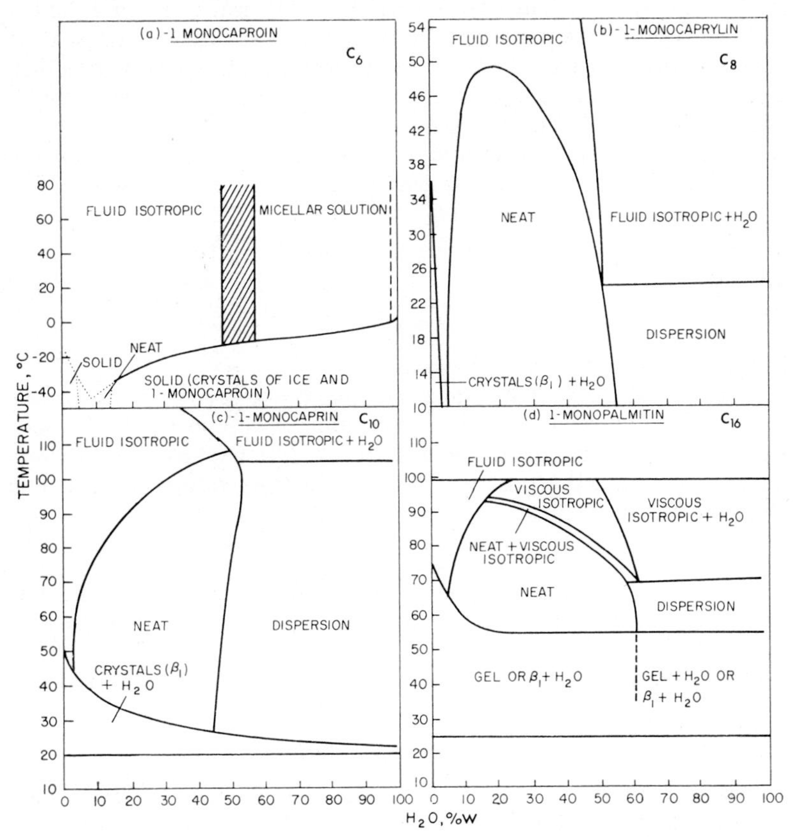

FIG. 5.34. Aqueous systems of the lower monoglycerides [51].

glyceride/water systems were published by Larsson [51] as well as a more detailed diagram for the monopalmitin/water system (Fig. 5.34(*d*)).

In X-ray diffraction studies Larsson found that the high concentration "fluid isotropic" phases of all the higher 1-monoglyceride/water systems show similar characteristics. The anhydrous melts show a well-defined intensity maximum which moves to higher spacings as water is added. The position of the maximum for these solutions is roughly the same as that of the first-order line of the neat phase obtained on cooling. This was interpreted as indicating that the fluid isotropic phase has locally a lamellar structure similar to that found in extended form in the neat phase.

The X-ray diffraction diagrams for the viscous optically isotropic and middle phases were taken to indicate a face-centred cubic and a two-dimensional hexagonal lattice for the two phases respectively. The cubic phase was assigned a structure of water spheres embedded in an amphiphilic continuum while the middle phase was believed to be the inverse form (M_2) rather than the hydrophilic form (M_1) found with the alkali metal soap/water systems. The change from the neat phase into the cubic and hexagonal mesophases on heating was visualized as caused by a "tendency for the chain directions to diverge from the polar region due to increased temperature movements". Within the neat phase itself an increase in the interfacial area per polar group was deduced from the X-ray measurements as occurring both on dilution or rise of temperature.

The question whether the sequence of fused mesophases observed in the monoglyceride/water systems is

$$G \rightarrow V_2 \rightarrow M_2$$

or

$$G \rightarrow V_1 \rightarrow M_1$$

is not easily answered unambiguously.

In non-ionic amphiphile/water systems, phase sequences observed on rise of temperature at constant composition usually proceed from left to right in Fig. 5.5 or 5.6 (cf. Fig. 5.32(*a*), and (*b*)). On this basis one would interpret the sequence observed with rise of temperature in the 1-monoglyceride water systems as

$$G \rightarrow V_2 \rightarrow M_2.$$

On the other hand, in binary amphiphile/water systems the phase sequences observed on dilution normally proceed from right to left in Figs. 5.5 and 5.6.

On this basis one would represent the phase sequence observed in the 1-monoglyceride/water systems on dilution as

$$G \rightarrow V_1 \rightarrow M_1.$$

It is possible that the anomaly could be accounted for if the monoglyceride samples contain free glycerol, either initially or formed by hydrolysis. In such a system it is conceivable that addition of water could give the reverse phase sequence.

$$G \rightarrow V_2 \rightarrow M_2$$

(cf. p. 268, Table 5.16, Method II*b* in reverse).

Dispersions of the neat phase in 1-monoglyceride/water systems

As noted, at higher water/monoglyceride ratios the mono-glyceride mesophases exist in equilibrium with a conjugate aqueous phase of very low monoglyceride content. The fluid neat mesophase may readily be dispersed by agitation in the excess aqueous phase S_1 giving very interesting "dispersions" shown by Larsson to have optical properties resembling those of a nematic (cf. p. 284).

Semi-crystalline mesophase: the "gel" phase

At lower temperatures the higher monoglycerides give "gels" or "coagels", somewhat similar to those formed by the soaps (p. 234). The structure of the "gel" phase in 1-monoglyceride/water systems consists, according to Larsson, of bimolecular lipid layers with the same structure as that of the α crystal [52], separated by water layers up to 20 Å thick. The "gel" phase formed by 1-monopalmitin (Fig. 5.34(*d*)) is metastable but can persist for a few days at 25°C before it transforms to a coagel containing the crystal form ($\beta 1$) stable at this temperature.

MULTICOMPONENT SYSTEMS

In the following sections only thermodynamically stable phases of the fused type are considered. Partly crystalline, often metastable, mesophases that probably play a part in grease technology are not included. Foundation work by the schools of McBain and of Harkins and others was largely directed to elucidating the processes of micelle formation and of solubilization in isotropic solutions although mesophases were incidentally encountered and studied. This work has already been covered in several books and reviews [3, 5, 53, 54].

Some Miscellaneous Multicomponent Systems

The phase sequences encountered in a large number of systems containing up to five components were studied by the writer and co-workers [2–5]. These systems included amphiphilic salts, non-ionic amphiphiles (alkanols, amines, carboxylic acids, polyethanoxy derivatives), hydrocarbons, water, water-soluble organic compounds (ethylene glycol, ethanol, etc.) and water-soluble inorganic salts (sodium sulphate, etc.). The regularities observed in this work formed the basis for the *R*-theory. In these systems only one liquid crystalline phase was identified, the lamellar phase, which from its appearance and consistency was originally termed "gel" G. The absence of other liquid crystalline phases provided a fortunate simplification in these earlier "orientative" studies and was due to the nature of the amphiphilic salts employed (usually branched-chain and of rather low molecular weight), the frequent presence of hydrocarbon and the particular concentration ranges investigated. Recent experimental re-investigation by the writer of a number of the earlier systems using polarization microscopy has confirmed that the liquid crystalline phase encountered was, as believed, the lamellar or G phase and that, at least in the majority of the systems studied, other mesophases were not involved.

The experimental method most frequently employed in this work is illustrated as follows.

To a mixture of hydrocarbon (e.g. n-octane) and an aqueous solution

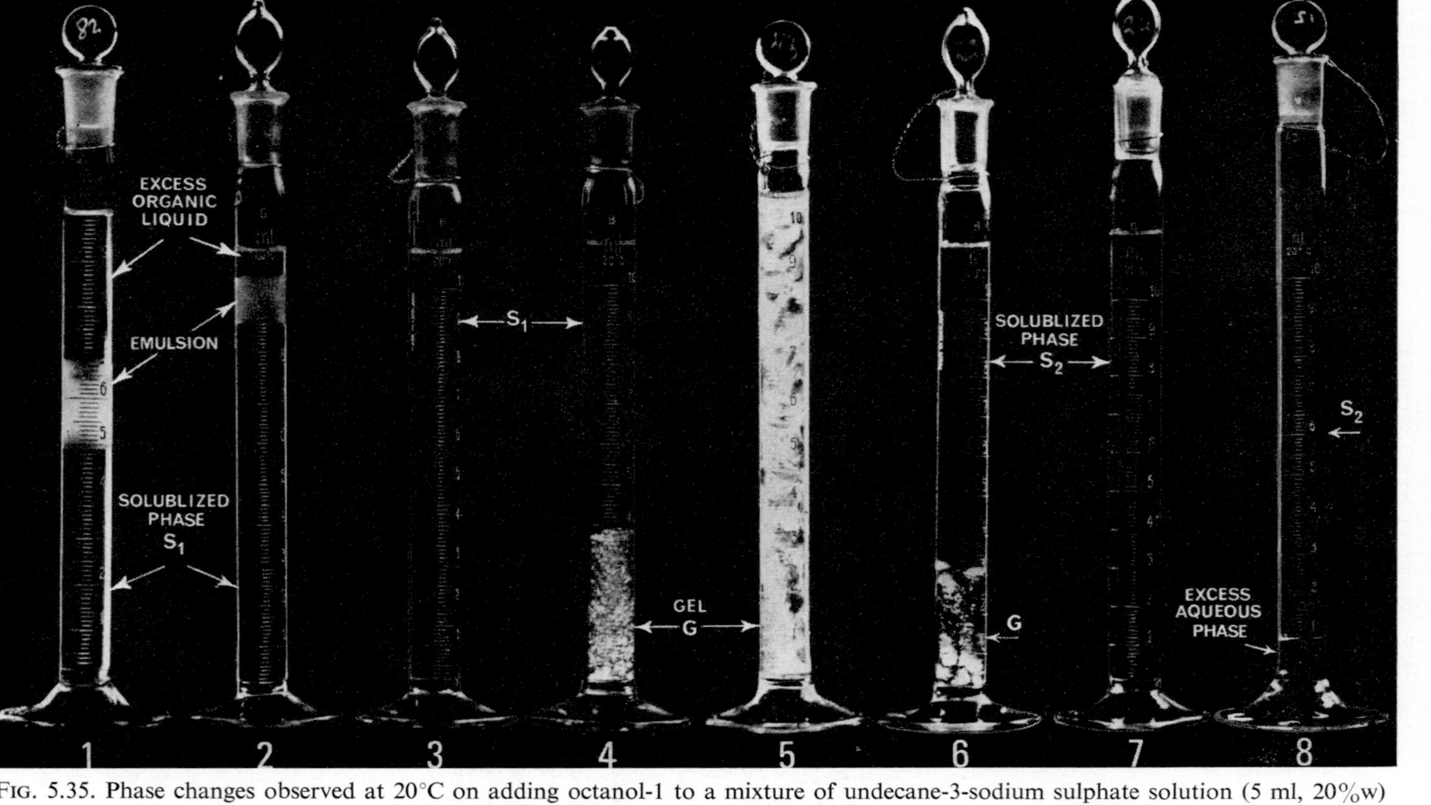

FIG. 5.35. Phase changes observed at 20°C on adding octanol-1 to a mixture of undecane-3-sodium sulphate solution (5 ml, 20%w) with aromatic-free hydrocarbon (b.p. 188° to 213°C) (5 ml); systems photographed between crossed polar sheets. Note that S_1 and S_2 solutions are clear and that there is persistent orientation in certain regions of G phases.

(1) No octanol added, (S_1 + excess hydrocarbon phase)
(2) 0·85 ml octanol-1 added
(3) 0·91 ml octanol-1 added, S_1 (isotropic)
(4) 1·00 ml octanol-1 added, (S_1 + G)
(5) 1·20 ml octanol-1 added, G (birefringent)
(6) 1·38 ml octanol-1 added, (G + S_2)
(7) 1·70 ml octanol-1 added, S_2 (isotropic)
(8) 2·13 ml octanol-1 added, (S_2 + excess aqueous phase)

of an amphiphilic salt (e.g. a 20% w solution of undecane-3-sodium sulphate) a second more lipophilic amphiphile (e.g. octanol-1) was gradually added so that the mixture of amphiphiles present was made overall progressively more lipophilic. The general type of phase sequence observed is illustrated by Fig. 5.35. These phase changes were followed visually (with or without crossed polars) and by observation of changes in consistency and electrical resistance.

The typical phase changes, exemplified in Fig. 5.35, were interpreted on the basis of the *R* concept as follows.

Fig. 5.35, 1

In the aqueous phase S_1, the undecane-3-sodium sulphate is dissolved as spherical micelles, S_1 ($R \ll 1$). A small amount of hydrocarbon enters the micelles by solution ("solubilization") in the hydrocarbon interiors of the micelles. The consequent swelling necessarily tends to reduce the convexity of the $\bar{C}$ layer (undecane-3-sodium sulphate) towards the aqueous region $\bar{W}$ (equilibrium aqueous intermicellar solution of undecane-3-sodium sulphate at a concentration somewhat above the cmc (about 1%) and containing a trace only of hydrocarbon). Because $R \ll 1$, the extent of the reduction in convexity that is possible is limited. Beyond this limit excess hydrocarbon (containing traces only of water and undecane-3-sodium sulphate) separates as a second phase.

Fig. 5.35, 2

The added octanol-1 distributes itself between the excess of hydrocarbon phase and the aqueous amphiphilic solution phase S_1 [55]. Within the S_1 phase itself, the octanol is distributed between the $\bar{W}$ region, the micellar interior $\bar{O}$, and the interfacial layer, $\bar{C}$ (p. 273), the largest proportion entering the $\bar{C}$ layer. Here the octanol molecules are interposed on the like-to-like principle (p. 199) between the undecane-3-sodium sulphate molecules (ion-pairs and ions). The highly hydrophilic $-SO_4Na$ groups on the polar face are thus diluted by the less polar groups of the octanol. $A_{\bar{C}\bar{W}}$ is thus diminished and *R* increased. This increase in *R* permits a reduction in the convexity of the $\bar{C}$ layer towards the $\bar{W}$ region and the entry of further hydrocarbon into the micellar interior, the excess of hydrocarbon phase thereby being diminished.

Fig. 5.35, 3

Further addition of octanol continues the processes described and results in complete solution of the excess hydrocarbon.

Fig. 5.35, 4

Progressive additional incorporation of octanol-1 into the $\bar{C}$ layer, displaces the intermicellar equilibrium within the S_1 solution (Fig. 5.6, lower sections) towards the right and *R* approaches unity. With thermal fluctuations, loci arise where $R = 1$ and transitory lamellar groupings are produced. Further incorporation of octanol-1 increases these groupings in extent and persistence until separation of the G phase ($R = 1$) occurs. This phase at first exists conjugately with residual S_1 phase.

FIG. 5.35, 5

The processes described above are completed and the system is wholly converted to the lamellar liquid crystalline solution phase G ($R = 1$ throughout).

FIG. 5.35, 6

With increasing octanol-1 content of the $\bar{C}$ layer, R tends to become >1 and the $\bar{C}$ layer to become convex towards $\bar{O}$. Up to a certain point this may be offset by dimensional changes within the lamellae (p. 205). Beyond this point a stage is reached at which the local convexities towards the $\bar{O}$ region produced by thermal fluctuations attain an amplitude sufficient to break down the G phase with separation of the S_2 phase within which the $\bar{C}$ region is statistically convex towards $\bar{O}$ ($R > 1$). The S_2 phase is at first conjugate with remaining G phase.

FIG. 5.35, 7

The processes described above are completed and the system is wholly converted to the amorphous solution phase, S_2.

FIG. 5.35, 8

With further increase of the octanol content of the $\bar{C}$ layer R tends to become much greater than unity and within the thermally fluctuating micellar system local regions (S_2 micelles) arise where the convexity of the $\bar{C}$ region becomes sufficient to cause extrusion of an excess aqueous phase containing only very low concentrations of hydrocarbon, octanol and undecane-3-sodium sulphate (cf. reference [3], Chap. 5). The mechanism of separation of this aqueous phase from S_2 solutions when $R \gg 1$ is thus complementary to the mechanism for the separation of the hydrocarbon phase (Fig. 5.35, 2) from S_1 solutions when $R \ll 1$. All the phase transitions described in the foregoing paragraphs are readily reversible.

Many series of broadly similar systems were examined and found to show similar phase sequences. By observing the effects of stepwise modifications in composition on the phase equilibria, the influence of each such modification on the ratio R could be inferred. The following examples are given.

(*a*) When the saturated hydrocarbon (boiling range, 188°–213°C) of the initial system (Fig. 5.35) is replaced with an equal volume of *n*-hexane, the various subsequent stages of the phase sequence are reached on the addition of smaller amounts of octanol-1 than were required with the original hydrocarbon. It may be inferred that replacement of the solubilized hydrocarbon of high molecular weight by an equal volume of a lower hydrocarbon increases R. Further, from the relationship

$$R \rightarrow [(A_{\bar{C}\bar{O}} - A_{\bar{O}\bar{O}})/(A_{\bar{C}\bar{W}} - A_{\bar{W}\bar{W}})] \qquad \text{(cf. p. 204)}$$

it is reasonable that the increase in R be associated with that decrease in

$A_{\overline{OO}}$ which is also responsible for the increased volatility of the lower hydrocarbon.

Replacement of the *n*-hexane (b.p. 60°C) with cyclohexane (b.p. 81°C), however, still further reduces the amount of octanol required to reach a given stage in the phase sequences. Here a further mechanism must operate, e.g. greater polarizability of the cyclic hydrocarbon could increase $A_{H\overline{CO}}$. These effects, as followed by specific resistance changes, are illustrated by Fig. 5.36.

(*b*) If increasing amounts of sodium sulphate are added to the initial system (Fig. 5.35) the amounts of added octanol-1 required to reach any

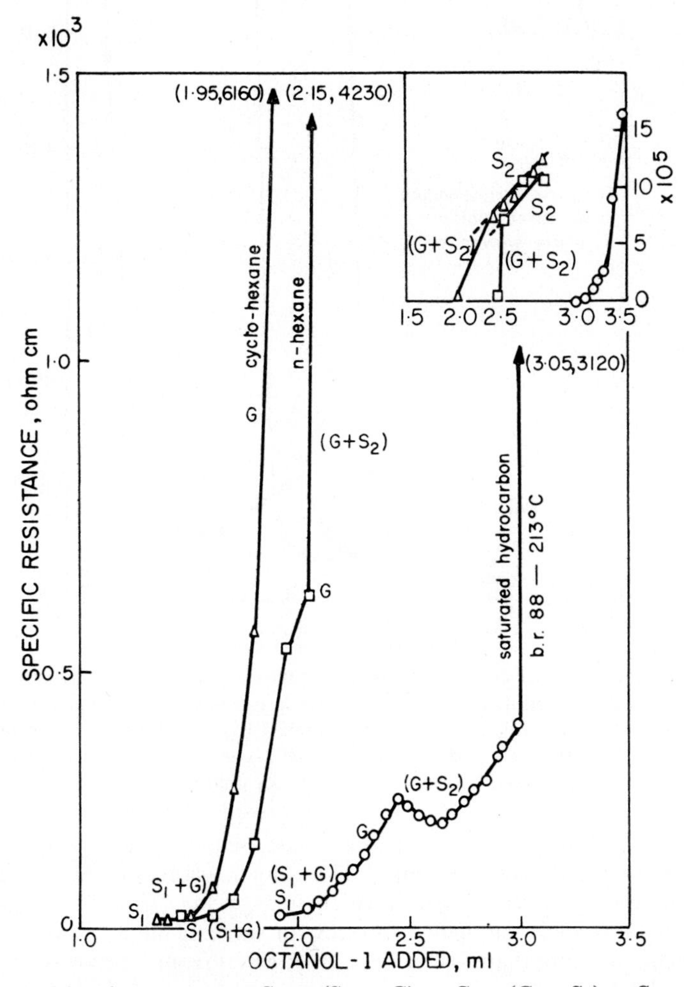

Fig. 5.36. The phase sequence, $S_1 \rightarrow (S_1 + G) \rightarrow G \rightarrow (G + S_2) \rightarrow S_2$, at 20°C produced on addition of octanol-1 to a mixture of undecane-3 sodium sulphate solution (10 ml, 20% w) and hydrocarbon (10 ml) as followed by electrical resistance charges [2*b*].

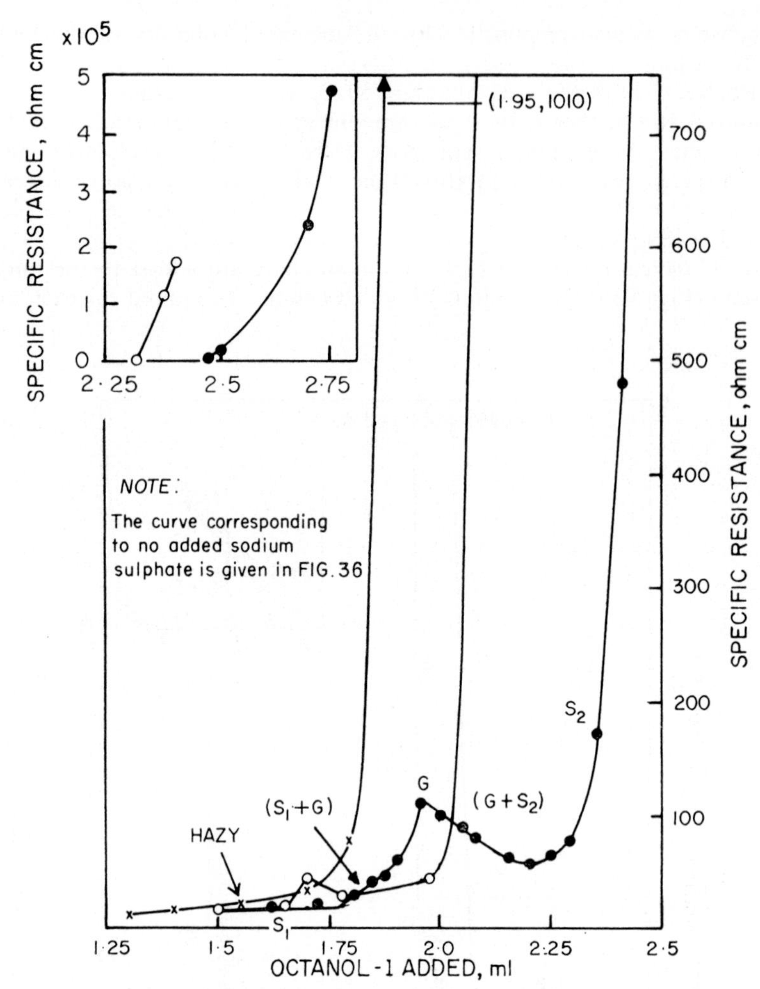

FIG. 5.37. The influence of added sodium sulphate in affecting the phase sequence at 20°C produced on adding octanol-1 to a mixture of undecane-3 sodium sulphate solution (10 ml, 20% w) and saturated hydrocarbon (10 ml, b.p. 188–213°C). Sodium sulphate: ● 0·1 g, or O 0·2 g or ×0·4 g. Phase sequence followed by electrical resistance changes [2b].

particular subsequent stage in the phase sequence are progressively reduced. This is illustrated by Fig. 5.37 in which, again, the phase sequence was followed by electrical resistance changes. It may be inferred that the addition of sodium sulphate produces an increase in R. Further, it is reasonable to think that this increase in R arises from a decrease in $A_{\mathrm{H\bar{C}\bar{W}}}$ in the relationship

$$R \rightarrow [(A_{\mathrm{H\bar{C}\bar{O}}} + A_{\mathrm{L\bar{C}\bar{O}}})/(A_{\mathrm{H\bar{C}\bar{W}}} + A_{\mathrm{L\bar{C}\bar{W}}})] \qquad \text{(cf. p. 204)}$$

The reduction in $A_{\mathrm{H\bar{C}\bar{W}}}$ will be due in part to reduced polarity of the $\bar{\mathrm{C}}$

layer owing to decreased ionization (common ion effect) and in part to effects of the added inorganic ions in interrupting water-to-amphiphile and water-to-water hydrogen bonding. The latter effects may also be responsible for the progressive decrease, on addition of sodium sulphate, in the range of compositions which give rise to the liquid crystalline G phase as indicated by the specific resistance peaks in Fig. 5.37.

(*c*) Figure 5·38 illustrates the phase sequence,

$$S_2 \rightarrow (S_2 + G) \rightarrow G \rightarrow (G + S_1) \rightarrow S_1,$$

observed on addition of further hydrocarbon to the S_2 solution produced as in Fig. 5.35, 7, by mixing octanol-1 (3·7 ml), saturated hydrocarbon (b.p.

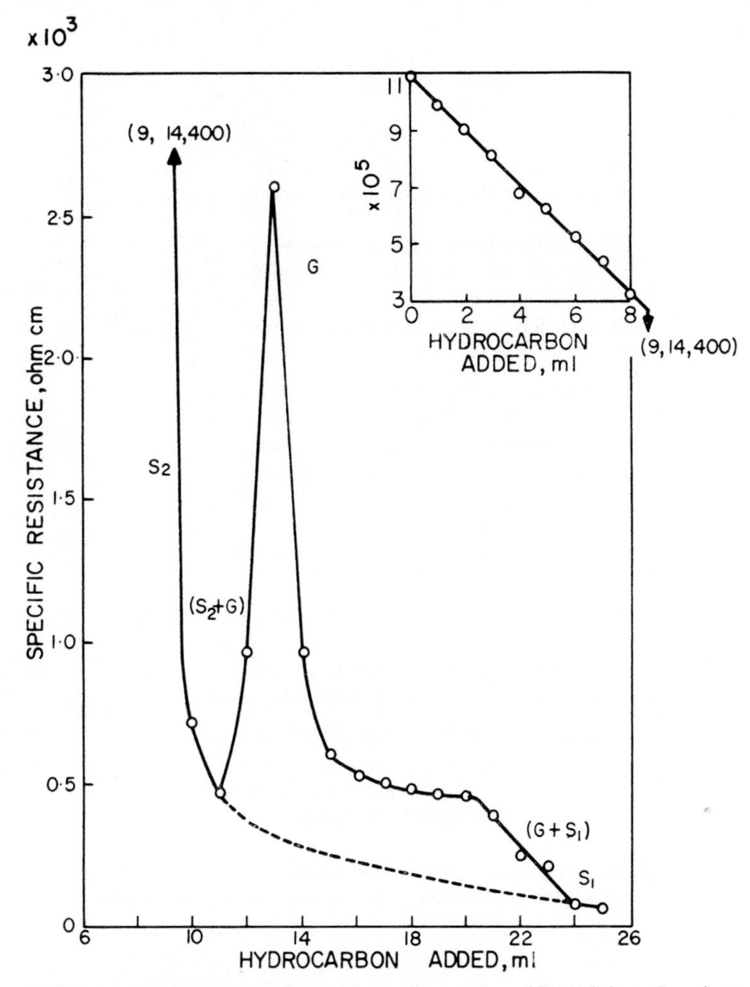

FIG. 5.38. The phase sequence, $S_2 \rightarrow (S_2 + G) \rightarrow G \rightarrow (G + S_1) \rightarrow S_1$, observed on addition of further hydrocarbon to the S_2 solution produced on mixing octanol-1 (3·7 ml), saturated hydrocarbon (b.p. 188–213°C) (10 ml) and aqueous undecane-3-sodium sulphate (10 ml, 20% w) [2b].

TABLE 5.16. *General methods for influencing R and for thus controlling the phase equilibrium in amphiphilic systems*

Note: R increases with $[A_{\overline{CO}}/A_{\overline{CW}}$, i.e. with $(A_{H\overline{CO}} + A_{L\overline{CO}})/(A_{H\overline{CW}} + A_{L\overline{CW}})]$ W_1, O_1 and C_1 respectively represent the initial more hydrophilic (water-soluble), lipophilic (oil-soluble) and amphiphilic parts of the system. In absence of water or organic liquid, W_1 and O_1 represent respectively the juxtaposed polar and hydrocarbon moities of C_1. O_1* and W_1* respectively represent excess non-solubilized organic or aqueous phases if present (cf. Fig. 5.35, 1 and 8)

Method classification	Details of method	Influence on R
I*a*	Reduction in the relative proportion of W_1	R tends to be increased
I*b*	The addition to O_1 of an oil-soluble compound of more polar (hydrophilic) character than O_1. Alternatively, this may be regarded as the addition to C_1 of an amphiphilic compound less polar than C_1	R tends to be increased
I*c*	Increase in the relative lipophilic character of C_1, e.g., by increase of molecular weight, substitution of an organic for an inorganic counter-ion or by certain changes in constitution†	R tends to be increased
I*d*	The addition to W_1 of an inorganic salt	R tends to be increased
II*a*	Reduction in the relative proportion of O_1	R tends to be decreased
II*b*	The addition to W_1 of a water-soluble organic liquid of more lipophilic character	R tends to be decreased
II*c*	Increase in the relative hydrophilic character of C_1, i.e., method I*c* in reverse, or the addition to C_1 of an amphiphile more polar than C_1, i.e., Method I*b* in reverse	R tends to be decreased
II*d*	The addition to O_1 of a less hydrophilic oil-soluble component, e.g., medicinal oil	R tends to be decreased

Probable mechanism of influence on R	Effect on the phase ratio in an initially heterogeneous system containing any two in the following succession of phases: O_1*, S_1, S_{1c}, M_1, V_1, G, V_2, M_2, S_2, W_1*	
$A_{\overline{CW}}$ diminished by mass action	Increase in proportion of the right-hand phase present	Where a single-phase system is initially involved, the progressive application of a particular method will finally tend to the precipitation from this phase of the neighbouring successional phase. With particular systems certain phases of the sequence do not appear, but the general order of succession with variation in R remains the same. Superimposed on the directional effects of methods I*d* and II*b* is the general tendency of $\bar{W}$-soluble materials, other than water, to inhibit the formation of liquid crystalline solutions
Reduction of $A_{H\overline{CW}}$ by incorporation of amphiphilic additive in $\bar{C}$ layer and/or increase of $A_{H\overline{CO}}$		
$A_{L\overline{CO}}$ increased and/or $A_{H\overline{CO}}$ diminished		
$A_{H\overline{CW}}$ diminished		
$A_{\overline{CO}}$ diminished by mass action	Increase in proportion of the left-hand phase present	
$A_{L\overline{CW}}$ increased		
$A_{H\overline{CW}}$ increased and/or $A_{L\overline{CO}}$ diminished		
Reverse of mechanism of method I*b*. Decrease of $A_{H\overline{CO}}$		

† E.g., by change in position of polar group from terminal towards medial in n-alkane sodium sulphates [2(*a*), 3].

188°–213°C) (10 ml) and aqueous undecane-3-sodium sulphate solution (10 ml, 20% w). In this system the further hydrocarbon added enters the $\bar{O}$ region and, since the octanol-1 is distributed between the $\bar{O}$ and $\bar{C}$ regions (p. 273), part of the octanol will be withdrawn from the $\bar{C}$ region into the $\bar{O}$ region. The relative concentration of the highly hydrophilic $-SO_4Na$ groups on the polar face of the $\bar{C}$ layer will therefore increase and R diminish, by the reverse of the processes outlined on p. 263, giving rise to the phase sequence observed. Note that the electrical resistance changes recorded in Figs. 5.36, 5.37 and 5.38 are all qualitatively in good accord with the structures assigned to the individual phases (Fig. 5.6). The great decrease in electrical resistance on addition of hydrocarbon indicated in Fig. 5.38, is, though theoretically expected, rather striking.

General Methods for Influencing R and for thus Controlling the Phase Equilibria in Amphiphilic Systems

As the result of many comparative experiments of the types just described and their interpretation on the basis of the relationships involving R, a tabulation and classification of general methods for influencing R, and for thus effecting particular phase sequences, was developed [2, 3]. This tabulation, extended to cover the results of later studies, is given in Table 5.16 [5]. The tabulation as originally compiled referred only to systems exhibiting the types of phase sequence illustrated by Fig. 5.35, 1 to 8. Its extension to systems exhibiting the additional mesophases in Figs. 5.5 and 5.6, although in many instances already justified by experiment, must in certain details be regarded as tentative and requiring further experimental verification (cf. below).

Although Fig. 5.6 is of general application, Table 5.16 applies more specifically to systems whose main amphiphile is ionic. In systems based on non-ionic amphiphiles derived from polyethylene glycols, Method II*b* of Table 5.16 produces phase changes indicating an increase rather than a decrease in R. This is probably because, whereas with an ionic amphiphile a water-soluble organic liquid such as ethanol almost exclusively enters the $\bar{W}$ region and increases $A_{L\bar{C}\bar{W}}$, with a polyethanoxy amphiphile, through the operation of the like-to-like principle, it enters the polyethanoxy zone of the $\bar{C}$ layer and, by itself forming hydrogen bonds with the ether linkages, competitively diminishes $A_{H\bar{C}\bar{W}}$ and thus increases R (cf. p. 273).

Little recent work has been published on the influence of inorganic salts on amphiphilic mesogenic systems. Method I*d* of Table 5.16 has thus largely been verified only with the series of systems investigated by the writer in which the G phase was the sole mesophase produced. However Larsson [56] found with certain 1-monoglyceride/water systems (p. 256) that addition of sodium chloride restricts the range of composition of the G phase while increasing that of the viscous optically isotropic phase (V_2 or V_1). Such restriction in the range of composition of the G phase on addition of inorganic salt has already been noted above (Fig. 5.37, p. 266). It is not to be accounted for on the basis of the effect of added sodium sulphate on the ratio R itself but may probably be related to an effect of inorganic salt in breaking down hydrogen-bonded structures within the $\overline{W}$ region and between the $\overline{W}$ region and the $\overline{C}$ layer, which probably play an essential part in contributing to the relative micellar immobility within the G phase, an immobility which is apparently not a feature of the structure of the V_1 or V_2 phases. Other water-soluble materials, e.g., ethylene glycol, ethanol, propanols, etc., that enter the $\overline{W}$ region likewise restrict formation of the G phase. A further effect of sodium sulphate in aqueous alkane-1-sodium sulphate/alkanol-1 systems is discussed on p. 284.

The influence of temperature on *R* has not been specified in Fig. 5.6 or

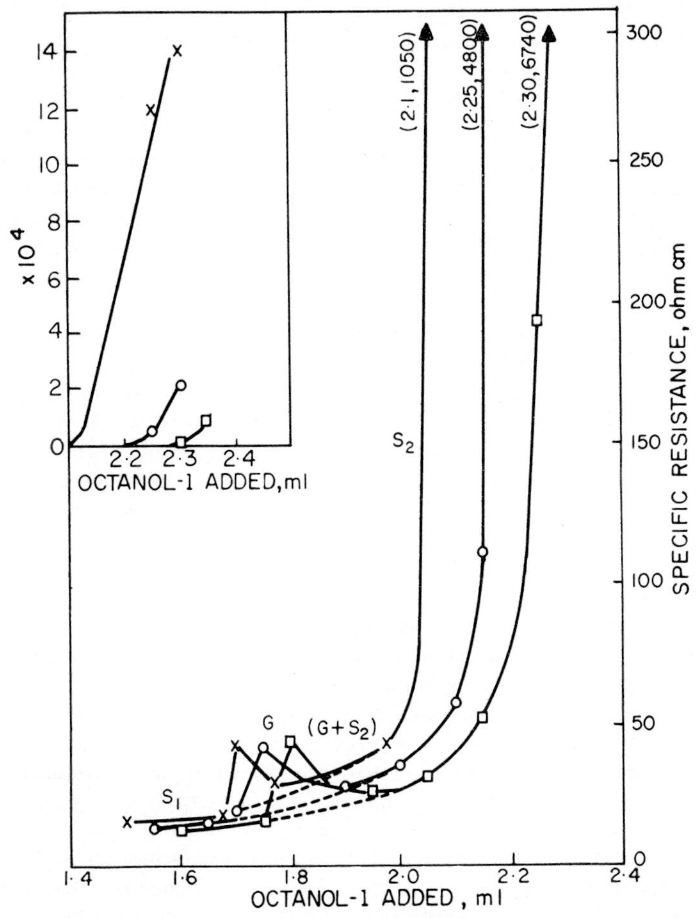

FIG. 5.39. The effect of temperature on the amount of octanol-1 required to reach corresponding stages in the phase sequence, $S_1 \rightarrow (S_1 + G) \rightarrow G \rightarrow (G + S_2) \rightarrow S_2$, in systems containing initially saturated hydrocarbon (b.p. 188–213°C) (10 ml), sodium sulphate (0·2 g) and aqueous undecane-3-sodium sulphate solution (10 ml, 20% w). Phase sequence followed by specific resistance changes: ×, at 20°C; ○, at 25°C; □, at 30°C [2*b*].

Table 5.16 since it varies from system to system. The general effect of rise of temperature in breaking down the organization of mesomorphous phases is, however, noted in Figs. 5.5 and 5.6.

It is often difficult to distinguish between this general effect of temperature and the effect of temperature exerted specifically through its influence on R. In the systems illustrated in Fig. 5.39, to reach corresponding stages in the phase sequence, $S_1 \rightarrow (S_1 + G) \rightarrow G \rightarrow (G + S_2) \rightarrow S_2$, requires progressively greater amounts of octanol-1 over the temperature range 20°C to 30°C. Here, then, rise of temperature decreases R.

In systems based on non-ionic amphiphiles, on the other hand, the effect of increase in temperature is usually to increase R. This is illustrated by the sequence, $M_1 \rightarrow (M_1 + V_1) \rightarrow V_1 \rightarrow (V_1 + G) \rightarrow G$, produced with suitably chosen compositions on rise of temperature in the systems represented in Fig. 5.32(*a*) and (*b*) (cf. also p. 260 and Figs. 5.33 and 5.34). In the non-ionic amphiphile systems it seems reasonable to attribute the increase in R with rise of temperature mainly to decrease in $A_{H\bar{C}W}$ through progressive thermal disruption of the dipole interactions between water and the non-ionic polar group of the amphiphile. A similar effect would be expected with ionic amphiphiles. However, in this case an opposing effect could arise through rise of temperature promoting ionic dissociation. This would make the $\bar{C}$ layer effectively more polar and tend to increase $A_{H\bar{C}W}$.

Extensions of the R-theory

Although Figs. 5.5 and 5.6 are probably the furthest schematic representations of the R-theory presently possible, it is expected that they will undergo additional development. For example, the details of the structures of the V_1 and V_2 phases, unspecified in Fig. 5.6, will, when elucidated need insertion as will also the structures of the other phases which arise intermediate between the G and M phases (p. 212).

It seems likely that systems will be found with which it is definitively established that the G phase exists in two or more forms. As pointed out (p. 208) the polar groups, at least in those G phases which succeed M_1 phases on increase in concentration, probably show mutual orientations of which, depending on concentration and temperature, more than one form may be possible. The lamellar D and B phases identified by Ekwall *et al.* in several ternary systems (Figs. 5.44, 5.45) may possess this polymorphic relationship which may also exist between the high-temperature high-soap content "neat soap" phase of the sodium soaps and the more dilute "soap boiler's neat phase", G (cf. p. 211 and Fig. 5.4).

The possibility of such polymorphism may also apply to the M_1 and M_2 phases. Indeed, Luzzati *et al.* have already characterized two forms of the phase, H (which is probably to be identified with the M_2 phase) among the high-temperature phases of the magnesium and cadmium soaps (p. 245).

In such fibrous phases it is also conceivable that an arrangement of the fibrous micelles other than the typical two-dimensional hexagonal array, might arise. It has been suggested that a two-dimensional square tetragonal arrangement occurs in Ekwall's C and K phases (p. 279) and Luzzati [6] earlier proposed an orthorhombic array in his "deformed middle" phase and a rectangular array in the rectangular phase R (cf. p. 212). The nature of the changes in the physical forces which could lead to such modifications in the micellar array is obscure.

If the fibrous micelles of the M_1 and M_2 phases are capable of local rotary Brownian displacements about their longitudinal axes so that all their constituent molecules are statistically indistinguishable, then in each of the S_1, M_1, G, M_2 and S_2 micellar forms of Fig. 5.6 the roles of all the individual amphiphile molecules of the same chemical type are statistically geometrically and energetically equivalent.

It would seem to the writer that this equivalence, also recently noted by Larsson [51], is probably of a fundamental character and that any proposed "fused" micellar form in which such statistical molecular equivalence is not evident should be treated with some reserve (cf. p. 252).

Distribution of the Components Between the $\bar{O}$, $\bar{W}$ and $\bar{C}$ Regions of Amphiphilic Phases

The distribution of the components between the $\bar{O}$, $\bar{W}$ and $\bar{C}$ regions of amphiphilic phases has been discussed by the writer [5, 58] and is not considered in detail here. Nevertheless, it is remarked that in the discussion of amphiphilic phases sufficient regard is rarely paid to the thermodynamic necessity that, at equilibrium, throughout a given phase and throughout any other phases in equilibrium with it, the activity of each individual chemical species must everywhere be the same (cf. Fig. 5.2(*a*)). If this were not so and the activity of the water, for example, were higher at one point than at another, then water would diffuse from the point of higher to the point of lower activity, i.e. the system would not be in equilibrium. Thus, in an aqueous lamellar G phase the activity of the water must equal the activity of the water throughout any other phases in equilibrium with the G phase, e.g., S_1, S_2, M_1, M_2 or vapour. The fact that with the G and M phases there is usually marked depression of the aqueous vapour pressure indicates that no part of the $\bar{W}$ region within such phases can consist of pure water in the random liquid state. N.m.r. and other studies indicate that in the G and M mesophases of highest concentration, all the water [46] is held ("oriented" or "bound") at the polar micellar surface where it no longer gives the n.m.r. high-resolution spectrum characteristic of water molecules in free thermal motion. At somewhat lower concentrations, in both the G and M_1 phases, the water is partly held and partly free, the free water giving a high resolution n.m.r. spectrum. Since, however, the vapour pressure is considerably depressed, this free water must contain a significant amount of amphiphile in solution. This will particularly be the case with amphiphiles of lower molecular weight and relatively high cmc, e.g., sodium caprylate.

The curve of vapour pressure against composition for aqueous S_1 solutions of this salt has recently been carefully redetermined [57]. When account is taken of the necessity that the vapour pressure of the intermicellar solution, $\bar{W}$, must be equal to that of the system as a whole, it can be inferred from the vapour pressure curve that the sodium caprylate concentration of the intermicellar solution, $\overline{W}$, rises from about 6% w at the cmc to at least three times this value at the overall concentration of 40% w. Even with the Aerosol OT/water system (c.m.c. ca. 0·2% w) it has been inferred from X-ray studies of the G phase [58] that the $\overline{W}$ regions contain a significant concentration (in the region of 5% w) of Aerosol OT. Further, such inclusion of a significant fraction of the amphiphile in a given system outside of the $\overline{C}$ layer and within the $\overline{W}$-region has a pronounced effect on the calculated micellar dimensions. In almost all X-ray studies of aqueous amphiphilic phases, however, the possibility of the presence of such significant concentrations of amphiphile within the $\overline{W}$ region, as well as the possibility that these concentrations may vary with temperature [58], have been tacitly disregarded (cf. p. 240). Nevertheless, in the complementary case of the $\overline{O}$ region, Fontell *et al.* [59] have recognized that to derive reasonable micellar dimensions for the $F(= M_2)$ phase in the sodium caprylate/decanol-1/water system, it is necessary to assume that a considerable proportion of the decanol-1 is located outside the $\overline{C}$ layer and within the $\overline{O}$ region.

The distribution of a particular species within an amphiphilic phase must depend on the variation in the activity coefficient* with respect to that species from location to location within the phase. This may be considerable even inside an individual region, for example, inside either the $\overline{O}$ or $\overline{W}$ regions at points close to or remote from the interface. Thus the writer found by direct chemical analysis that in certain conjugate G and S phases containing sodium chloride, the sodium chloride/water ratio is considerably higher (up to four times) in the S phase than in the G phase (reference [3], p. 77). The average activity coefficient for sodium chloride in the $\overline{W}$ region within the G phase must therefore be considerably greater than that in the $\overline{W}$ region within the equilibrium S phase. This relative intolerance of the $\overline{W}$ region of the G phase for inorganic ions is possibly related to the effect of such ions in restricting the stability of the G phase (p. 267).

TERNARY AQUEOUS SYSTEMS: PHASE DIAGRAMS

On account of the large number of components present, the miscellaneous multicomponent systems discussed above were studied only over limited ranges of composition and temperature chosen to provide a broad preliminary phenomenological characterization of the phases and processes involved. A number of studies on ternary aqueous systems cover either the whole range of compositions or a broad selected range, e.g., the complete range excepting that where solid crystalline phases may be expected [61]. The results have frequently been expressed on triangular isothermal phase diagrams, the temperature in the majority of cases being close to room temperature.

Many of the systems have contained two amphiphiles, one relatively hydrophilic (A) and the other relatively lipophilic (B). Examples are

* Activity = activity coefficient × concentration.

potassium n-butyrate with butanol-1, hexanol-1 or octanol-1; potassium n-octanoate (caprylate) with butanol-1 or octanol-1; sodium n-butyrate with octanol-1 (at 50°C and 75°C); potassium isobutyrate with octanol-1 [61]; sodium caprylate with pentanol-1, octanol-1, nonanol-1, decanol-1, caprylic acid, methyl caprylate, caprylnitrile or caprylaldehyde; sodium octylsulphate with decanol-1; sodium octylsulphonate with decanol-1; octylammonium chloride with decanol-1; cetyltrimethylammonium bromide with hexanol-1; decaethylene glycol monolauryl ether with caprylic acid or oleic acid; sodium cholate with decanol-1; sodium desoxycholate with decanol-1; 1-monocaprylin with decanol-1 or caprylic acid [62]. In addition, ternary aqueous systems have been studied containing sodium caprylate with *p*-xylene, octane, 1-chloro-octane or tetrachloromethane; Aerosol OT with *p*-xylene [62]. Further, the ternary aqueous systems containing sodium caprylate with ethylene glycol, glycerol or tetraethylene glycol have been examined [63].

With the exceptions of potassium isobutyrate, sterol derivatives and Aerosol OT, all the amphiphiles, both relatively hydrophilic and relatively lipophilic, have been "straight-chain" compounds. This contrasts with the work with the miscellaneous multi-component systems considered in the previous section in which many branched-chain and cyclic compounds were also examined.

Most prominent among studies of ternary systems are those of Ekwall *et al.* who recently reviewed their work and conclusions in detail [62]. Here only a limited number of illustrative systems will be considered. It should be noted that a mixture of amphiphiles A and B within a single amphiphilic phase behaves largely as an individual amphiphile. Where two phases exist conjugately, however, the two amphiphiles are distributed between them [55, 61, 62]. If in the schematic triangular diagram of Fig. 5.40, X, Y and Z are compositions represented on the sides opposite to the A, B and water vertices respectively, any sequence of "fused" phases encountered along a line Y–B may be considered as exemplifying Table 5.16, Method I*b*; a sequence along water–Z exemplifies Method I*a*, while a sequence along X–A exemplifies Method II*c*. Examination of the phase sequences actually encountered along such lines in systems such as those considered below shows that they all conform with the generalizations expressed in Figs. 5.5 and 5.6 and Table 5.16.

The Ternary Aqueous Systems at 25°C Containing: (a) Potassium n-Butyrate and Butanol-1; (b) Potassium n-Butyrate and Hexanol-1; (c) Potassium n-Butyrate and Octanol-1; (d) Potassium Isobutyrate and Octanol-1

Systems (*a*), (*b*) and (*c*) (Fig. 5.41) [61] illustrate the effect of increase in the molecular weight of amphiphile B in the presence of the very short chain amphiphile, potassium n-butyrate, A. In these diagrams "gel" represents the lamellar phase, G (Ekwall's D). Systems (*c*) and (*d*) illustrate the profound effect of replacement of the straight chain n-butyrate anion by the branched chain isobutyrate anion on the formation of liquid crystalline

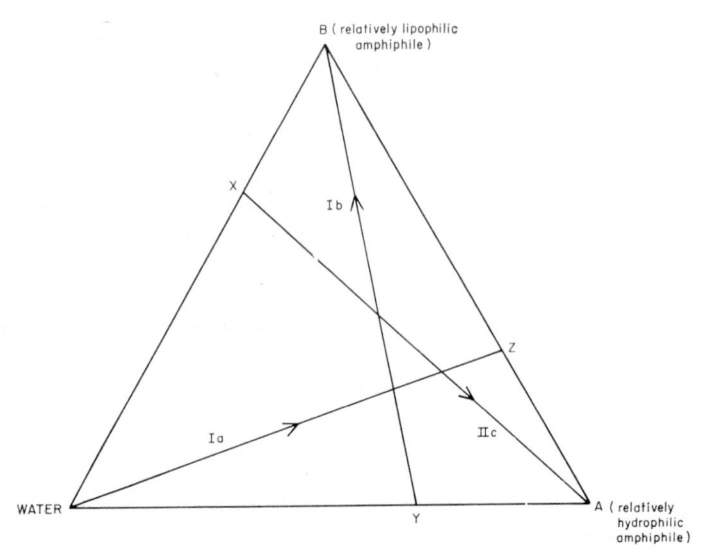

FIG. 5.40. Diagram illustrating the composition changes corresponding to Methods Ia, Ib and IIc of Table 5.16 as applied to a ternary aqueous system containing two amphiphiles.

solutions. The aqueous sodium cholate/decanol-1 and sodium desoxycholate/decanol-1 systems also give phase diagrams qualitatively similar to Fig. 5.41(*d*) [62].

The Ternary Aqueous Systems Containing: (a) Potassium Caprylate and Butanol-1 at 25°C [61]; (b) Sodium Caprylate and Butanol-1 at 20°C [62]

Systems (*a*) and (*b*) (Fig. 5.42) show only very small changes on replacement of potassium with sodium in contrast to the profound change on replacement of n-butyrate with isobutyrate (Fig. 5.41(*c*) (*d*)). Comparison of Fig. 5.41(*a*), with Figs. 5.42(*a*) and 5.45, shows how replacement of potassium butyrate with potassium caprylate results in the appearance of two additional liquid crystalline phases, the middle phase M_1 (Ekwall's E) and the cubic phase V_1 (Ekwall's J). The V_1 phase (Fig. 5.45) was not recorded in Fig. 5.42(*a*) because a sufficiently high concentration of potassium caprylate was not investigated.

The Ternary Aqueous System Containing Sodium n-butyrate and Octanol-1 at 50°C and 75°C

Since replacement of potassium with sodium exerts little effect on the phase equilibria in the systems under consideration, comparison of Fig. 5.41(*c*) with Fig. 5.43(*a*) and (*b*) illustrates the marked effect of rise of temperature over the range 25–75°C on liquid crystal (G) formation. It is seen that at 75°C the behaviour of the potassium n-butyrate/octanol-1 system becomes similar to that of the potassium isobutyrate/octanol-1 system at room temperature.

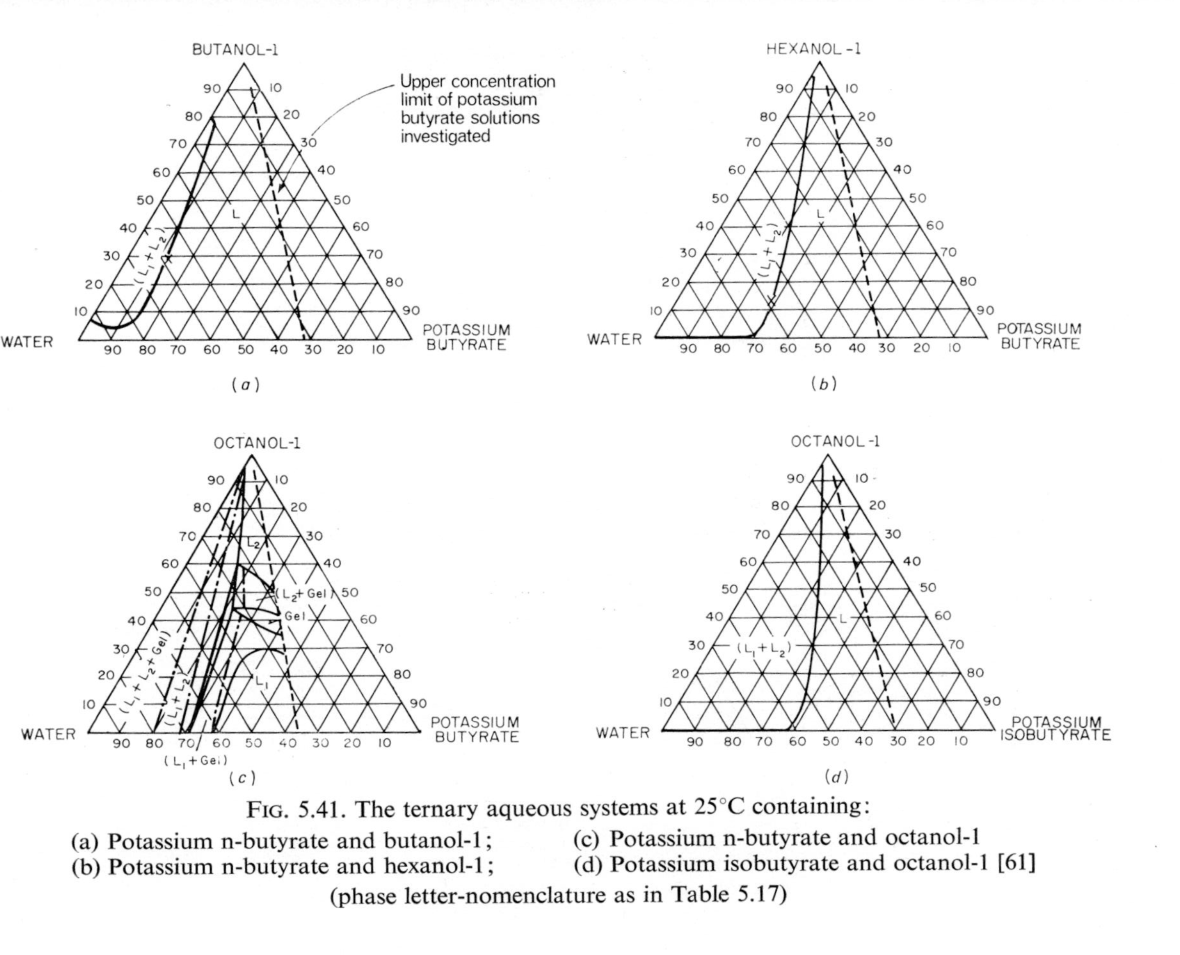

FIG. 5.41. The ternary aqueous systems at 25°C containing:

(a) Potassium n-butyrate and butanol-1; (c) Potassium n-butyrate and octanol-1
(b) Potassium n-butyrate and hexanol-1; (d) Potassium isobutyrate and octanol-1 [61]

(phase letter-nomenclature as in Table 5.17)

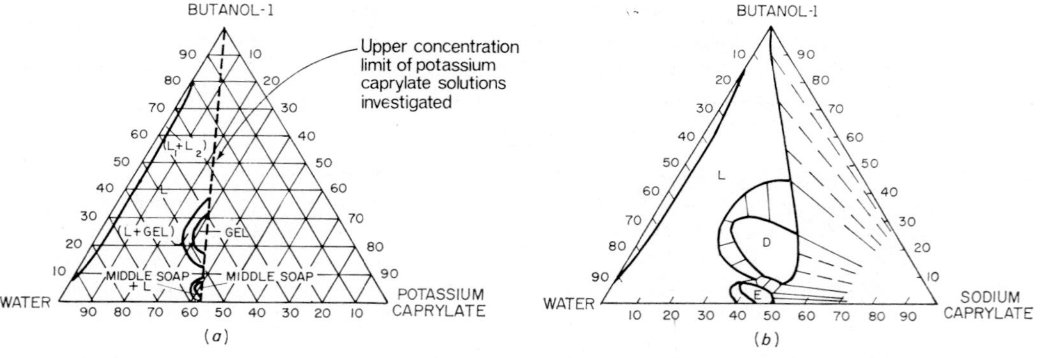

FIG. 5.42. The ternary aqueous systems containing: (a) potassium caprylate and butanol-1 at 20°C [61]; (b) sodium caprylate and butanol-1 at 25°C [62].

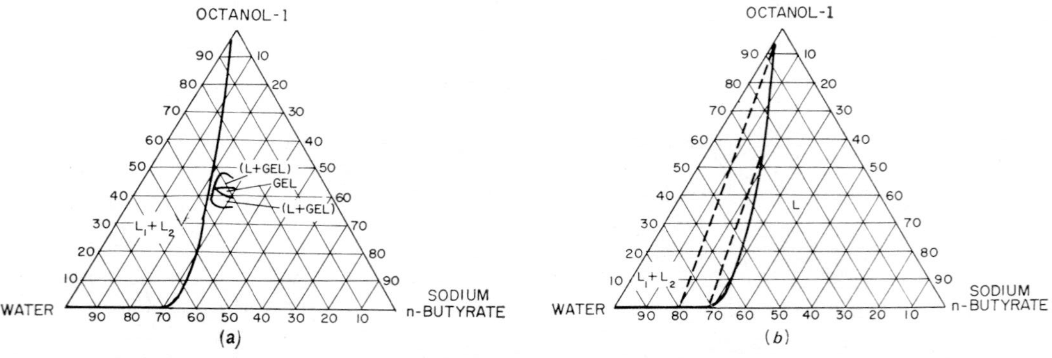

FIG. 5.43. The ternary aqueous system containing sodium n-butyrate and octanol-1 at (a) 50°C and (b) 75°C [61].

(In Figs. 5.43–5.44 phase letter-nomenclature as in Table 5.17, p. 280).

The Ternary Aqueous System Containing Sodium Caprylate and Decanol-1 at 20°C

This system (Fig. 5.44) has been thoroughly investigated by Ekwall *et al.* In addition to establishing the phase diagram, principally by centrifugal separation methods, these authors also examined the individual phases using X-ray diffraction, polarization microscopy, n.m.r. spectroscopy, optical and other methods. To facilitate the correlation of the results expressed in Fig. 5.44 and related diagrams with those of other workers,

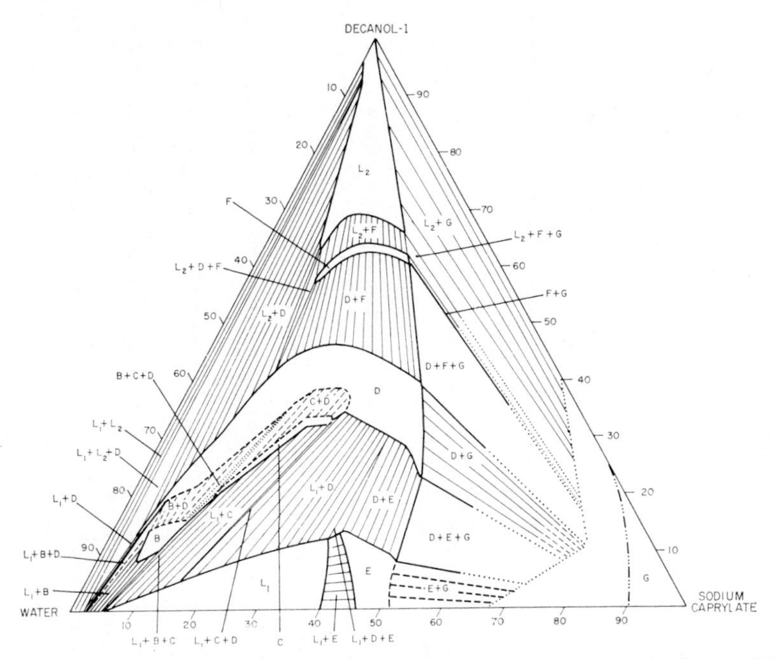

FIG. 5.44. The ternary aqueous system containing water and decanol-1 at 20°C. Phase letter-nomenclature as indicated in Table 5.17, p. 280, except that, in Fig. 5.44, G indicates region with solid substances [64].

and with the discussion in Chapter 2.3, a list of equivalent phase nomenclatures is given in Table 5.17.

Examples of phases analogous to L_1, E, D, F and L_2 of Fig. 5.44 have already been considered; B, and C, represent additional phases.

Phase B is closely related to the D phase (G) and likewise has been identified as of the lamellar type [64]. The two phases probably differ in the relative orientations of the polar ends.

Phase C, or the "white phase", has been definitively characterized as an individual phase. Its structure [65] however remains problematic. A phase (K) which apparently has an "inverse" structure complementary to that of phase C, has been found in the potassium caprate/octanol-1/water system [66]; cf. also Chap. 4 of reference [70].

TABLE 5.17. *Equivalent nomenclatures used by different authors for the individual fused amphiphilic phases*

Name of phase	Proposed micellar constitution	Letter nomenclature
Fluid isotropic;* fluid amorphous; micellar	Equilibrium of successive micellar forms without long-range order; hydrophilic forms predominate (p. 217)	$L(L_1)$, $S(S_1)$
Viscous isotropic;† cubic; normal cubic; hydrophilic cubic	Hydrophilic spherical micelles in body-centred cubic lattice (p. 212)	I, J_1, S_{1c}
Hexagonal; normal hexagonal; hydrophilic hexagonal; middle	Parallel indefinite hydrophilic fibrous micelles in two-dimensional hexagonal array (p. 207)	E, H, H_I, M_1
White phase; normal tetragonal	Parallel indefinite hydrophilic fibrous micelles in two-dimensional square tetragonal array (p. 279)	C
'Intermediate phases'		
(1) deformed hexagonal	Birefringent mesophases; structures not unambiguously determined. Possibly 'hybrid' (p. 212)	
(2) rectangular		R
(3) complex hexagonal		H_c
(4) cubic;† viscous isotropic; normal cubic	Micellar units in cubic lattice, structures not definitively determined (p. 212)	$I^1, J, I, I_{f1}, I_{b1}, Q_I, V_1$
Neat; superneat; soap boilers' neat; lisse; lamellar; lamellar labile; gel	Indefinite parallel lamellae (p. 201)	D, B, L, L_α, L.L., G
Reversed tetragonal	Parallel indefinite lipophilic fibrous micelles in two-dimensional tetragonal array (p. 279)	K
Viscous isotropic;† cubic; reversed cubic; lipophilic cubic	Micellar units in cubic lattice, possibly several types; structure not definitively determined (p. 211)	$I^1_2, J, I, I_2, I_{f2}, Q_{II}, V_2$
Hexagonal; reversed hexagonal; lipophilic hexagonal; inverse middle	Parallel indefinite lipophilic fibrous micelles in two-dimensional hexagonal array (p. 210)	$F, H_{II}(H_1, H_2), M_2$
Fluid isotropic;* fluid amorphous; micellar	Equilibrium of successive micellar forms without long-range order; lipophilic forms predominate (p. 217)	$L(L_2), S(S_2)$

* These phases, being amorphous, are isotropic with respect to all their properties.

† These phases are isotropic with respect to their optical properties but not necessarily with respect to other properties, e.g. the tendency to develop crystal faces.

The Ternary Aqueous System Containing Potassium Caprylate and Decanol-1 at 20°C

As in Fig. 5.42(*a*) and (*b*), replacement of the sodium ion (Fig. 5.44) with the potassium ion (Fig. 5.45) exerts only a small influence over most of the phase diagram. The solubility of the solid potassium salt is greater than that of the sodium salt and, with increase in concentration, the E phase, along the water–potassium caprylate axis is succeeded by a cubic phase,

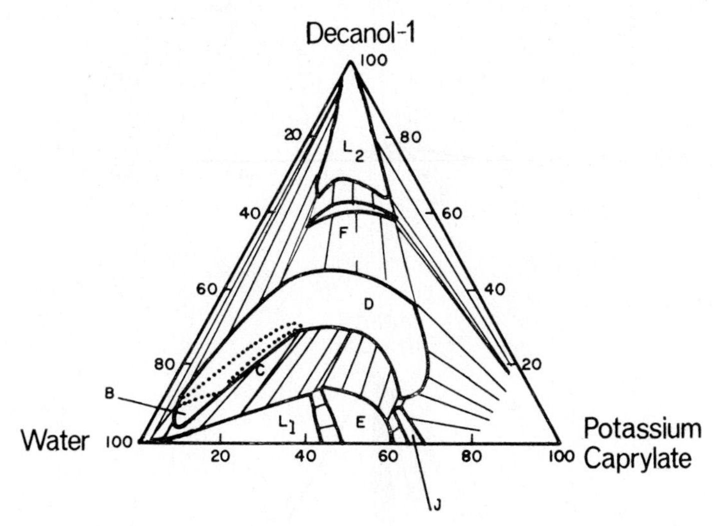

FIG. 5.45. The ternary aqueous system containing potassium caprylate and decanol-1 at 20°C [62].

(In Fig. 5.45 phase letter-nomenclature as in Table 5.17, p. 280).

J (= V_1). The formation of this phase by the higher potassium soaps at more elevated temperatures was noted by Luzzati. Although the V_1 phase does not arise with the sodium soaps of the fatty acids it is formed by sodium 2-ethylhexyl sulphate and by sodium undecane-3-sulphate etc. (p. 253, Table 5.15).

Ternary Aqueous Systems Containing Sodium Caprylate with Non-polar Organic Compounds

The systems represented in Fig. 5.46 provide further examples of the viscous optically isotropic cubic phase, S_{1c}(I). Incorporation of a non-polar additive into the spherical micelles of sodium caprylate at concentrations of the L_1 phase not too closely approaching the point of separation of the E(M_1) phase, would be expected to occur by entry of the additive into the non-polar micellar interior without markedly influencing the ratio *R* (p. 263). The micelles would be expected to swell to a certain limit but to

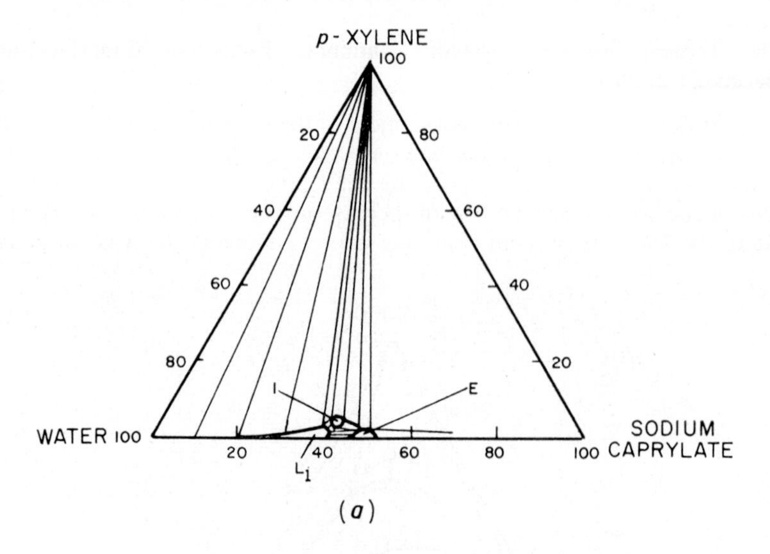

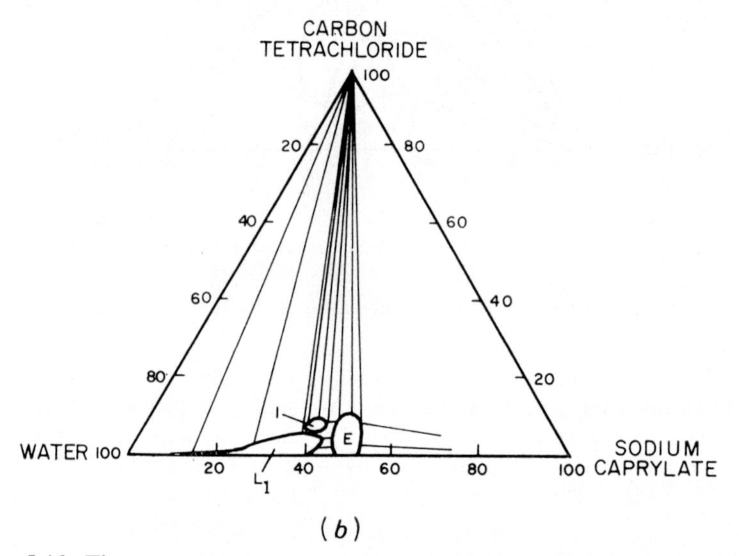

FIG. 5.46. The ternary aqueous systems containing sodium caprylate with (a) p-xylene and (b) carbon tetrachloride, at 20°C [62].

retain their spherical form. Under favourable conditions of composition and temperature this could lead to mutual ordering of the micelles to give the S_{1c} cubic phase. If, to a phase so constituted, additional non-polar organic liquid were added, the ratio R would remain unaffected, and at some limit, one would expect separation of excess non-polar liquid containing little amphiphile or water, just as in the case of the amorphous S_1 phase (p. 263). This behaviour is represented for both the L_1 (S_1) and I

(S_{1c}) phases in Fig. 5.46(*a*) and (*b*). On the other hand, if to a phase constituted as S_{1c} one added additional sodium caprylate one would produce an increase in *R* (Table 5.16, Method I*a*) which could lead to the phase sequence, $S_{1c} \longrightarrow (S_{1c} + E) \longrightarrow E$, shown in Fig. 5.46(*a*) and (*b*).

Ternary Aqueous Systems Containing an Amphiphile and Water-soluble Organic Liquid

Ternary aqueous systems containing sodium caprylate with ethylene glycol, glycerol or tetraethylene glycol have been examined by Ekwall, Mandell and Fontell [63]. Figure 5.47 illustrates the type of diagram obtained in all three systems. The general effect of water-soluble substances in reducing the range of liquid crystalline compositions (p. 267) is evident.

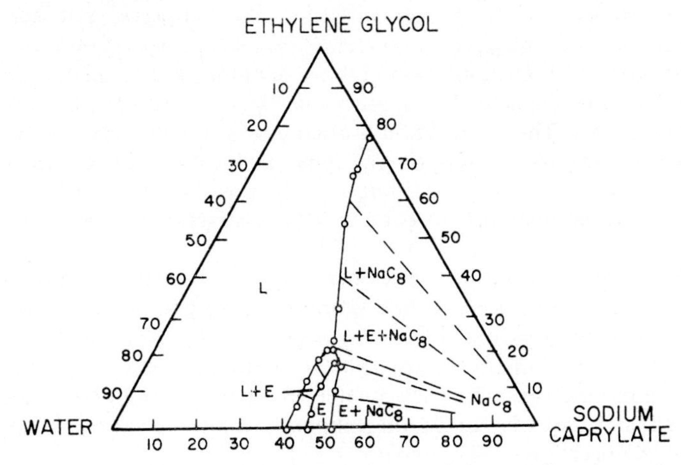

FIG. 5.47. The ternary aqueous system containing sodium caprylate and ethylene glycol at 20°C [63].

Further it will be seen that in proceeding from the 50% w sodium caprylate solution (E phase) towards the ethylene glycol vertex one encounters the phase sequence, $M_1 \longrightarrow (M_1 + S_1) \longrightarrow S_1$. This provides an example of Method II*b* of Table 5.16 for decrease in *R*. Further, the decrease in *R* produced is attributed in Table 5.16 to an increase in $A_{L\overline{CW}}$. In support of this it was found that the incorporation of ethylene glycol produced an apparent increase in the interfacial area per polar group. A similar increase was noted by Gilg, François and Skoulios on additions of water-soluble organic liquids to aqueous solutions of potassium laurate [67] and has been discussed by the present writer (reference [5] p. 34). When the ionic amphiphile is replaced by a polyethanoxy amphiphile, the interfacial area per polar group (in this case defined as the effective cross-sectional area per amphiphile molecule at the $R \cdots O\ (CH_2{\cdot}CH_2O)_nH$ link) was found to be little dependent on the water/organic liquid ratio with both G and M_1 phases [67]. This difference is probably due to the different distribution of the organic liquid between the $\bar{C}$ and $\bar{W}$ regions in the case of the ionic and non-ionic amphiphile systems respectively

The Formation of Systems Having the Properties of Nematic Phases on the Addition of Sodium Sulphate to Ternary Aqueous Systems Containing a Sodium n-Alkyl Sulphate with the Corresponding n-Alkanol

Lawson and Flautt [49(*a*)] found that a mixture of C_8 or C_{10} alkyl sulphates, the corresponding alcohol, sodium sulphate and water in the approximate proportions 40, 5, 5, 50 (by weight) respectively behaves over the temperature range 10°C–75°C as a nematic phase in that (*a*) it shows a typical nematic thread-like texture under the polarizing microscope (*b*) it can be oriented by a magnetic field and (*c*) it can confer this orientation on dissolved molecules. It further possesses the property, not generally shown by nematic mesophases, of retaining its induced orientation when spun at normal high-resolution n.m.r. spinning speeds about an axis perpendicular to the field. This implies that the oriented units lie perpendicular to the field. The preparation is viscous and orients only slowly in the field but retains its orientation on removal of the field. Its molecular structure was not determined but it may possibly be constitutionally analogous to the nematic "dispersions" of the 1-monoglycerides (p. 261) studied by Larsson.

Certain aqueous solution phases that appear to be true nematic phases are known. Thus 9-bromophenanthrene 3-sulphonic acid is said to give with progressively increasing amounts of water, first a smectic then a nematic and finally an amorphous solution. Whether the behaviour of this compound is more closely related to the non-amphiphilic mesogens discussed in Chap. 4.1 or to the amphiphilic mesogens of the present chapter, is uncertain.

The Formation of Mesophases in Amphiphilic Preparations of Biological Origin

X-Ray studies of mesophases found in amphiphilic preparations of biological origin have been carried out, particularly by Luzzati *et al.* Luzzati has recently reviewed this field [11]. The preparations studied included monoglycerides, phospholipids, sphingolipids, brain lipids and mitochondria lipids. Semi-crystalline mesophases of the "gel" type and fused mesophases, including the M_1, G, V_2 and M_2 phases, have been encountered dependent on the nature, water content and temperature of the systems.

The Formation of Mesophases by Amphiphilic Polymers

Mesophases arise in solutions of certain polymers of amphiphilic character, for example, with polysoaps, e.g., the acid, under differing conditions of neutralization [68], prepared by hydrolysis of an alternate copolymer (molecular weight about 10^5), n of 260, of maleic anhydride and hexadecyl vinyl ether, and with block-copolymers in which successive blocks are of contrasting solvent character, e.g., polystyrene-polyethylene oxide, polypropylene-oxide-polyethylene oxide and polystyrene-polyvinyl-2-pyridine [69]. When to such block-copolymers, X–Y, a solvent A is added which is preferentially miscible with block X, mesophases arise, under favourable conditions of temperature and composition, in which there is a regular alternation of Y regions and regions of X containing solvent A.

REFERENCES

[1] Stosick, A. J. *J. chem. Phys.*, **18,** 1035 (1950).

[2] Winsor, P. A. (*a*) *Trans. Faraday Soc.* **44,** 376 (1948); (*b*) *Trans. Faraday Soc.* **46,** 762 (1950).

[3] Winsor, P. A. *Solvent Properties of Amphiphilic Compounds*, Butterworths, London (1954).

[4] Winsor, P. A. *Chem. Ind.* (*London*) p. 632 (1960).

[5] Winsor, P. A. *Chem. Rev.* **68,** 1 (1968).

[6] (*a*) Luzzati, V., Mustacchi, C., Skoulios, A. and Husson, F. *Acta Cryst.* **13,** 660 (1960);
(*b*) Luzzati, V. and Husson, F. *J. Cell Biology* **12,** 207 (1962);
(*c*) François, J., Gilg, B., Spegt, P. A. and Skoulios, A. E., *J. Colloid Interface Sci.* **21,** 293 (1966).

[7] Rogers, J. and Winsor, P. A. *J. Colloid Interface Sci.* **30,** 247 (1969).

[8] E.g., Parsegian, V. A. (*a*) *Trans. Faraday Soc.* **62,** 848 (1962); (*b*) *Science*, **156,** 939 (1967).

[9] McBain, J. W. and Lee, W. W. *Oil and Soap* **20,** 17 (1943).

[10] E.g., Reiss-Husson, F. and Luzzati, V. *J. Colloid Interface Sci.* **21,** 534 (1966); also Reference [5], p. 15.

[11] Cf. Luzzati, V. *Biological Membranes*, Academic Press, London (1968), Chap. 3, p. 71.

[12] (*a*) Fontell, K., Mandell, L. and Ekwall, P. *Acta Chem. Scand.* **22,** 3209 (1968);
(*b*) Balmbra, R. R., Clunie, J. S. and Goodman, J. F. *Nature* **222,** 1159 (1969);
(*c*) Tardieu, A. and Luzzati, V. *Biochim. biophys. Acta* **219,** 11 (1970).

[13] Rogers, J. and Winsor, P. A., *Vth International Congress on Surface Active Substances, Barcelona*, 1968, Vol. 2, p. 933, Ediciones Unidas, Barcelona (1969).

[14] Rogers, J. and Winsor, P. A. *J. Colloid Interface Sci.* **30,** 500 (1969).

[15] Friedel, G. *Annls Phys.* **18,** 273 (1922).

[16] Friedel, G. and Friedel, E. *Z. Kristallogr.* **79,** 1 (1931).

[17] Vold, R. D. and Vold, M. J. *J. Am. chem. Soc.* **61,** 808 (1939).

[18] Balmbra, R. R., Clunie, J. S. and Goodman, J. F. *Proc. R. Soc.* (*London*) **A285,** 534 (1965).

[19] Vold, M. J., Macomber, M. and Vold, R. D. *J. Am. chem. Soc.* **63,** 168 (1941).

[20] GRANT, R. F. and DUNNELL, B. A. *Can. J. Chem.* **38,** 1951, 2395 (1960) (and references).
[21] LAWSON, K. D. and FLAUTT, T. J. *Molec. Crystals* **1,** 241 (1966) (and references).
[22] SKOULIOS, A. E. and LUZZATI, V. *Acta Cryst.* **14,** 278 (1961).
[23] LAWSON, K. D. and FLAUTT, T. J. *J. phys. Chem.* **69,** 4256 (1965).
[24] VOLD, R. D. and HELDMAN, M. J. *J. phys. Chem.* **52,** 148 (1948).
[25] GALLOT, B. and SKOULIOS, A. E. *Kolloid-Z. u. Z. Polymere* **213,** 143 (1966) (discusses results for caesium soaps and gives a recapitulation of earlier work with references).
[26] GALLOT, B. and SKOULIOS, A. E. *Kolloid-Z. u. Z. Polymere* **210,** 143 (1966).
[27] GALLOT, B. and SKOULIOS, A. E. *Molec. Crystals* **1,** 263 (1966).
[28] GALLOT, B. and SKOULIOS, A. E. *Kolloid-Z. u. Z. Polymere* **222,** 51 (1968).
[29] VINCENT, J. M. and SKOULIOS, A. E. *Acta Cryst.* **20,** 432, 441, 447 (1966).
[30] VINCENT, J. M. and SKOULIOS, A. E. *Acta Cryst.* **20,** 441 (1966).
[31] A recent review of the structures of the fused mesophases in soap/water systems has been given by Skoulios, A. E., *Advances in Colloid and Interface Science* **1,** 79 (1967).
[32] GALLOT, B. and SKOULIOS, A. E. *Kolloid-Z. u. Z. Polymere* **210,** 143 (1966).
[33] HATTIANGDI, G. S., VOLD, M. J. and VOLD, R. D. *Ind. Eng. Chem.* **41,** 2320 (1949).
[34] VOLD, R. D. and HATTIANGDI, G. S. *Ind. Eng. Chem.* **41,** 2311 (1949).
[35] SPEGT, P. and SKOULIOS, A. E. *Acta Cryst.* **21,** 892 (1966) (discusses the strontium soaps and gives citations and summary of work on the soaps of the other metals of Group II).
[36] LUZZATI, V. and SPEGT, P. A. *Nature* **215,** 701 (1967).
[37] LUZZATI, V., TARDIEU, A. and GULIK-KRZYWICKI, T. *Nature* **217,** 1028 (1968).
[38] LUZZATI, V., GULIK-KRZYWICKI, T. and TARDIEU, A. *Nature* **218,** 1031 (1968).
[39] SPEGT, P. A. and SKOULIOS, A. E. *J. Chimie Physique*, p. 418 (1965).
[40] SPEGT, P. A. and SKOULIOS, A. *C.r. Lebd. Séanc Acad. Sci., Paris* **254,** 4316 (1962); SPEGT, P. A. and SKOULIOS, A. E. *Acta Cryst.* **16,** 301 (1963).
[41] SPEGT, P. A. and SKOULIOS, A. E. *Acta Cryst.* **17,** 198 (1964).
[42] SPEGT, P. A. and SKOULIOS, A. E. *J. Chimie Physique*, p. 377 (1965).
[43] LUZZATI, V. and REISS-HUSSON, F. *Nature* **210,** 1351 (1966).
[44] CLUNIE, J. S., CORKHILL, J. M. and GOODMAN, J. F. *Proc. R. Soc., (London)* **A285,** 529 (1965).
[45] BALMBRA, R. R., CLUNIE, J. S. and Goodman, J. F. *Molecular Crystals* **3,** 218 (1967).
[46] GILCHRIST, C. A., ROGERS, J., STEEL, G., VAAL, E. G., and WINSOR, P. A. *J. Colloid Interface Sci.* **25,** 409 (1967).

[47] LUTTON, E. S. *J. Amer. Oil Chemists Soc.* **43,** 28 (1966).
[48] LAWSON, K. D., MABIS, A. J. and FLAUTT, T. J. *J. phys. Chem.* **72,** 2958 (1968).
[49] (*a*) LAWSON, K. D. and FLAUTT, T. J. *J. phys. Chem.* **72,** 2066 (1969); cf. also (*b*) HANSEN, J. R. and LAWSON, K. D. *Nature* **225,** 542 (1970).
[50] LUTTON, E. S. *J. Am. Oil Chemists Soc.* **42,** 1068 (1969).
[51] LARSSON, K. *Z. phys. Chem.*, Neue Folge **56,** 173 (1967).
[52] LARSSON, K. *Ark. Kemi* **29,** 35 (1964).
[53] McBAIN, M. E. L. and HUTCHINSON, E. *Solubilization and Related Phenomena*, Academic Press, New York (1955).
[54] KLEVENS, H. B. *Chem. Rev.* **47,** 1 (1950).
[55] LUMB, E. C. and WINSOR, P. A. *Ind. Eng. Chem.* (*Indust.*) **45,** 1086 (1953), and reference [3], Chap. 5.
[56] LARSSON, K. *Chemistry and Physics of Lipids* **2,** 129 (1968).
[57] EKWALL, P., LEMSTROM, K., EIKREM, H. and HOLMBERG, P. *Acta Chem. Scand.* **21,** 1401 (1967).
[58] PARK, D., ROGERS, J., TOFT, R. W. and WINSOR, P. A. *J. Colloid Interface Sci.* **32,** 81 (1970).
[59] FONTELL, K., MANDELL, L., LEHTINEN, H. and EKWALL, P. *Acta Polytechnica Scandinavica* **No. 74, III,** p. 19 (1968).
[60] SKOULIOS, A. E. *Acta Cryst.* **14,** 419 (1961).
[61] LUMB, E. C. *Trans. Faraday Soc.* **47,** 1049 (1951), and unpublished results (cf. reference [3], pp. 90–92, 107).
[62] Most of these studies have been carried out by Ekwall and collaborators. A comprehensive review with detailed references is given by EKWALL, P., MANDELL, L. and FONTELL, K. in *Molec. Crystals Liq. Crystals* **8,** 157 (1969).
[63] EKWALL, P., MANDELL, L. and FONTELL, K. *J. Colloid Interface Sci.* **28,** 219 (1968).
[64] MANDELL, L. and EKWALL, P. *Acta Polytechnica Scandinavica* **No. 74, I,** 92 (1968).
[65] FONTELL, K., MANDELL, L., LEHTINEN, H. and EKWALL, P. *Acta Polytechnica Scandinavica* **No. 74, III,** p. 32 (1968).
[66] EKWALL, P., MANDELL, L. and FONTELL, K. *Acta Chem. Scand.* **22,** 697 (1968).
[67] GILG, B., FRANÇOIS, J. and SKOULIOS, A. E. *Kolloid-Z.u. Z. Polymere* **205,** 139 (1966).
[68] SCHMITT, A., VAROQUI, R. and SKOULIOS, A. E. *C.r. Lebd. Séanc Acad. Sci., Paris* **268,** 1469 (1969).
[69] GROSIUS, P., GALLOT, Y. and SKOULIOS, A. *Die Makromolekulaire Chemie* **127,** 94 (1969).
[70] *Liquid Crystals and Plastic Crystals*, Vol. 2 (edited by G. W. Gray and P. A. Winsor), Ellis Horwood Ltd., Chichester (1974).

6

Significance of Liquid Crystals in Biology

6.1 LIQUID CRYSTALS AND CELL MEMBRANES

D. Chapman

Introduction

An important topic of modern molecular biology is the structure and function of cell membranes. This topic is now the subject of much discussion [1]. The components of cell membranes include lipids (e.g., phospholipids), proteins, metal ions, sometimes cholesterol and, of course, water. One question which is presently receiving a great deal of attention is how these components are organized and mutually interact with each other. A model often suggested is that of a lipid bilayer sandwiched between protein and interacting electrostatically with protein, i.e., the Danielli–Davson model. Other models placing greater emphasis on the rôle of the protein and on hydrophobic interactions have also been put forward.

A considerable amount of analysis of the lipid components has now been carried out and it has been shown that various classes of lipids are present, e.g., lecithins, phosphatidylethanolamines and phosphatidylserines. Typical structures of some phospholipids are shown in Fig. 6.1.1.

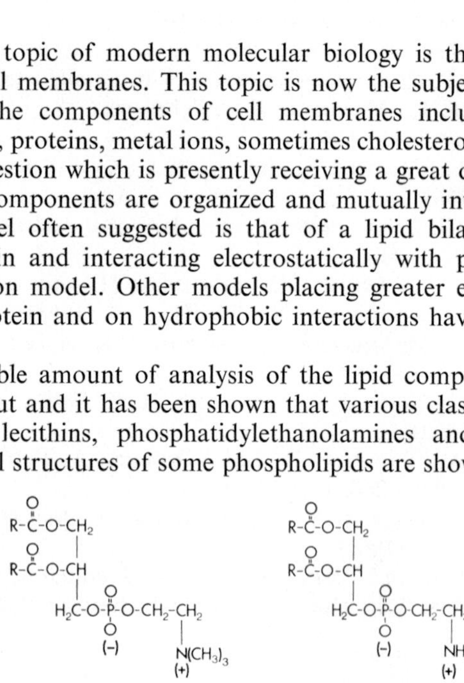

Phosphatidyl choline

Phosphatidyl ethanolamine

FIG. 6.1.1. Molecular structures of some phospholipids.

Associated with each lipid class are a variety of fatty acids of differing chain length and unsaturation. At present much work is being carried out to analyse the protein components of membranes. This will enable studies to be made of the mutual effect of the lipid and protein components.

Important to the subject of this book is the fact that the lipids which occur in membranes have liquid crystalline properties. They appear to exhibit long range order with a certain amount of short range disorder. Let us now examine these properties and their relevance to membrane structure.

BEHAVIOUR OF INDIVIDUAL LIPIDS IN THE SOLID AND MESOMORPHIC STATES

Solid State Behaviour

No complete X-ray structure of a phospholipid fully esterified and containing fatty acid chains has been published. Preliminary X-ray studies have, however, been reported for a number of related derivatives, e.g., the crystal structure of L-1-glycerylphosphorylcholine cadmium chloride trihydrate has been obtained [2]. This has shown that this molecule exists with the positive charge on the quaternary nitrogen neutralized by the negative charge on the phosphoric acid residue. The choline residue exists in the *gauche* conformation rather than the more extended form. The crystal structure of L-1-glycerylphosphorylcholine has also been reported [3].

It is too early to say just how relevant these structures may be to the lipid organization in membranes. The most significant points seem to be (*a*) that the choline residue exists in the *gauche* conformation rather than the extended zig-zag form and (*b*) that the conformation of the glycerol residue can be *gauche gauche*. The molecular configuration of a deacylated plant sulpholipid has been studied *via* the anhydrous rubidium salt [4] and the crystal structure of 2-aminoethanol phosphate has also been reported [5].

Single crystal X-ray studies of some diacyl-DL-phosphatidylethanolamines, e.g., 1,2-dilauroylphosphatidylethanolamine, are at present in progress.* This study of 1,2-dilauroylphosphatidylethanolamine is, as yet, incomplete, but it suggests the following important structural features: (*a*) the two hydrocarbon chains of the fatty acid residues present in the lipid molecule are pointing in the same direction and are parallel to each other and (*b*) the polar portion of the molecule containing the phosphate and amine groups appears to be in line with the hydrocarbon chains.

The suggestion that the hydrocarbon chains in this structure are parallel to each other is not surprising. Long chain fatty acids, esters and glycerides all pack in this way [6]. This is because, beyond a certain chain length, the long hydrocarbon chains tend to dominate the crystallization processes. This in turn is related to the favourable dispersion interaction forces which arise from parallel packed chains.

In related lipid molecules, while the hydrocarbon chains within the

* Hitherto unpublished studies.

molecule lie parallel to each other, the chains do not always point in the same direction. The single crystal X-ray studies of triglycerides, e.g., trilaurin [7] and tricaprin [8], show that the two hydrocarbon chains of the ester groups on adjacent positions, i.e., at the 1- and 2-positions on the glycerol molecule, point in opposite directions. The different situation which appears to exist with the diacylphosphatidylethanolamines is important because the solid state structure for these can be regarded as built up from infinite sheets of phospholipid bilayers with the polar groups organized in sheets and with all the hydrocarbon chains of adjacent lipid molecules lying parallel, i.e., it has a somewhat similar arrangement to the organization of the lipid which, it has been suggested, occurs in many biological membrane systems.

Other attempts to deduce information about the solid state properties have been made using electron diffraction and X-ray powder diffraction methods. Studies using electron diffraction are few and, as yet, have given only a limited amount of information [9]. Studies using X-ray powder diffraction methods have provided more information, e.g., the way in which the hydrocarbon chains are packed in the crystal lattice, and the angle of tilt of the chains to the basal planes has been determined. The diffraction data also provide information about the long and short X-ray spacings. The values of the X-ray short spacings suggest that the hydrocarbon chains are packed in similar arrangements to those adopted by many other long chain molecules [10]. The long spacings of an homologous series of phospholipids are shown in Fig. 6.1.2.

Some impression of the polar group organization in phospholipid systems can be obtained by a consideration of their capillary melting points.* The capillary melting points of a number of pure phospholipids have been determined and are shown to be quite high, e.g., the values of the capillary melting points for the diacylphosphatidylethanolamines are about 200°C, while for the phosphatidylcholines these capillary melting points are about 230°C. It is important to note that these values are independent of both the chain length and the degree of unsaturation of the fatty acid residues associated with the phospholipid. These capillary melting points can be compared with those of other compounds having long chains. The capillary melting points of fatty acids containing the same length of hydrocarbon chain are much lower, e.g., stearic acid melts at 69·7°C. On the other hand, the capillary melting point of sodium stearate is ~300°C. The high values of the capillary melting points of phospholipids are, therefore, consistent with the occurrence of ionic linkages associated with the polar groups of the phospholipid. The higher capillary melting points of the sodium soaps, compared with the phospholipids, suggest that there is possibly greater ionic character associated with the polar groups of these molecules than with the diacylphosphatidylcholines and phosphatidylethanolamines.

* For the purposes of this section, capillary melting point is defined as the temperature at which the amorphous isotropic liquid is formed either directly from the crystal (in the case of fatty acids) or from an intermediate liquid crystal (in the case of phospholipids or sodium stearate).

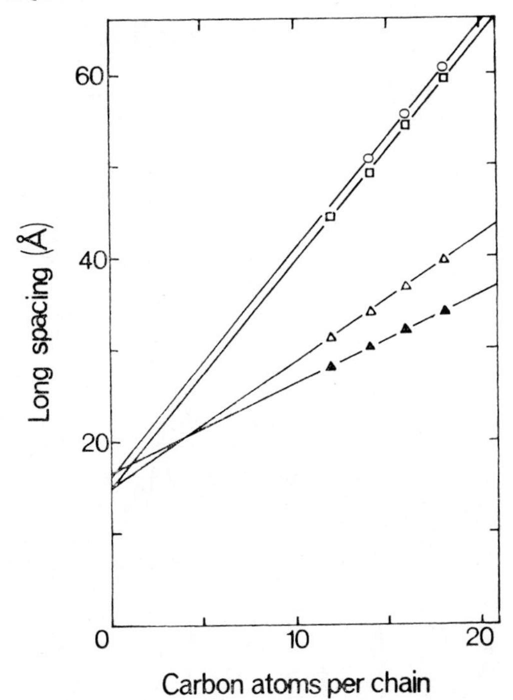

FIG. 6.1.2. The long spacings of a number of diacylphosphatidylethanolamines in various polymorphic forms (from Chapman, *et. al.* [10])
○, A form; □, A′ form; △, B form; ▲, C form.

Liquid Crystals Formed by the Action of Heat (Thermotropic Mesomorphism of Lipids)

In addition to the capillary melting point, other phase changes have been shown to occur with phospholipids at lower temperatures; for example, when a pure phospholipid, dimyristoylphosphatidylethanolamine, containing two fully saturated chains is heated from room temperature up to the capillary melting point, a number of thermotropic phase changes occur (i.e., phase changes caused by the effect of heat). This was first shown by infrared spectroscopy [11], then by thermal analysis [12] and has now been studied by a variety of other physical techniques [10]. Optical studies of thin films show that dimyristoylphosphatidylethanolamine at room temperature is birefringent when viewed between crossed polars. On heating, three processes occur: first some loss of birefringence at the first transition temperature at ~120°C, then a small increase at 135°C and a pronounced overall loss of birefringence near the melting point of 200°C. Above 120°C, pressure on a coverglass with a needle causes the material to flow. When the temperature of the phospholipid reaches ~120°C, the infrared (i.r.) absorption spectrum undergoes a remarkable change and loses all the fine structure and detail which was present at lower temperatures. The spectrum is then similar to that obtained with a phospholipid

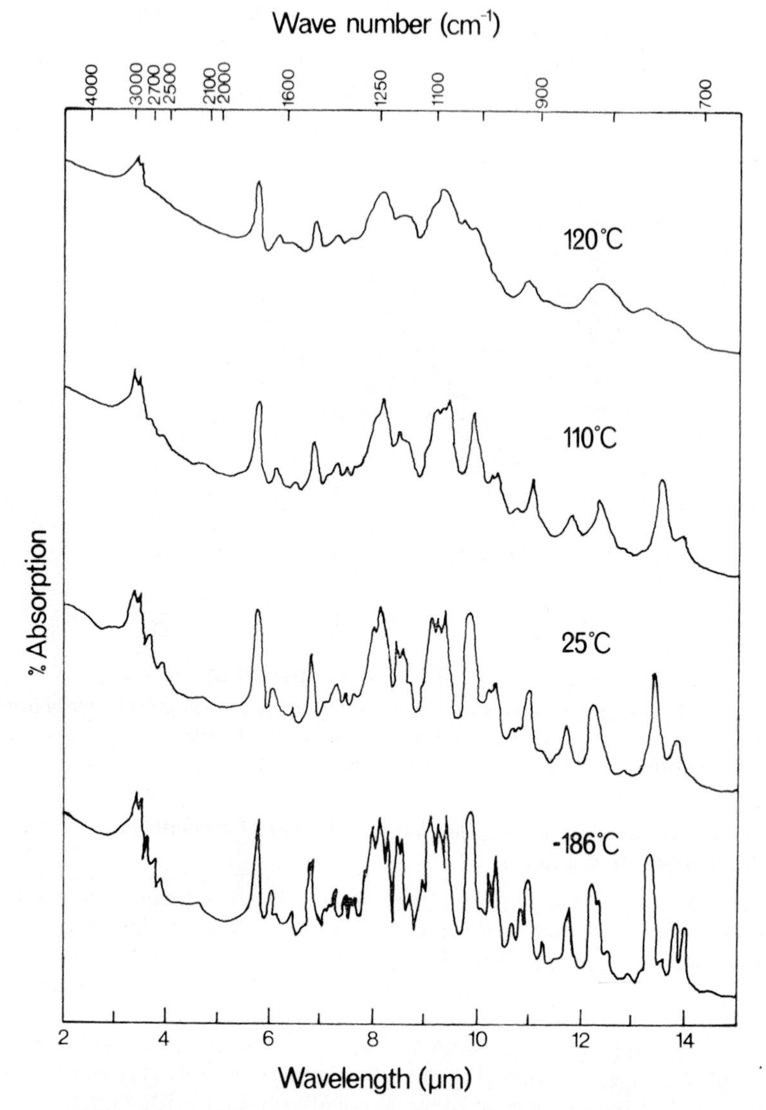

FIG. 6.1.3. The infrared spectra of 1,2-dilauroyl-DL-phosphatidylethanolamine at different temperatures (from Chapman, *et. al.* [10]).

dissolved in a solvent such as chloroform. The i.r. spectra of a phospholipid at different temperatures are shown in Fig. 6.1.3.

Differential thermal analysis (d.t.a.) shows that a marked endothermic transition (absorption of heat) occurs at about 120°C [12]. An additional heat change occurs at ~135°C and only a small heat change is involved near the capillary melting point of the lipid (see Fig. 6.1.4). This behaviour is similar to that which occurs with compounds which form liquid crystals,

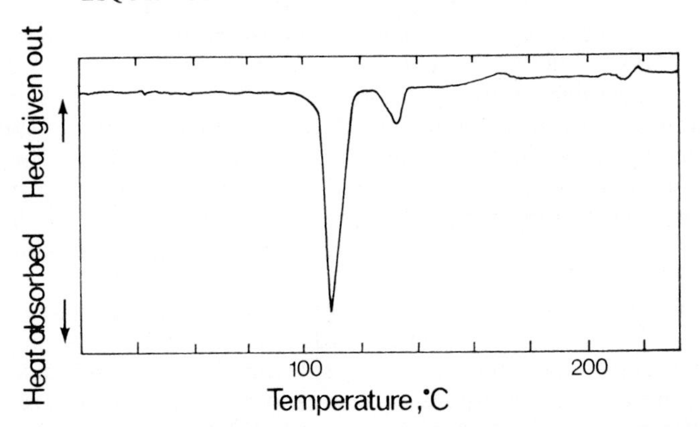

FIG. 6.1.4. Differential thermal analysis (d.t.a.) heating curve for 1,2-dimyristoyl-phosphatidylethanolamine (from Chapman and Collin [12]).

e.g., *p*-azoxyanisole or cholesteryl propionate, which give nematic and cholesteric liquid crystalline phases, respectively.

The X-ray long spacings show a dramatic reduction to some two-thirds of their original value at the first transition temperature with a further small reduction at the second transition temperature. The X-ray short spacings change at the first transition point from sharp diffraction lines to

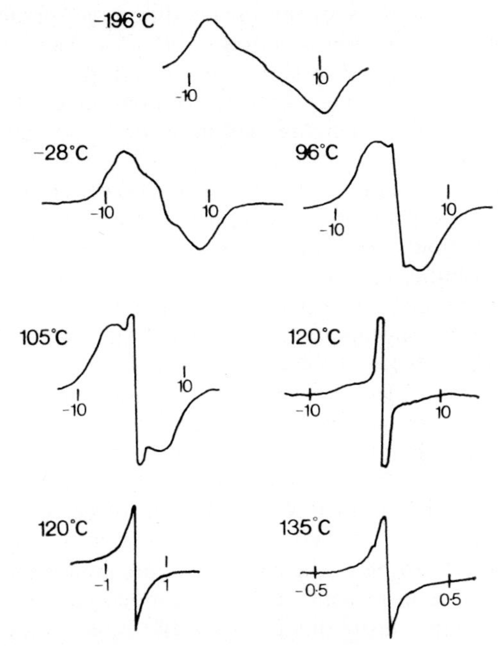

FIG. 6.1.5. P.m.r. derivative absorption spectra of 1,2-dimyristoyl-DL-phosphatidylethanolamine at various temperatures. The abscissae are marked in gauss. (Chapman and Salsbury [13]).

a diffuse spacing at ~4·6 Å. Nuclear (proton) magnetic resonance (p.m.r.) studies show a gradual reduction in line width from about 15 gauss at liquid nitrogen temperature until, at the first transition temperature, there is a sudden reduction in the line width (to ~0·09 gauss). P.m.r. spectra of dimyristoylphosphatidylethanolamine are shown in Fig. 6.1.5. These show that molecular motion increases gradually as the temperature increases until, at the transition temperature, a considerable increase in the molecular motion takes place [13], as revealed by an appreciable reduction in line width.

The main conclusions from these various studies are that (*a*) even with the fully saturated phospholipid at room temperature, some molecular motion occurs in the solid. This is evident from the p.m.r. spectra and from the i.r. spectra taken at liquid nitrogen and at room temperatures. (Note the difference between the i.r. spectra at −186°C and room temperature shown in Fig. 6.1.3.) (*b*) When the phospholipid is heated to a higher temperature, it reaches a transition point at ~120°C, a marked endothermic change occurs and the hydrocarbon chains in the lipid "melt" and exhibit a very high degree of molecular motion. This is evident both in the appearance of the i.r. spectrum and also in the narrow p.m.r. line width at temperatures of 120°C and higher. On the one hand the broad diffuse appearance of the i.r. spectrum is consistent with the chains flexing and twisting and with a "break-up" of the all-planar *trans* configuration of the methylene groups of the chains. (The i.r. spectrum can be regarded as giving a "rapid snapshot" showing the condition and organization of the phospholipid molecule within a period of 10^{-15} s. The p.m.r. spectrum, on the other hand, is affected by slower motions. It provides an impression of any molecular motion, i.e., rotation, translation or diffusion, of the phospholipid molecule occurring within a much longer time period, some 10^{-7} s.)

The fact that the phase transition is concerned primarily with the hydrocarbon chains of the phospholipid is confirmed by the X-ray data.

When phospholipids contain shorter chains, or unsaturated bonds, those marked endothermic phase transitions occur at lower temperatures. The changes in the temperatures at which these transitions occur parallel the trends in the melting points of the related fatty acids. The transition temperatures are high for the fully saturated long chain phospholipids. They are lower when there is a *trans* double bond present in one of the chains and lower still when there is a *cis* double bond present. This trend in the transition temperatures also confirms that this phase transition is primarily associated with a "melting" of the hydrocarbon chains of the phospholipid, while this in turn is related to the dispersion forces between the chains.

Only one main "melting of the chains" occurs even when there are two different types of chain present in the phospholipid. The transition temperatures for different phospholipid classes differ even though they contain exactly the same fatty acid residues. The difference between the transition temperatures for a diacylphosphatidylcholine and a diacylphosphatidylethanolamine are shown in Fig. 6.1.6. While natural phospholipids from

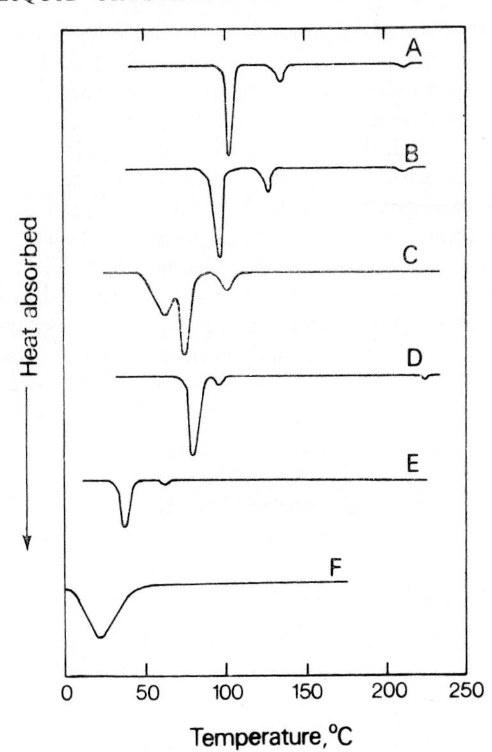

FIG. 6.1.6. Differential thermal analysis heating curves for different phospholipids (from Chapman [27]).

(A) 1,2-dimyristoyl-DL-phosphatidylethanolamine;
(B) 1,2-dielaidoyl-DL-phosphatidylethanolamine;
(C) 1-stearoyl-2-oleoyl-DL-phosphatidylethanolamine;
(D) 1,2-distearoyl-DL-phosphatidylcholine;
(E) 1-stearoyl-2-oleoyl-DL-phosphatidylcholine;
(F) egg yolk lecithin (from Chapman [27]).

erythrocyte or mitochondrial membranes contain large numbers of unsaturated *cis* double bonds and, therefore, in the dry condition have endothermic transition temperatures either near or below room temperatures, the highly saturated derivatives exhibit transition temperatures much higher than room temperature.

With mixtures of phospholipids we might expect the transition temperatures to occur over a wider range of temperature than with a single phospholipid. This is illustrated in the d.t.a. heating curve for egg yolk phosphatidylcholine (see Fig. 6.1.6).

When a crystalline phospholipid approaches the first thermotropic transition temperature we can imagine that the molecular motion of the chains increases until, at the first transition point, lateral expansion of the crystal lattice is forced to take place. This expansion allows even greater chain mobility possibly involving cooperative motion of the chains with rotation about C–C bonds occurring. Such an effect could explain the observed reduction in long spacing. A similar effect probably takes place with polar long chain molecules in a mono-layer at an air-water interface dependent upon the density of the chain packing.

A very simple theoretical system has been devised [14] to provide some insight for such situations. The model system is restricted to a set of simple chains in two dimensions and with the chains confined to lie on a two-dimensional hexagonal lattice. The end-to-end distance of each chain (corresponding in a real system to the distance from the polar group to the methyl group) and other properties were determined as a function of the density of packing of the chains using two simple potential functions and the Monte Carlo computational method. Typical configurations of sets of simple chains on a two-dimensional hexagonal lattice are shown in Fig. 6.1.7. At the highest density only the fully extended configuration

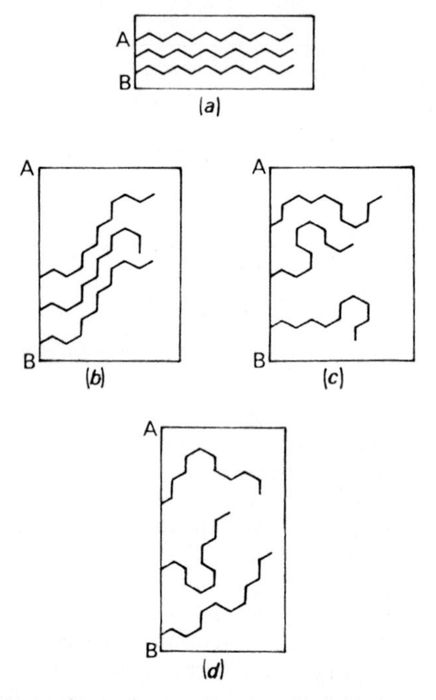

Fig. 6.1.7. Typical configurations of sets of chains on a two-dimensional hexagonal lattice as the interchain separation is increased. Separation increases from (*a*) to (*d*) (from Whittington and Chapman [14]).

occurs. At lower densities other configurations are allowed. As the density of the chains decreases, the end-to-end distance of the chains suddenly falls, consistent with the occurrence of a co-operative phase transition.

An important conclusion which we obtain from these phase transition studies is that, near to the endothermic transition temperature, a given phospholipid can be in a highly mobile condition with its hydrocarbon chains flexing and twisting. The more unsaturated the chain, the lower the temperature will be at which this occurs. This is a fundamental property of the phospholipid and we can expect this chain mobility in whatever situation the phospholipid occurs unless, for special reasons, this motion is somehow inhibited. We can envisage that inhibition of chain motion by interaction with other molecules would provide one of these special reasons, e.g., cholesterol or protein (see later discussion). In other circumstances, due to less perfect packing arrangements, we might expect, at a particular temperature, even greater mobility of the chains of the lipid and indeed of the whole lipid molecules themselves.

Liquid Crystals Formed in the Presence of Water (Lyotropic Mesomorphism of Lipids)

Small amounts of water can have important effects upon the mesomorphic behaviour. Thus the diacylphosphatidylcholines (lecithins) exhibit an additional liquid crystalline form between the first transition

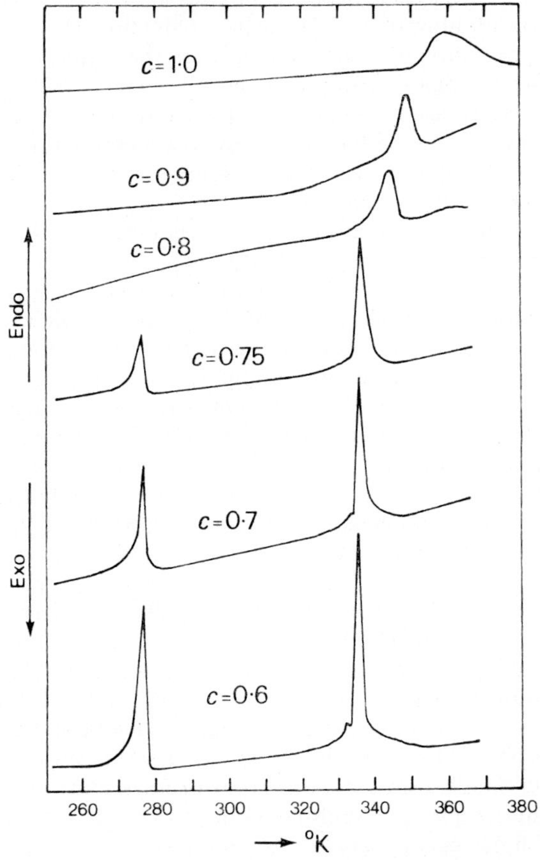

FIG. 6.1.8. D.s.c. heating curves for 1,2-distearoylphosphatidylcholine in increasing amounts of water; c is the concentration of the phospholipid.

temperature and the capillary melting point [15]. The intermediate liquid crystalline form is found to exhibit X-ray spacings consistent with a cubic phase organization.

When phospholipids are examined in the presence of increasing amounts of water, the various physical techniques, such as microscopy, p.m.r. spectroscopy or differential thermal analysis, show that as the amount of water increases, the marked endothermic transition temperature for a given phospholipid falls. The transition temperature does not fall indefinitely; it reaches a limiting value independent of the water concentration (Fig. 6.1.8). We can understand this if we regard the effect of water as leading first to a "loosening" of the ionic structure of the phospholipid crystals. This in turn affects the whole crystal structure and causes a reduction, up to a certain limit, of the dispersion forces between the hydrocarbon chains. Large amounts of energy are still required to counteract the dispersion forces between the chains and quite high temperatures are still required to cause the chains to melt. These limiting transition temperatures parallel the melting point behaviour of the analogous fatty acids, becoming lower with increasing unsaturation. This further reduction of the endothermic transition temperature by water means that the natural phospholipids extracted from biological membranes usually exhibit this crystalline to liquid crystalline transition many degrees below the biological environmental temperature. At the biological environmental temperature we can expect the phospholipids which contain highly unsaturated chains to be in a highly mobile and fluid condition.

There are a number of important features associated with the transition temperature for the lipid when it is in the presence of water. The first of these is that the ability to disperse the lipid in water increases markedly above this transition temperature. Only those phospholipids which give transitions when placed in water below or near to room temperature, spontaneously form myelin figures. Fully saturated phospholipids which have high transition temperatures do not form myelin figures at room temperature. However, if the temperature is raised to the transition point, these phospholipids form myelin figures. In the presence of an excess of water, phospholipids, such as 1,2-dipalmitoylphosphatidylcholine, spontaneously form myelin figures at 42°C. The behaviour of lipids in water can be conveniently summarized by means of the phase diagram shown in Fig. 6.1.9.

A further feature of lipid/water systems is their monolayer behaviour. Monolayer studies of phospholipids have been carried out for a number of years, but usually this work has been performed with natural phospholipid mixtures and, in the vast majority of cases, with egg yolk phosphatidylcholine. In recent years however, a few studies have been made with pure synthetic phospholipids. These show that the fully saturated phospholipids form, at room temperature, monolayers which are more condensed than are the unsaturated phospholipids containing *cis* hydrocarbon chains, i.e., the saturated lipids occupy less area at low surface pressures than do the unsaturated derivatives.

Monolayers obtained with phosphatidylcholines are observed to be

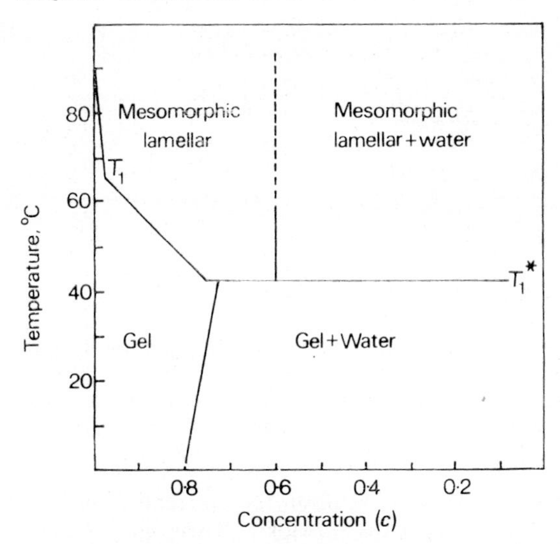

FIG. 6.1.9. Phase diagram of the 1,2-dipalmitoyl-L-phosphatidycholine/water system. (Chapman, *et al.* [15]).

In the gel phase water separates the double layers of lipid, and the hydrocarbon chains are packed in an organized crystalline manner—probably in an hexagonal arrangement.

The cubic phase occurs at *high* temperatures and in the presence of only small amounts of water. It therefore occurs close to the left hand side of the phase diagram.

much more expanded than are the corresponding phosphatidylethanolamines containing the same acyl chains. These results can be compared with the d.t.a. results discussed earlier. A high transition temperature for liquid crystal formation is, in general, correlated with a condensed type monolayer and a low transition temperature with an expanded film [10]. The d.t.a. transition temperatures are higher for the phosphatidylethanolamines than for the corresponding phosphatidylcholines. Phospholipids containing *trans* (elaidoyl) unsaturated chains have higher transition temperatures than those containing *cis* (oleoyl) chains. Phospholipids containing one fully saturated chain and one *trans* unsaturated chain give condensed monolayers similar to those observed with completely saturated phospholipids [16].

Types of Lyotropic Phase

Above the first transition temperature, when sufficient water is present, phospholipids can form several different types of phase organization, e.g., some phospholipids give lamellar and hexagonal type structures. Recent studies have shown that the lecithins appear to exhibit only a lamellar type structure over a large range of water concentration. Other phospholipids, such as phosphatidylethanolamines and samples of brain lipid, appear to be able to exist in both the hexagonal and lamellar types of

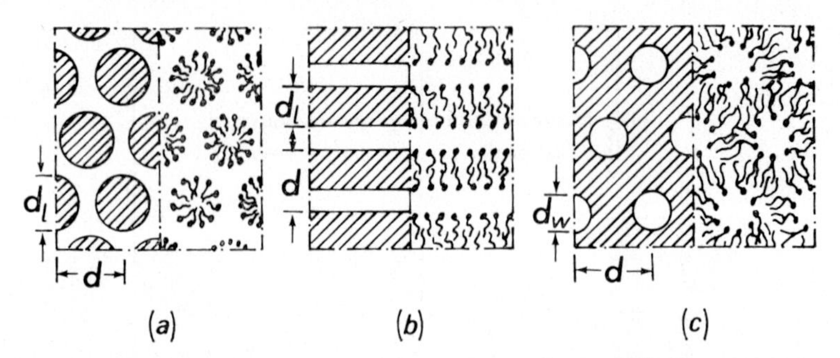

FIG. 6.1.10. Structures of some high-temperature phases of lipid-water systems; schematic representation of a cross-section of the rods and of the lamellae. The hydrophilic group is represented by a dot, the paraffin chains by a wriggle: (*a*) hexagonal I; (*b*) lamellar; (*c*) hexagonal II. (From Luzzati, *et al.* [17a]).

organization, [17a] depending upon the concentration in water (see Fig. 6.1.10). Studies by Gulik-Krzywicki, Rivas and Luzzati [17b] have shown that four phases can be distinguished with a mitochondrial lipid extract. The composition of the extract from beef heart mitochondria was 34% lecithin, 29% phosphatidylethanolamine, 10% phosphatidylinositol, 20% cardiolipin, 2% cholesterol and 5% neutral lipids. The phases observed were hexagonal as well as various lamellar phases. The phase diagram is shown in Fig. 6.1.11. As yet no relationship between these phases and their biological function has been made. Nevertheless, these phase properties have to be considered and held in mind in discussions of lipid behaviour.

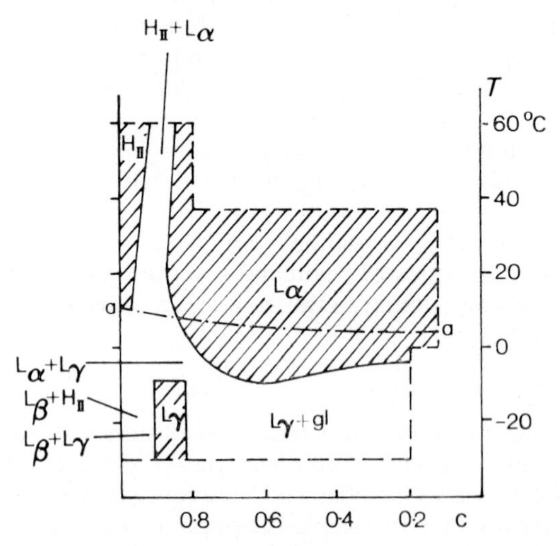

FIG. 6.1.11. Phase diagram of mitochondrial lipid-water system. Five phases are observed; one hexagonal (H_{II}), three lamellar (L_α, L_β, L_γ) and ice (gl). (From Gulik-Krzywicki, *et. al.* [17b]).

EFFECTS DUE TO CHOLESTEROL

An interaction which may be of considerable biological importance is that of cholesterol with phospholipids. Cholesterol occurs in many membranes, particularly in the myelin sheath and red blood cell membranes, but its precise function and arrangement are not understood. The biochemical importance of the solubilization of cholesterol by phospholipid and its vehicular possibilities have often been discussed [18].

As long ago as 1925 Leathes [19] showed that cholesterol, when mixed with certain fatty acids in monomolecular films on water, caused a diminution of the area occupied by the fatty acids. The study of this condensing effect of cholesterol on phospholipids has continued up to the present day using monolayer techniques. This has led to a variety of models and discussions of the interaction between the phospholipid and cholesterol.

The interaction of egg yolk lecithin and cholesterol has also been studied

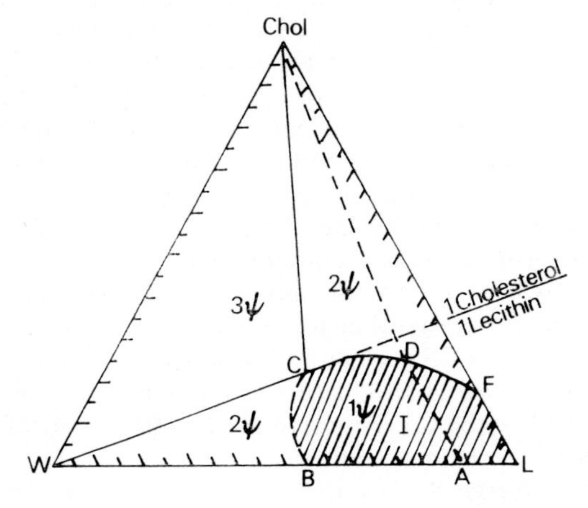

FIG. 6.1.12. Egg-yolk lecithin (L)-cholesterol (Chol)-water (W) system at 25°C. The number of phases present is indicated (from Small and Bourges [20]).

in bulk systems in the presence of water using microscopy and X-ray techniques [20]. The three component phase-diagram of this system is shown in Fig. 6.1.12. Along the water-cholesterol side of the triangle one notes that the cholesterol is totally insoluble in water at 25°C. However, along the lecithin-water side, from about 12–45% water (A–B), lecithin forms with water a lamellar liquid crystalline phase. At $<12\%$ water, the water is taken up to form "bound" water associated with the polar group of the lipid. Mixtures containing more than 45% water separate into this lamellar phase suspended in the excess of water in the form of myelin figures and anisotropic droplets.

The microscopic appearance of lecithin or of lecithin–cholesterol mixtures is similar to that of a neat soap and only one sharp long spacing is observed at 48 Å by X-ray examination. There is no second or third spacing to show whether it is in fact truly lamellar. Below 5% of water, several spacings are observed. From the phase diagram it appears that cholesterol can be added to lecithin-water mixtures up to a molecular lecithin-cholesterol ratio of 1:1 to give the same lamellar liquid crystalline phase already given by lecithin and water. Any excess of cholesterol above this proportion separates as cholesterol crystals. X-ray analysis shows that, in general, the addition of cholesterol to lecithin at a constant water concentration tends to make the lipid layer slightly thicker. This slight increase of X-ray spacings, according to these authors, does not seem sufficient to account for the cholesterol being positioned between the ends of the hydrocarbon chains of the lecithin in the lamellar layer. They suggest that it is more probable that the cholesterol is interdigitated between the lecithin molecules and that its hydroxyl group is lying in the water layer in a similar manner to mixtures of lecithin and cholesterol spread on the surface of water.

Studies of lecithin–cholesterol–water interactions have also been made using differential scanning calorimetry (d.s.c.) [21]. This work has shown that addition of cholesterol to dipalmitoyl-L-lecithin in water lowers the transition temperature between the gel (see legend to Fig. 6.1.9) and liquid crystalline phase and decreases the heat absorbed at the transition. No transition is observed with an equimolar ratio of the lecithin with cholesterol. Unsaturated lecithins and the lipid extract of human erythrocyte ghosts exhibit similar behaviour.

The d.s.c. curves between 280 and 360°K for a series of 1,2-dipalmitoyl-L-lecithin/cholesterol mixtures each containing 50% by weight of water and varying ratios of lecithin to cholesterol are shown in Fig. 6.1.13. As the concentration of cholesterol increases the main endothermic transition remains sharp while a small peak at 35°C disappears. This is followed by a profound change in which the main transition becomes broad and decreases in area. When the concentration reaches 50 m% of cholesterol, no endothermic peak can be observed.

The effect of cholesterol is to disrupt the ordered array of the hydrocarbon chains of the lipid in the gel phase and, when cholesterol and lecithin molecules are present in equimolar proportions, all the chains are in a fluid condition. The presence of equimolar amounts of cholesterol with lipid causes the phospholipid/cholesterol mixtures to be dispersible in water over a much wider temperature range than occurs with the individual phospholipid.

High resolution p.m.r. spectra are usually observed when molecules are dissolved in a solvent and the molecular motion of the solute is sufficient so that all dipole interactions are averaged out. The fact that phospholipids are not readily soluble in water might be expected to be a limitation on the possibility of obtaining such high resolution. However, if natural phospholipids such as egg yolk lecithin or red blood cell lipids are dispersed by means of sonication in water, a high resolution p.m.r. spectrum can indeed be obtained [22, 23]. This shows the phospholipid in this dispersion has considerable diffusional freedom. The spectrum is similar to that which is observed when a phospholipid is dissolved in a solvent such as chloroform [24]. Peaks associated with different proton groupings are

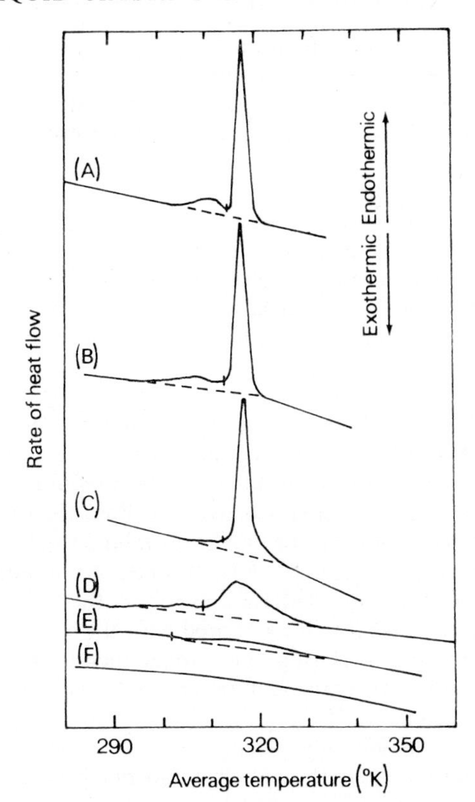

FIG. 6.1.13. D.s.c. curves of 50 wt % dispersions in water of 1,2-dipalmitoyl-L-lecithin/cholesterol mixtures containing: (A) 0·0 m%, (B) 5·0 m%(C) 12·5 m%, (D) 20·0 m%, (E) 32·0 m% and (F) 50·0 m% cholesterol (from Ladbrooke, *et al.* [21]).

observed. Peaks in the spectrum can be assigned to the $[CH_2]_n$ protons of the hydrocarbon chain of the lipid, the nine protons of the choline group $N(CH_3)_3$ and so on. This discovery led to the idea that this technique could provide a new method for investigating the interaction of molecules such as cholesterol with phospholipids. This p.m.r. spectroscopic method is important because the organization of lipid and cholesterol should be analogous to the situation which is likely to occur *in vivo*, particularly in its vehicular situations. It therefore has advantages over the more classical monolayer type investigation.

As a result of the cholesterol interaction, the peak arising from protons in the hydrocarbon chain $[CH_2]_n$ signal is broadened out and is absent from the spectrum.

These various results seem to show that, at a particular temperature, the presence of cholesterol causes the hydrocarbon chains of differing phospholipid molecules to be in an "intermediate fluid" condition. Those lipids which would normally be above their limiting transition temperature may have a certain amount of inhibition of chain motion, while the hydrocarbon chains of those lipids which would normally be in a gel condition are given greater fluidity. These results are supported by other more recent studies using spin label probes and e.s.r. spectroscopy.

Relevance to Membranes

When we consider the relevance of the liquid crystalline behaviour of phospholipids to membrane structure and function we have to remember that the interaction with protein has still to be explored. However, even with this limitation we can make a few general comments.

Model Membranes*

The spontaneous formation by phospholipids in water, above their transition temperature, of globules containing bilayers of lipid separated by aqueous layers has led to their use as model membranes. The argument is that if real membranes contain bilayers of lipid then here is a natural system whose constitution can be chosen to mimic the lipid characteristic of a real membrane. Ions can then be trapped in the aqueous compartments and the permeability of these ions across the lipid bilayer studied. Admittedly, there is no protein associated with the system but this can be included in later studies. Using this model membrane system various effects which occur with real membranes can be mimicked, e.g., anaesthetic and drug effects.

A second model membrane system also uses the liquid crystalline behaviour of phospholipids. In this system a phospholipid above its transition temperature is painted with a brush across an orifice separating two aqueous compartments. It spontaneously thins down and forms a single bilayer. This model membrane can be used to study ion permeability, water flow, electrical resistance and excitability characteristics. The model membrane is fairly stable and some of this stability is probably associated with the flow characteristics of the phospholipid in its liquid crystalline phase.

Natural Membranes

Despite their apparent complexity, natural membranes have been successfully examined using the same techniques which have been applied to simple lipid/water systems.

Calorimetric studies

We have seen that lipids in water may show a marked thermal transition associated with a phase change from gel to liquid crystal, but that the presence of cholesterol can modify or remove this transition. It is important to enquire whether thermal transitions occur with cell membranes, similar to those observed with the phospholipids themselves.

Studies by Steim, *et al.* [25] have been with a membrane which does

* An excellent review of the two model membrane systems referred to below is given by A. D. Bangham in *Progress in Biophysics*, vol. 18 (1968).

not contain cholesterol. This is the membrane of *Mycoplasma laidlawii*. With this membrane, a strong endothermic transition is indeed observed. The transition temperature corresponds with the transition temperature of the extracted lipids dispersed in water. Comparison of the heat changes associated with the transitions observed with the membrane and with the lipid extract indicate that appreciable amounts of lipid contribute to this membrane transition. The existence of such a transition may be consistent with the presence of lipid bilayer regions in the membrane. These authors consider that this evidence is in support of the Danielli model of lipid sandwiched between protein. They do not appear to consider the effects of electrostatic interaction with the protein on the lipid transition temperatures. A *simpler* explanation is that there are in the membrane extensive regions of lipid bilayer linking lipid–protein regions.

Studies have also been made with membranes which contain cholesterol, e.g., myelin and red blood cell membranes. The studies with the myelin membrane isolated from white matter of ox brain show [26] that:

(*a*) With wet myelin, thermal transitions are not detectable. In this case the cholesterol and other lipids appear to be organized in a single phase. The organization of cholesterol in the membrane appears to prevent the lipids from crystallizing.

(*b*) To maintain the organization of the lipid in myelin a critical amount of water appears to be required. This water is unfreezable at 0°C and may correspond to "bound" water.

(*c*) On drying the myelin, the cholesterol and other lipid crystallize and precipitate. Endothermic transitions associated with the cholesterol and other lipid can then be observed.

(*d*) The total lipid extract in water does *not* show a detectable endothermic transition but the cholesterol-free lipid does. In the absence of cholesterol, part of the myelin lipid is crystalline at body temperature.

Studies of the red blood cell membrane also show no detectable thermal transitions associated with the lipid component.

These results are understandable if there are regions of lipid bilayer present in the membrane with the cholesterol preventing lipid chain crystallization.

The fluidity of membranes may be related to the transition temperatures of their components. Thus membranes which contain phospholipids having little unsaturation will have less fluidity than those having phospholipids with much unsaturation. This control of fluidity may then be related to diffusional characteristics of molecules passing into and out of the cell.

In line with this idea it is interesting that poikilothermic organisms, i.e., those which alter their body temperature to correspond to their environment, apparently alter the unsaturation in their phospholipids [27]. The alteration may provide a more or less constant fluidity of the membrane structure. At high temperatures the chains are more saturated; at low temperatures, they become more unsaturated.

Some cell membranes, e.g., in some bacterial systems, contain lipids having saturated branched chains. In this case, the presence of the methyl groups reduces the dispersion forces, and hence the transition temperatures.

It may be the case that in these biological structures it is more "convenient" to have branching rather than unsaturated double bonds in the membrane system whilst at the same time having the appropriate fluidity of structure.

The effects of cholesterol on the simple lipid water systems suggest that a possible rôle for cholesterol in membranes is to control the fluidity of the hydrocarbon chains of the phospholipids providing a coherent structure stable over a wide temperature range and permitting some latitude in the fatty acid content of the component lipids.

REFERENCES

[1] Chapman, D. in *Biological Membranes*, Academic Press, London and New York (1968).

[2] Sundaralingam, M. and Jensen, L. H. *Science, N.Y.* **150,** 1035 (1965).

[3] Abrahamsson, S. and Pascher, I. *Acta Crystallogr.* **21,** 79 (1966).

[4] Okaya, U. *Acta Crystallogr.* **17,** 1276 (1964).

[5] Kraut, J. *Acta Crystallogr.* **14,** 1146 (1961).

[6] Chapman, D. in *The Structure of Lipids*, Methuen, London (1965).

[7] Larsson, K. *Ark. Kemi.* **23,** 1 (1964).

[8] Jensen, L. H. and Mabis, A. J. *Nature, Lond.* **197,** 681 (1964).

[9] Parsons, D. F. and Nyberg, S. L. *J. appl. Phys.* **37,** 3920 (1966).

[10] Chapman, D., Byrne, P. and Shipley, G. G. *Proc. R. Soc.* **A290,** 115 (1966).

[11] Byrne, P. and Chapman, D. *Nature, Lond.* **202,** 987 (1964).

[12] Chapman, D. and Collin, D. T. *Nature, Lond.* **206,** 189 (1965).

[13] Chapman, D. and Salsbury, N. J., *Trans. Faraday Soc.* **62,** 2607 (1966).

[14] Whittington, S. G. and Chapman, D. *Trans. Faraday Soc.* **62,** 3319 (1966).

[15] Chapman, D., Williams, R. M. and Ladbrooke, B. D. *Chem. Phys. Lipids* **1,** 445 (1967).

[16] Chapman, D., Owens, N. F. and Walker, D. A. *Biochim. biophys. Acta* **120,** 148 (1966).

[17a] Luzzati, V. and Reiss-Husson, F. *Nature, Lond.* **210,** 1351 (1966).

[17b] Gulik-Krzywicki, T., Rivas, E. and Luzzati, V. *J. molec. Biol.* **27,** 8 (1967).

[18] Fleischer, S. and Brierley, G. *Biochem. biophys. Res. Commun.* **5,** 367 (1961).

[19] Leathes, J. B. *Lancet* **208,** 853 (1925).

[20] Small, D. M. and Bourges, M. C. *Molec. Crystals* **1,** 541 (1966).

[21] Ladbrooke, B. D., Williams, R. M. and Chapman, D. *Biochim. biophys. Acta* **150,** 333 (1968).

[22] Chapman, D. and Penkett, S. A. *Nature, Lond.* **211,** 1304 (1966).

[23] Chapman, D., Kamat, V. B., de Gier, J. and Penkett, S. A. *J. molec. Biol.* **31,** 101 (1968).

[24] Chapman, D. and Morrison, A. *J. biol. Chem.* **241,** 5044 (1966).

[25] Steim, J. M., Tourtellotte, M. E., Reinert, J., McElhaney, R. and Rader, R. L. *Proc. natn. Acad. Sci. U.S.A.* **63,** 104 (1969).
[26] Ladbrooke, B. D., Jenkinson, T. J., Kamat, V. B. and Chapman, D. *Biochim. biophys. Acta* **164,** 101 (1968).
[27] Chapman, D. in *Thermobiology* (edited by A. H. Rose), Academic Press, London and New York (1967), pp. 123–145.

6.2 THE ROLE OF LIQUID CRYSTALS IN LIFE PROCESSES

G. T. Stewart

In looking at the rôle of liquid crystals in living tissue and in systems arising from living matter, we should first consider what, in physical and chemical terms, are the prerequisites of animate matter.

Organic matter must be regarded as a product of terrestrial cooling. The inorganic world had no room for ordered structures except crystals. Cooling, however, brought possibilities of reaction between carbon and other elements forming compounds, consisting of linear, helical and cyclic molecules, some of which, as we know from physical studies, behave as liquid crystals under appropriate thermal conditions. The properties of the thermotropic mesophase have been extensively studied and it is now possible to recognize or predict [1, 2, 3] which compounds or classes of compounds are most likely to enter it. Such compounds often remain stable at extremes of temperature and may indeed be products of condensations and cyclizations proceeding at high temperatures. Apart from their ability to display one or more mesophases, compounds in this category conform with other non-liquid-crystalline substances in their thermodynamic behaviour and interatomic bonding.

In terms of chemical palaeontology, cooling of the earth brought carbonaceous solids, semi-solids and solvents into contact with water and electrolytes. It can be assumed that this contact produced a variety of liquid systems, from true solutions to dispersions of one compound in another. In the millennia before life appeared, comparative stabilization of temperature of the earth's surface must have permitted evolution of molecules of long chain and polycyclic hydrocarbons, and, eventually, carbohydrates, lipids and amino acids. From and within this chemically richer brew, or primeval soup, there developed what has been nicely conceptualized by Teilhard de Chardin [4] as a Biosphere, whose chemical and physical state, over an incalculable period of time, provided conditions appropriate to the creation of a thin layer of animate from inanimate matter around the earth's organic and mineral crust.

It is a curious fact that most theories of evolution take Life for granted, skipping the extraordinary processes which permitted or produced the miraculous state known as vitality. Looking at chemical evolution mechanistically, we can identify substances—carbohydrates, amino acids, simple aliphatic compounds—which are likely and probably necessary precursors of living matter: but even the simplest forms of protoplasm contain self-replicating complex molecules and we are very far from understanding, let alone reproducing, this essential difference between a chemical and a biochemical process. It is however likely that, in the rich primeval liquid of the non-living organic world, there existed special forms of liquid crystalline matter, lyotropic mesophases [5] which, as Bernal [6] predicted in 1933, have now been shown [7, 8] to possess properties compatible with the organization of precursor molecules into the macromolecules, coacervates and other complexes which are characteristic of and peculiar to living matter. The relationship between life, as we know it, and the properties of lyotropic mesomorphic systems becomes clearer if we define that living matter requires the following set of collective physico-chemical characteristics for its existence.

(1) The ability to replicate essential chemical structures indefinitely.

(2) The ability to derive energy from its environment and to store energy during inactive periods.

(3) Convertibility of energy from chemical to kinetic form for movement and division of protoplasm.

(4) Utilization of small jumps in energy levels for all protoplasmic reactions.

(5) Formation of polymers from identical or similar sub-units.

(6) Formation of macromolecules from different sub-units.

(7) Combination of invariable with variable chemical and physical structures to form colloids, dispersions and coacervates.

(8) Orderly growth and differentiation of chemical and physical constituents into continuous (syncytial) or cellular forms of protoplasm.

(9) Arrangement of molecules and macromolecules of varying complexity in a manner permitting, simultaneously, the retention of order and individuality in a state of kinetic and thermal mobility.

(10) Preservation of physical and chemical individuality (i.e., of Bernard's *milieu interieur*) during changes in external environment.

To these physico-chemical properties and processes, there must be added, for completeness, a set of imponderables, such as the direction and purpose of the Life Force, which are therefore far beyond the terms of reference of this chapter though not of the natural philosophy which is or should be the inquisitive *alpha* and intellectual *omega* of all science. Thus we can recognize in the change in birefringence of myosin the change in ordered structure which accompanies muscular contraction [9, 10]. We can also envisage [11–15] and to some extent imitate [16–19] by experiment the ordered bimolecular layers of polar lipid and water which are the primary components of a cell membrane, but we cannot yet tell or even glimpse how such a structure becomes differentiated into a cellular covering which not only contains protoplasm but regulates its electron transport, osmosis and various other internal reactions, as well as its relationships with other cells and with the external environment [20, 21]. The garment is alive, as well as the body which it clothes; and yet, the membrane, whether conceived as a bimolecular leaflet [14, 15, 22], a Danielli–Davson model [20], or even as a wide range of lipid structures [23, 24] is one of the simplest of all living structures. If we cannot understand how it came about, we cannot understand the fundamental mystery of Life—and that is where we stand.

To this mystery and to this vital problem, the study of lyotropic liquid crystals and of ordered macromolecules brings more light in the physicochemical sense than any other branch of science. In the first part of this chapter, Chapman describes the essential components of membranes of living cells and connective tissue. In this part of the chapter, we shall therefore confine ourselves to substances and structures other than membranes, though necessarily in less detail, because membranes are convenient to handle and have therefore received detailed attention by a variety of morphological and biophysical techniques [16–20, 24–29] which should be studied closely because they are applicable to other biological structures and processes. The studies of Finean [14], Stoeckenius [12], Robertson [15] and others on myelin, for instance, have contributed immensely to knowledge, not only of membranes in general, but also of the nature and function of dispersed phases and the organization of lipid in nerve tissues. Other workers [17–20, 22–28] have developed improved methods for fractionation, separation and viewing of membranes and associated structures, but there are nevertheless many wide gaps in knowledge and between disciplines about the state in which membranes and other vital components exist in the living cell. There is a need to apply, biologically, physical measurements such as infrared spectrometry [23], proton resonance spectroscopy, polarization microscopy and micrometry [7, 30, 31] and light-scattering [32] and, complementarily, for those who make elaborate measurements to employ simple optical or even visual methods on whole-cell or fresh tissue preparations (see pages 314 and 319). In this respect, the uses of the simple microscope with polarizing or interference optical attachments and calibrations furnish an abundance of direct visual information on cells, tissues and tissue-cultures; the electron microscope introduces artefacts which have to be themselves identified but, when that is done, the results are highly informative. The introduction more recently of scanning electron microscopy holds promise of major advances.

Phase-dependent States of Living Matter

Living tissue cannot exist without the formation, presence and exact replication of certain macromolecules, notably of lipids, nucleic acids and proteins, which possess biochemical reactivity only when water is present. Elementary or first order biological reactions, such as enzymatic hydrolysis and synthesis, gaseous or ion exchange and diffusion, can proceed in simple or colloidal amorphous isotropic solutions of these macromolecules, but all the structural conditions necessary for the formation and individual function of discrete cells and tissue cannot be met in amorphous isotropic solutions or, indeed, in any other single phase of matter except the liquid crystal. Where there are boundaries between cells or tissues, and more than one kind of subcellular component, a second order of biological reactions has to be defined for reactants in different phases and this is where the liquid crystal enters the picture as a distinctive birefringent optical entity (Fig. 6.2.1) exemplified by the myelinic structures formed by amphiphilic lipids like lecithin in contact with water.

FIG. 6.2.1. Myelinic tubes of phospholipid (lecithin) at the interface with water.

When substances in different physical states or phases react, the number of molecules reacting is a function of the intimacy of contact or mixing. This is well shown by the action of a hydrolytic enzyme, such as an esterase, upon a water insoluble substrate. Dispersion of the substrate, by increasing interfacial contact, accelerates hydrolysis by several orders of

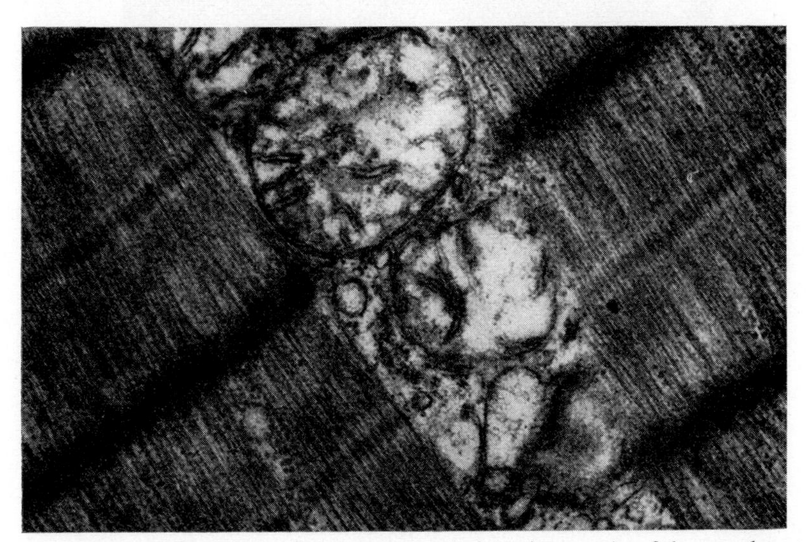

FIG. 6.2.2. Longitudinal and transverse striations in muscle of human heart, with lipid and collagen between the fibres. (Electron micrograph $\times 39{,}200$—(OsO_4).)

magnitude [33]. For such reactions to proceed intracellularly, the dispersion of reactants must occur in what might be called a third order of biological reactions, the essential condition of which is the establishment of order and dynamic equilibrium between the phases of the reactants, as in the distribution of subcellular particles like ribosomes and lysosomes. Myoglobin in its natural state is an example of the relationship between phase and biofunction: excitation, mediated by electrical change and high-energy phosphates, alters the arrangement of elongated molecules with corresponding increase in viscosity leading to alteration in shape, and to movement. This system is, in fact, a comparatively simple lyotropic mesophase in which the typical elongated protein fibrils of striped muscle provide the contractile chemical structure (Fig. 6.2.2).

The fourth order of reaction occurs when more than one type of molecule is included in one or more of the phases in a preferred orientation, as in cell membranes in which phospholipid and protein are the essential components. Experimentally, this can be demonstrated by bringing lecithin into contact with water to form a mobile smectic mesophase, with water molecules between ordered bimolecular layers of phospholipid [8]. Biophysical models of this situation have been studied in detail by Lawrence [5], Dervichian [11], Luzzati [16] and others. In such systems, phase transition within lipid-water dispersions must affect the formation and function of even the simplest biostructures, especially in poikilothermic forms of life which assume the temperature of their ambient

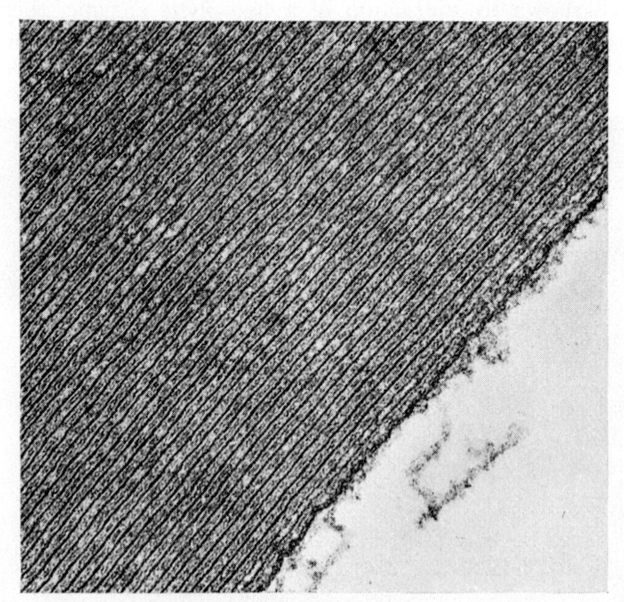

FIG. 6.2.3. Lamellar lipid in myelin from a human nerve removed surgically. Note the black layers of lipid coagulated by osmic acid. (Photographed by courtesy of Dr. James C. Harkin, New Orleans, La.) (Electron micrograph $\times$110,000).

environment. Long chain saturated palmitoyl and stearoyl phosphatides are mesomorphic when anhydrous, or in the presence of water at temperatures up to about 41°C, a temperature above which organized cellular function ceases in all metazoa and most protozoa. It would seem possible that, biophysically, the limits of vitality in a cell or tissue are the limits of the phase-dependent reactions of that cell or tissue.

Phase-equilibria and phase transitions have to be kept in mind in the investigation, especially by morphological methods, of living cells or tissues, which are aggregates of membranes and subcellular elements. As such, they require fluid and oxygen. They have limited biothermal ranges and usually perish if subjected to mechanical disintegration, high vacuum, dehydration or embedding. Living cells and tissues do not normally encounter substances such as potassium permanganate, glutaraldehyde, formalin, osmium tetroxide, lead acetate or uranium acetate. Yet most histological and ultrastructural information is obtained from cells or cellular components processed under these conditions with scant regard

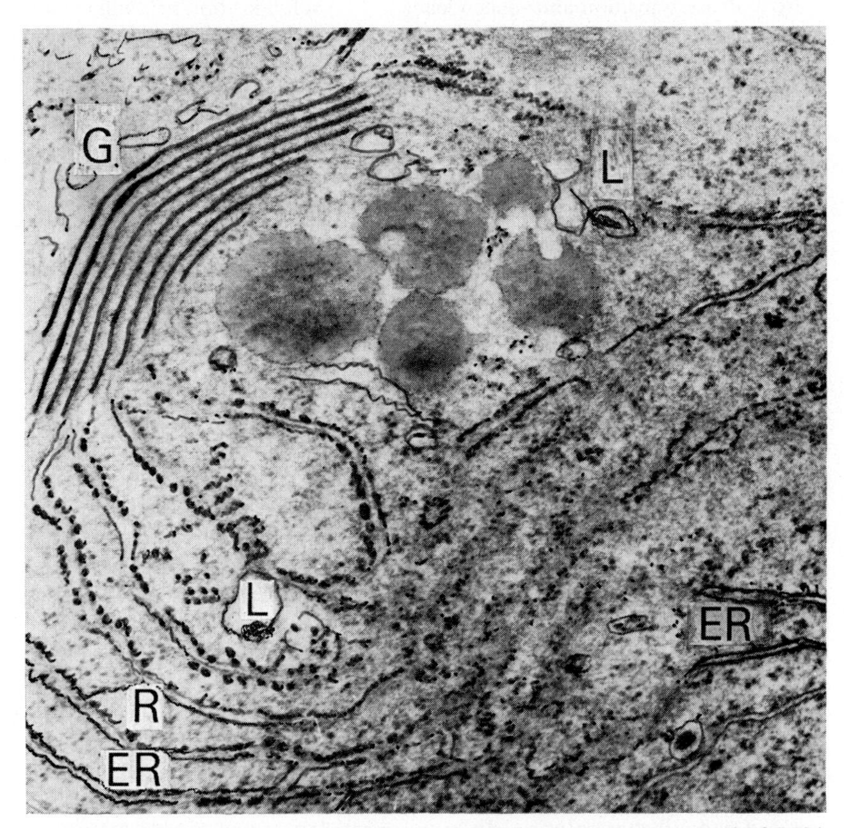

FIG. 6.2.4. Ordered structures within a cell. Ribosomes (R), composed of RNA and protein, are distributed in rows of black dots along the endoplasmic reticulum (ER). Other ordered structures include the Golgi apparatus (G) and lysosomes (L). (Photograph touched-up.) (Electron micrograph ×63,840.)

for the thought that a living cell is very much greater than the sum of its parts and that, if fractionated or fixed, it cannot be revitalized. To study structure, and especially to study relationship between structure and function in living systems, it is essential to look at living cells or, at the very least, to be aware of the alterations and artefacts introduced by processing and extracting biological matter (Fig. 6.2.3 and Fig. 6.2.4).

For reference purposes, important points relating to the four orders of biological reactions discussed above are summarized in Table 6.2.1.

TABLE 6.2.1. *Orders of biological reactions*

Order	Conditions	Examples
1st	Hydrophilic molecules in simple or colloidal solution	Transport, simple exchanges, hydrolysis.
2nd	Interface between reactants	Selective exchanges of ions and crystalloids; localization of charges.
3rd	Phase-transition and -dependence	Solubilization of lipids.
4th	Preferred orientation of molecules in layers, chains, ribbons, sheets, coils and more complex configurations	Synthesis of steroids, nucleic acids and proteins.

Change of Phase

Dehydration, extremes of temperature, radiation and other insults to the phase-equilibria and energy relationships of living tissue cause damage or death, even when chemical change is not detectable. Change of phase and bioenergetic transformation in living systems must therefore proceed within relatively narrow limits which, thermodynamically, would seem to limit reactivity but, biophysically, present an infinitude of possibilities, since viscosity, conductivity, hydrophilia, detergence, osmosis, flow, shape, consistency, dispersion and numerous other properties of matter are easily altered by small changes in phase and temperature. In complex, fourth-order reactions (see page 312 and Table 6.2.1) involving macromolecules, steric and conformational factors also contribute to the potential for mobile structural change and therefore for biophysical reactivity [21]. In addition to changes in aqueous sols and gels, the boundaries and dispersions between lipids and water furnish many examples of the extent and complexity of phase-change in animal tissues and body fluids (see Plates). The absorption of solid fat requires a phase change from a semi-solid to an emulsion or micellar solution in a non-ordered or low-order form. Absorption of fat in animals and man *via* intestinal lymphatic capillaries (lacteals) into lymph and thence into the blood stream requires another phase change to form the finely dispersed chylomicrons in which newly absorbed fats circulate (alimentary lipemia) in an easily reversible non-ordered micellar solution. In humans, chylomicrons consist mainly of triglycerides which are either dispersed as such directly into tissue cells or hydrolysed by "clearing factor" comprising a lipase and its activator [34] whereby the bulk of the expendible circulating fat is removed for anabolic storage or is consumed metabolically. The need for conversion to an ordered system arises only when lipid has to be incorporated integrally into the structure of cells and tissues; abnormally, low-density lipoprotein molecules in the ordered phase can occur or be produced [35] in the blood stream or in the wrong tissues and thereby lead to atherosclerosis and degenerative disease [8]. Indeed, in many ways, the growth and vitality of a cell depends upon regulation of the ordered phase, while degeneration and senescence are represented by the reverse, i.e., lack of this regulation which, to some extent, appears to be a purely physico-chemical process. If the phase equilibrium is disturbed by a simple physical agency, sufficient to affect the balance of the weak forces maintaining the ordered liquid crystal, the result physically is the same as alteration in phase but, biologically, it is degeneration, disaster or death of that cell or tissue. In physico-chemical terms, it may be said that death of a cell or tissue can be caused simply by a change in weak intermolecular forces in the ordered state.

Fat therefore has a dual biological rôle in vital intracellular ordered structures, from the membrane inwards, including integral components like mitochondria, Golgi apparatus, and various functional micelles. The liquid crystals concerned in these subcellular particles are all lyotropic [36]. The variability of lipid phases according to temperature indicates that there must be differences in the types of lipid needed in cold-blooded and warm-blooded species. Cold-blooded animals in general use fats which are semi-solid or liquid in their environment; lowering the temperature slows transportation and utilization of their cellular and depot fat, and hence in winter they have to cut down oxygen consumption or perish. The range of lipids eligible for cellular function under these restricted conditions is limited, and hence the potential for metabolism, environmental adaptation, cellular chemistry and diversification of behaviour of cold-blooded species is correspondingly limited. In warm-blooded species a much greater range of complex fats can be accommodated and utilized. Cerebral function requires a variety of lipid-mediated intracellular reactions and it is therefore no surprise to find that the higher levels of cerebration occur only in species such as primates, canines and porpoises which are not only warm-blooded but have stable, highly regulated temperature maintenance and an omnivorous drive to consume naturally a wide range of glycerides which provide side chains for phosphatides, cholesteryl esters, cerebrosides, and other polar fats whose orientation and reactivity come from the hydrophilic polar groups of their molecules.

The Lipid Mesophase in Neural Tissue

Most substances known to cause alterations in the function of nerve cells and tissues are lipid-soluble. The Overton–Meyer theory of narcosis [37] postulates that the effect of a drug upon a nerve cell is proportional to its fat solubility; the process is influenced by molecular size and free energy of absorption of the chemical in the lipid phase [38]. Pauling [39] and Miller [40] propose, alternatively, that narcotic and anaesthetic effects depend upon the ability of the drug to form hydrates stabilized on the side-chains of proteins in the aqueous phase of the cell-substance. Changes in the intracellular cations K^+ and Ca^{++} are also relevant. Chloroform and ether cause immobilization or loss of K^+ ions through the interface for an artificial phospholipid mesophase [41]. A similar effect is observed for n-alkanols with four to eight carbon atoms, and, since it is known that anaesthetic and narcotic effects are proportionately associated with loss of K^+ ions, this suggests that a change in the function of the cell might arise in the first place by the amphiphilic phospholipid in the lamellar membrane and in the cell taking up the fat soluble substance and, in so doing, losing electrolyte from the intermediate aqueous layers [42, 43] into the surrounding liquid. New knowledge about cation-permeability, charge-fluctuation and phase-solubility has therefore modified the old idea that anaesthetics like ether, chloroform and cyclopropane owe their action to their fat solubility, i.e., to their ability to penetrate the amphiphilic monolayer, enter the lipid mesophase of neural cells and somehow disturb the state of metabolic and physical equilibrium associated with consciousness. Anaesthetic gases like Xenon [44], neural hormones like acetyl choline and the hallucinogenic drug lysergic acid diethylamide (LSD) have a pronounced effect upon the dispersion of lipid, aqueous protein and ions in a three-phase dispersion resembling neural tissue, by causing in relatively low concentrations what has been well described as a "three-body interaction" [38].

Such changes in phase-dispersion and permeability provide a reasonable physico-chemical basis for the disturbances in nervous activity and consciousness which follow the use of drugs like anaesthetics, narcotics, certain tranquillizers and possibly hallucinogens. In the ordered structure of lipid, lipoprotein and nucleoprotein as well as in phase-dispersion within cells and fibres in the nervous system, we have a molecular basis for all levels of nervous activity from simple reflexes to intellectual creativity and we have, also, a neurochemical concept worthy of exploration in its relevance to the action of drugs and neurotropic substances.

Ordered Integration of Lipids and Lipoproteins in Tissues and Body Fluids

Complex, water insoluble substances are useful biologically only when they are solubilized or dispersed as water-continuous emulsions. In this respect, attention has been focused upon cholesterol, because it is an essential component of many cells and body fluids. Cholesterol is hydrophobic, but only moderately lipophilic, and it illustrates well the problems of physiologic phase transformations.

In the solid state, cholesterol is not readily dispersed in water or electrolyte solutions under conditions which could occur naturally. In the microcrystalline state, dispersion is practicable but not under physiological conditions. Yet, since cholesterol is integrated into all cells and most body fluids in non-particulate form, some very efficient methods of microdispersion and phase-change must be operative biologically. It is therefore necessary to find how cholesterol initially in the solid state can be dispersed or solubilized in aqueous model systems, to study the form in which cholesterol and its esters exist in cells, tissue and body fluids and, by experimental methods, to see if a cholesterol-containing mesophase can be produced in tissue from the solid state or non-mesomorphic precursor dispersions.

Solubilization of cholesterol from the solid state has been studied with a variety of amphiphiles and surfactants. Complete solubilization under semi-physiological conditions occurs in quaternary aqueous systems containing bile salt and lecithin [45], and also ternary systems with lecithin as the amphiphile [11], in which case a complex, stable mesophase is formed at certain critical concentrations [46]. In aqueous systems, long chain aliphatic compounds such as polyoxyethylene derivatives of partially esterified sorbitol (Tweens) exhibit a pseudo-membranous smectic mesophase [8]. Addition of crystalline cholesterol results in formation of a more complex mesophase (Plate 6.2.1 (*a*) and (*b*)). This transition occurs within a limited thermal range on warming but, on cooling, the mesophase remains stable at a lower temperature [47], which explains the ability of the natural mesophase containing cholesterol to maintain its ordered state at temperatures below the freezing point of water.* In these amphiphilic systems, cholesterol enters the mesophase only as an unsubstituted molecule: cholesteryl esters, irrespective of the chain length of the fatty acid tail, do not form ordered smectic layers in Tweens, like the corresponding phosphatide esters; on the other hand, the fatty acid chain length of the polyglycol ester used as amphiphile does influence the order and stability of unsubstituted cholesterol in the mesophase, supporting the view [24, 38] that the packing of cholesterol molecules occurs in a steric pattern different from that of phospholipids in bilayers. Different cholesteryl esters appear to enter or somehow influence the mesophase in membranes and bilayers differently, as shown by permeability studies [48] and by nuclear magnetic resonance [49], though it is not clear whether, in living systems, cholesteryl esters actually form a lyotropic mesophase as easily as cholesterol. The presence of cholesterol in a lamellar mesophase has a stabilizing effect so that, despite the sensitivity of surfactants, a cholesterol-containing mesophase can withstand high energy radiation [8].

Cholesterol in Physiological and Pathological States

The solubility of cholesterol in human bile is determined by the relative concentration of bile salts (cholates) and lecithin in the quaternary aqueous system [45]. In some pathological states, the concentration of cholesterol is increased beyond the point of solubilization, leading to the deposition of cholesterol and the formation of gall stones [50]. This might well explain the occurrence of some forms of gall bladder disease, especially when there is stasis of bile in the gall bladder.

Despite the relatively high solubility of cholesterol in lipids, certain physiological processes involving cholesterol depend upon aqueous or partially aqueous phases, notably those processes involving transport of

* This depends upon rate of cooling. Instant cooling to $-20°C$ can damage membranes and kill cells which may survive when cooled slowly to $-70°C$. Hygroscopic liquids like glycerol can also stabilize lyotropic liquid crystals.

PLATE 6.2.1. Conversion of slab crystals of cholesterol into liquid crystals by polyoxyethylene derivatives of partially esterified sorbitol (Tweens)

(*a*) Solid crystals (lower left) at the interface with the polyoxyethylene sorbitol/water system, forming liquid crystals (upper right, interference figures) in a three-phase system of the type discussed in Chapter 5. (Polarised light ×100.)

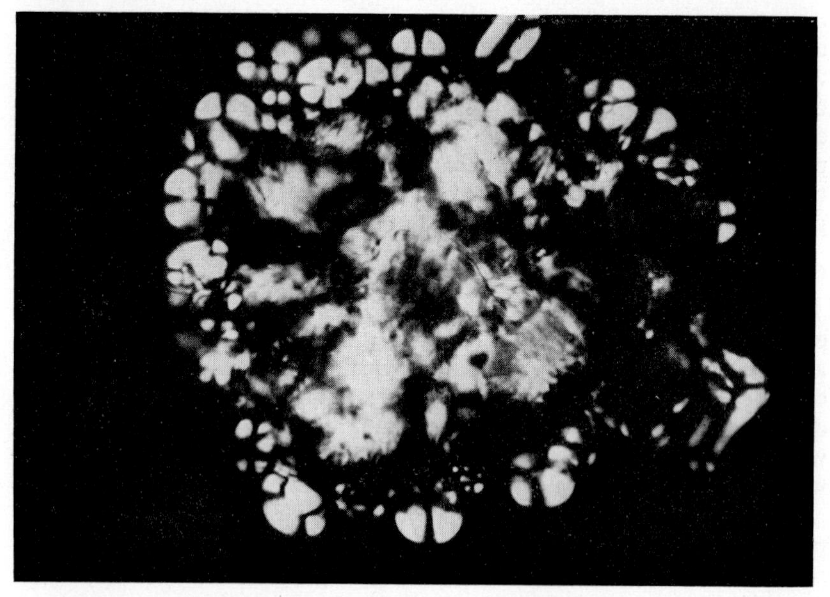

(*b*) Effect of amphiphile (polyoxyethylene sorbitol) in water at the edge of a cholesterol crystal (centre) converting it from the solid to the liquid crystalline state (cruciate spheroids at margin). (Polarised light ×400.)

ions, or protein or enzymatic hydrolysis. Thus cholesterol in erythrocytes, serum and lipoprotein complexes of tissue is either in a dispersed or in the liquid crystal state, in which form it can be identified intra- or extra-cellularly as anisotropic particles in a variety of situations, including leucocytes, nervous tissue, adrenal cortex, corpus luteum, low density β-lipoprotein fraction of plasma [7, 8] and atherosclerotic arteries [51, 52] (see Plate 6.2.2).

Of particular interest in relation to the ageing process in humans is the partitioning of cholesterol in plasma lipoproteins [53]. The total concentration tends to increase in middle age, but it is the distribution rather than

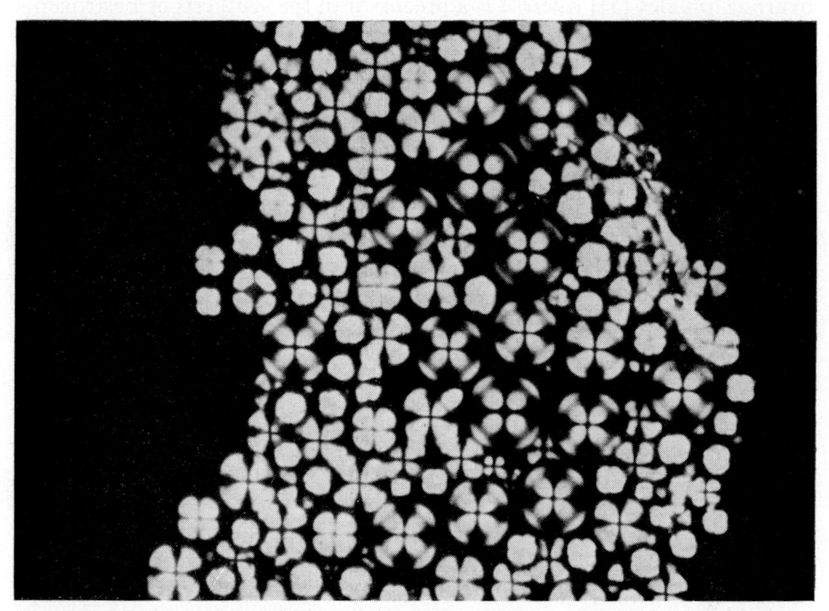

PLATE 6.2.2. Spheroids in the cholesterol containing mesophase from atheromatous deposit in human artery excised surgically, photographed using polarized light, at room temperature. Note secondary interference figures. (Magnification $\times 400$.)

the total amount which shows the most striking change. In youth, cholesterol and its esters are carried in the α-lipoprotein fractions which are, in general, of higher density when they are hydrated. In men, from the mid-twenties onward, and in women after the menopause, there is a sharp increase in β-lipoprotein which carries a relatively higher concentration of cholesterol, triglycerides and other lipids, and is therefore of lower density. About 50% of the cholesterol is esterified in the lipid phase; the remainder forms an ordered macromolecular complex with protein. Normally, the excess of lipid is split hydrolytically by lipases (activated by a "clearing factor") into water soluble products [54], but this physiological lipolysis is less effective against the low-density β-lipoproteins in which alteration of phase-equilibrium leads to deposition of cholesterol as its characteristic

β-lipoprotein mesophase [51]. Thus the clearing mechanism fails when it is most needed and cholesterol remains in the blood stream or enters the tissues. Cholesterol in its β-lipoprotein mesophase can be deposited in the intimal lining of aortas from young rabbits by pulsation at normal rabbit blood pressure [55, 56], which supports the viewpoint that disturbance of phase-equilibria is a strong biophysical factor in the pathogenesis of atherosclerosis and therefore of coronary heart disease and ageing. In similar fashion, the striking immunity of women to coronary heart disease during their reproductive years finds a biophysical explanation in the fact that cholesterol is diverted to or held in the luteinized cells of expended ovarian follicles [51] where it is a precursor in the synthesis of oestrogen, a hormone which is known to be effective *per se* in lowering excessive cholesterol in the β-lipoprotein phase in men [57], more or less in proportion to its capacity to cause femininization. One of the regulatory mechanisms of ageing appears therefore to be linked to reproductive function, an observation not entirely inconsistent with the natural, philosophic assumption that, at the organic as well as at the behavioural level, ageing intensifies as reproductive energy decreases. Consistent with this hypothesis is the finding that, in diseases associated with signs of premature ageing like untreated diabetes, hypothyroidism and some forms of chronic nephritis, the low-density β-lipoprotein phase is markedly increased and deposition of a cholesterol-containing mesophase in arteries and elsewhere is very conspicuous [51, 52].

Cellular and Intercellular Contacts

The structure and function of all tissues depend in large part upon local contacts and activity at the surfaces of the constituent cells. The surface layers, including the membrane, contain protein, polysaccharide and lipid in varying proportions, of which the phospholipid alone tends to conform to the intracellular composition [58]. The membrane is essentially an ordered, sterol-stabilized bimolecular (52 Å) phospholipid micelle, probably produced from intracellular phospholipid coated (28 Å) with protein, which is permeable to lipid-soluble substances, relatively impermeable to contained ions, insoluble in water and of high electrical resistance [59]. The configuration and composition of the membrane, and its relationship to other structures, have been well defined by numerous chemical, electron micrographic and X-ray diffraction studies. The arrangement depends upon the packing of steroid molecules which are probably accommodated between charged hydrophilic polar groups of phosphatidylethanolamine, lecithin or sphingomyelin [60], according to the nature of the phospholipid in the cell. However Ambrose [59] and others have raised theoretical objections to this hypothesis based upon the calculation of dipole moments of surface molecules, while other workers [16, 62, 63, 64] have shown that many different structural arrangements—leaflets, micelles, globules, particles and hexagonal columns—are likely to occur in the lipid-water systems of different types of cells. From the biofunctional viewpoint, the presence of "pores" of about 4 Å radius [65] and the amphiphilic nature of the cell membrane provide for the transportation of hydrophilic and lipophilic molecules in colloidal form, while the charge on the surface influences adhesion and intercellular contacts [59]. It is of interest to note that the cells of invasive forms of cancer have a decreased ability in comparison with normal cells of the same tissue to form firm adhesions, even in solid tissues [66]. Various enzyme molecules like phosphatase are also located in the cell surface, but the manner in which enzyme-substrate fit occurs at the cellular level has not yet been demonstrated with any certainty. Cholesterol and possibly other steroids are packed in monolayers with the type of phospholipid appropriate to the cell [61], and esterases may lie between these monolayers, which suggests a unique cellular site and mode of action for steroid hormones in sites of

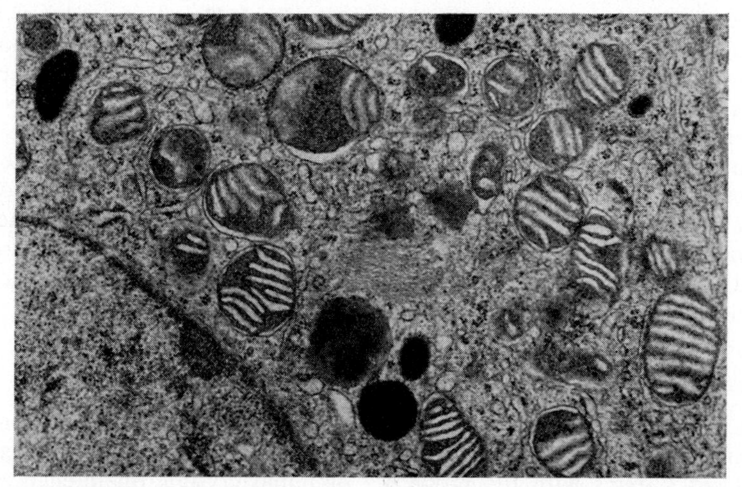

FIG. 6.2.5. Layering in lipid particles in ovarian cells in a mouse luteoma. (Photograph by courtesy of Dr. Lawrence M. Roth, New Orleans, La.). Note also the black, unlayered lipid droplets beside the nuclear membrane. (Electron micrograph $\times 27{,}360$ (OsO_4).)

active steroid metabolism and hormone synthesis such as the reproductive follicles in ovaries [51, 61] (Fig. 6.2.5).

Cells on absorptive surfaces e.g., intestinal mucosae, are usually different in structure, on their free edges at least. The outer layer may possess secretory areas, villi or other structures which may or may not be ordered. The membrane, in intestinal villi, is trilaminar [64] and the whole structure—unlike non-absorptive cell surfaces—is designed to "screen" or reject substances as well as to absorb them. This vital and fascinating property of cell surfaces is obviously a function of charge, solubility, molecular size and conformation, enzymatic activity and many other identifiable factors which from the biofunctional viewpoint are well worthy of study by a variety of available techniques.

Molecular Fit and Linkage (Antigens, Antibodies and Antibiotics)

There are also a number of situations in which highly-ordered molecules exist and interact independently of cells, though they originate in cells or are liberated by breakdown of cells or tissue, like the lipoproteins and enzymes described above. Of special interest in this respect, because of the need for very precisely ordered molecular tailoring and specificity of action, are the substances known as antigens, antibodies and antibiotics.

Antibodies formed in response to antigenic stimuli are highly specific in that they are functionally inert unless they fit the given antigen, or part of it, structurally. The antigen-antibody complex is to that extent ordered, in so far as combination cannot occur unless the relevant parts of their molecules fit exactly on a cell-surface or in a colloid. Hence the order of the antigen-antibody combination and the message to the immunologic information mechanism may depend upon morphology rather than upon chemical specificity. In the immunological situation, the specificity of interaction between antigen and antibody is very exact. In bacterial cells, antibiotics such as penicillins and cephalosporins appear to simulate or interfere with metabolic processes by a more approximate conformational resemblance to natural constituents, possibly in accordance with Quastell's signature principle [67]. In either case, the activity is of a very high order and depends less on chemical reactivity than upon molecular or macromolecular structure, order and fit.

Macromolecules in Colloids and Gels

Characteristically, the lyotropic mesophase is formed by interaction of components originally present in two or more phases [5, 11, 13]. Highly ordered systems with the properties of liquid crystals (maintenance of orientation in flow, birefringence, *etc.*) can however exist in aqueous colloids and gels [21]. This is explained by the fact that a stable, ordered state can be developed and maintained, even between large molecules, when their polar groups are kept in position by mutually-attractive forces, with van der Waals forces holding together the non-polar groups [68]. It follows that peptides, proteins, polysaccharides, nucleic acids and other macromolecules may aggregate or loosely conjugate, without loss of their individual identities, to form ordered complexes bonded with water, electrolytes or dissimilar molecules in colloidal solution.

Of particular importance and endless fascination in this respect are the nucleic acids and the brilliant concept of a genetic code [72]. The details of hypotheses, experimentation and deduction in this forum are now a luminous chapter in creative scientific literature [69]. What concerns us here is the discovery, not only of a supremely and permanently ordered, adaptable, molecular system, but also of a configuration which must, by analogy with the work of Crick and Watson [71], Robinson [73], Wilkins [74] and Luzzati [75] with their colleagues, be liquid crystalline in nature. Of the existence of the mechanisms appropriate to the initiation of an ordered structure there can be no doubt. Actual proof of liquid crystalline structure *in vivo* rests on a slender, almost tenuous basis: the deductions of Perutz [76] about the structure of haemoglobin S, of Wilkins *et al.* about transfer RNA [74] and of Luzzati *et al.* [75] about DNA. Yet we are dealing here with a subject where the origins, not only of life but also of the genetic mainspring of the pattern of life are discernible on the periphery of our scientific view.

The vital intracytoplasmic process of macromolecular aggregation and the building of tissue—or the intraneuronal building of information—is a biophysical process which clearly depends upon the replication of ordered macromolecular structures. With the primary structure coded, the subsequent configuration of the larger protein molecules is a physical process "akin to crystallization" (Perutz)—or to liquid crystallization—according to the nature and function of the molecule.

In the foregoing review, I have concentrated upon a relatively small range of biological structures and products, chosen because of their functional importance, practical utility or availability for study. Many other cellular and subcellular components, including mitochondria, lysosomes, germinal and embryonic tissue merit description and fuller study of their ordered structure by newer techniques which could be applied very appropriately to several earlier, incomplete but promising studies [10, 77] or to more recent hypothetical projections [67, 78] aimed at penetrating the technological relationships between living and non-living matter. The immense leap in knowledge of protein and nucleic acid structure which came from chromatography [79], X-ray diffraction analysis [69], molecular models [70], electronic absorption spectroscopy [80], fluorescent probes [81] and electron microscopy have not been matched by comparable innovations in the direct visualization of living tissue or of macromolecules in their natural state. It is clear however that the liquid crystalline form is universally present in and required by living matter, especially where complex macromolecules and phase dispersions are concerned or where, in general terms, biofunctional requirements are such that orientation and low-energy positional reactivity must be preserved in states of flow, stress and mobility. Finally we remind the reader of the stability displayed [53] by a cholesterol-containing mesophase, resembling that of some membranous and subcellular structures, to levels of γ-irradiation lethal to intact cells and destructive even to solids such as glass. On the other hand, relatively small, non-lethal doses of X-rays affect DNA readily, producing lasting chromosomal damage [82], and "soft" irradiation of this nature is well known as a means of inducing mutations [83]. A mesophase with a potential for membranous and micellar growth might therefore have appeared—or might survive—in a world with a high level of radioactivity, whereas genetic coding, protein synthesis and cellular growth could only proceed—or survive—at much lower levels of radioactivity. In terms of liquid crystalline structure, we might therefore view the entire Biosphere as having two functionally distinctive lyotropic mesophases composed, respectively, of lipid/water and protein or nucleoprotein, which may develop separately but have to act jointly for Life, as we define it, to appear, develop and persist. The rôle of liquid crystals in life processes might therefore be summarized by stating that they represent, not components, but physico-chemical conditions for Life itself.

REFERENCES

[1] GRAY, G. W. *Molec. Crystals Liqu. Crystals* **7,** 127 (1969).
[2] GRAY, G. W. *Molecular Structure and the Properties of Liquid Crystals*, Academic Press, London and New York (1962).
[3] BROWN, G. H. and SHAW, W. G. *Chem. Rev.* **57,** 1049 (1957).
[4] THEILHARD DE CHARDIN, *The Phenomenon of Man*, Harper and Row (1959).
[5] LAWRENCE, A. S. C. *Trans. Faraday Soc.* **29,** 1008 (1933).
[6] BERNAL, J. D. *Trans. Faraday Soc.* **29,** 1082 (1933).
[7] STEWART, G. T. *Nature, Lond.* **183,** 873 (1959).
[8] STEWART, G. T. *Adv. Chem. Ser. Am. Chem. Soc.* No. **63,** 141 (1967).
[9] LAWRENCE, A. S. C. and NEEDHAM, J. *Nature, Lond.* **147,** 766 (1941).
[10] NEEDHAM, J. in *Biochemistry and Morphogenesis*, Cambridge University Press (1942), p. 663–665.
[11] DERVICHIAN, D. G. *Trans. Faraday Soc.* **42,** 180 (1946).
[12] STOECKENIUS, W. *J. biophys. biochem. Cytol.* **5,** 491 (1959).
[13] LUZZATI, V., MUSTACCHI, H., SKOULIOS, A. and HUSSON, F. *Acta. Crystallogr.* **13,** 600 (1960).
[14] FINEAN, J. B. *J. biophys. biochem. Cytol.* **8,** 31 (1960).
[15] ROBERTSON, D. G. *J. Cell Biol.* **19,** 201 (1963).
[16] LUZZATI, V. and HUSSON, F. *J. Cell Biol.* **12,** 207 (1962).
[17] BANGHAM, A. D. and HORNE, R. W. *Nature, Lond.* **196,** 952 (1962).
[18] GLAVERT, A. M., DINGLE, J. T. and LUCY, J. A. *Nature, Lond.* **196,** 953 (1962).
[19] CASS, A. and FINKELSTEIN, A. *J. gen. Physiol.* **50,** 1765 (1967).
[20] WHITTAKER, V. P. *Br. med. Bull.* **24,** 101 (1968).
[21] STEWART, G. T. *Molec. Crystals Liqu. Crystals* **7,** 75 (1969).
[22] NAPOLITANO, L., LEBASON, F. and SCALETTI, J. *J. Cell Biol.* **34,** 817 (1967).
[23] CHAPMAN, D. *The Structure of Lipids*, Methuen & Co., London (1965).
[24] VANDENHEUVEL, F. A. *J. Am. Oil Chem. Soc.* **40,** 455 (1963).
[25] EICHEL, B. and SHARIK, H. A. *Science* **166,** 1424 (1969).
[26] NEVILLE, D. M. *J. biophys. biochem. Cytol.* **8,** 413 (1960).
[27] PALADE, G. E. *J. Am. med. Ass.* **198,** 815 (1966).
[28] WALLACH, D. F. and ZAHLER, P. H. *Proc. natn. Acad. Sci., U.S.A.* **56,** 1552 (1966).
[29] *Surface Chemistry of Biological Systems* (edited by M. Blank), *Adv. exp. Med.* and *Biol.* Vol. 7, Plenum Press, New York (1969).
[30] HARTSHORNE, N. H. and STUART, A. *Crystals and the Polarising Microscope* (4th edition), Arnold, London (1970).
[31] BREWER, D. R. *J. Path. Bact.* **74,** 371 (1957).

[32] RHODES, M. B., PORTER, R. S., CHU, W. and STEIN, R. S. *Molec. Crystals Liqu. Crystals* **10,** 295 (1970).
[33] SAGGERS, B. A. and STEWART, G. T. *J. Bact.* **96,** 1006 (1968).
[34] KORN, E. D. *J. biol. Chem.* **226,** 827 (1957).
[35] STEWART, G. T. *Br. J. exp. Path.* **58,** 109 (1958).
[36] MANDELL, L., FONTELL, K. and EKWALL, P. *Adv. Chem. Ser. Am. Chem. Soc.* No. **63,** 89 (1967).
[37] MEYER, K. H. *Trans. Faraday Soc.* **33,** 1062 (1957).
[38] MISHRA, R. K. *Molec. Crystals Liqu. Crystals* **10,** 85 (1970).
[39] PAULING, L. *Science* **134,** 15 (1961).
[40] MILLER, S. L. *Proc. natn. Acad. Sci. U.S.A.* **47,** 1515 (1961).
[41] DANIELLI, J. F. *Proc. 7th Symp. Colston Res. Soc., Univ. Bristol.* 1 (1954).
[42] EISENMAN, G. *Fedn. Proc. Fedn. Am. Socs. exp. Biol.* **27,** 1249 (1968).
[43] WALLACH, D. H. F. and GORDON, A. *Fedn. Proc. Fedn. Am. Socs. exp. Biol.* **27,** 1263 (1968).
[44] CULLEN, S. C. and GROSS, E. G. *Science* **113,** 580 (1951).
[45] SMALL, D. M., BOURGES, M. and DERVICHIAN, D. G. *Nature, Lond.* **211,** 816 (1966).
[46] STEWART, G. T. *Molec. Crystals* **1,** 563 (1966).
[47] STEWART, G. T. *Molec. Crystals Liqu. Crystals* **7,** 75 (1969).
[48] FINKELSTEIN, A. and CASS, A. *Nature, Lond.* **216,** 717 (1967).
[49] CUTLER, D. *Molec. Crystals Liqu. Crystals* **8,** 85 (1969).
[50] ADMIRAND, W. H. and SMALL, D. M. *J. clin. Invest.* **47,** 1043 (1968).
[51] STEWART, G. T. *J. Path. Bact.* **81,** 385 (1961).
[52] SMALL, D. M. In *Surface Chemistry of Biological Systems*, Plenum Press, New York (1969), pp. 55–83.
[53] STEWART, G. T. *Nature, Lond.* **192,** 624 (1961).
[54] ROBINSON, D. S., FRENCH, J. E. and FLOREY, H. W. *Q. J. exp. Psychol.* **38,** 101 (1953).
[55] STEWART, G. T. *Br. J. exp. Path.* **41,** 389 (1960).
[56] STEWART, G. T. *Br. J. exp. Path.* **43,** 345 (1962).
[57] OLIVER, M. F. and BOYD, G. *Lancet* **2,** 499 (1961).
[58] VEERKAMP, J. H., MULDER, I. and VAN CEENEN, L. L. M. *Z. Krebsforsch.* **64,** 137 (1961).
[59] AMBROSE, E. J. In *Recent Progress in Surface Science* (edited by J. F. Danielli, K. G. A. Pankhurst and A. C. Riddiford), Academic Press, New York and London (1964), pp. 338–358.
[60] FINEAN, J. B. *Expl Cell Res.* **5,** 202 (1953).
[61] WILMER, E. N. *Biol. Rev.* **36,** 368 (1961).
[62] STOECKENIUS, W. *J. Cell Biol.* **12,** 221 (1962).
[63] LUCY, J. A. *J. theroret. Biol.* **7,** 360 (1964).
[64] REVEL, J. P. and ITO, S. In *The Specificity of Cell Surfaces* (edited by B. D. Davis and L. Warren), Prentice-Hall, Englewood Cliffs, New Jersey (1967), pp. 211–234.
[65] LINDEMAN, B. and SOLOMON, A. K. *J. gen. Physiol.* **45,** 801 (1961).
[66] ABERCROMBIE, M. and AMBROSE, E. J. *Expl Cell Res.* **15,** 332 (1958).

[67] QUASTELL, H. *Emergence of Biological Organization.* Yale University Press (1964).
[68] BERNAL, J. D. *Discuss. Faraday Soc.* **25,** 7 (1958).
[69] PERUTZ, M. *Proteins and Nucleic Acids*, Elsevier Publishing Co., Amsterdam (1962).
[70] KENDREW, J. C. *Brookhaven Symposia in Biol. No.* **15** (1962).
[71] CRICK, F. H. C. and WATSON, J. D. *Proc. R. Soc.* **A223,** 80 (1954).
[72] CRICK, F. H. C., GRIFFITH, J. S. and ORGEL, L. E. *Proc. natn. Acad. Sci. U.S.A.* **43,** 416 (1957).
[73] ROBINSON, C. *Trans. Faraday Soc.* **52,** 571 (1956).
[74] SPENCER, M., FULLER, W., WILKINS, H. H. and BROWN, G. L. *Nature, Lond.* **194,** 1014 (1962).
[75] LUZZATI, V., NICOLAIEFF, A. and MASSON, F. *J. molec. Biol.* **3,** 185 (1961).
[76] PERUTZ, M. F., LIGNORI, A. M. and EIRICH, S. *Nature, Lond.* **167,** 929 (1951).
[77] NAGEOTTE, J. *Actualités Scientifique et Industrielles* **9,** Hermann et Cie, Paris (1936), pp. 431–434.
[78] SCHRODINGER, E. *What is Life?*, Cambridge University Press, (1944).
[79] SANGER, F. and TUPPY, H. *Biochem. J.* **49,** 481 (1951).
[80] *Spectroscopic Approaches to Biomolecular Conformation*, (edited by D. W. Urry). Am. Med. Assoc., Chicago (1970).
[81] SMEKAL, E., TING, H. P., ANGENSTEIN, L. G. and TIEN, H. T. *Science* **168,** 1108 (1970).
[82] CHU, E., GILES, N. and PASSANO, K. *Proc. natn. Acad. Sci. U.S.A.* **47,** 830 (1961).
[83] RUSSELL, W. L. *Proc. Am. phil. Soc.* **107,** 11 (1963).

7

Scientific and Technological Applications of Liquid Crystals

7.1 CHOLESTERIC LIQUID CRYSTALS IN TEMPERATURE MEASUREMENT AND VAPOUR DETECTION

G. W. Gray

Abstract

Important uses for cholesteric liquid crystals arise from the unique optical properties of thin films of these mesophases in the Grandjean plane texture. Such films rapidly and directly indicate by their colour the temperature of a surface on which they are coated. In addition to their uses for temperature measurement, cholesteric liquid crystals provide a means for measuring microwave energy, infrared and visible light and laser output, and they are reported to have potential applications as detectors of small amounts of contaminants in the atmosphere.

As mentioned in earlier Chapters, the Grandjean plane texture of the cholesteric mesophase has a helical structure. A film of cholesteric having this texture is therefore optically negative, and since it is not symmetrical about any axis or point, it is optically active. The optical rotatory power of such a film may be very large; optical rotations as high as 1,800° mm^{-1} have been reported [1].

The complex optical properties of films of cholesteric mesophase having the Grandjean plane texture have been discussed in many publications—see for example articles by Friedel [2], Cano and Chatelain [3], de Vries [4], Dreher, Meier and Saupe [5] and Keating [6]. From the standpoint of this section, it is necessary to mention only those features relating to the coloured light reflected from such films. Although this property resembles that shown by a film of oil on a water surface, there are marked differences. Whereas reflection of coloured light from an oil film occurs when the thickness of the film is an exact multiple of half wavelengths, the behaviour of a plane film of cholesteric is unaffected by the thickness of the film; the wavelength maximum of the light reflected remains the same for different thicknesses. It is concluded therefore that the helical ordering of the mesophase is a bulk property and not one which is *confined* to the surfaces, although it is of course at the supporting surface that the helix is *initiated.* Moreover, the light reflected from a cholesteric film, unlike that from an oil film, is circularly polarized, either to the right or left, dependent upon the cholesterogen. This property, and the fact that the selective reflection is independent of whether the illuminating light is polarized or not gives cholesteric films their characteristic iridescence, reminiscent of a peacock's feathers.

Most colourless materials do not give *selective* reflection of light and the light which they transmit is in the same wavelength band as the light which they reflect. With a film of material giving selective reflection, the light reflected is complementary to the transmitted light. Selectively reflecting cholesteric films give [7] very little absorption of light, and the peak associated with the reflection is of narrow bandwidth.

Thus, if P is the pitch of the rotation and n is the refractive index, and the cholesteric film is illuminated with ordinary light directed parallel to the axis of rotation of the helix, then, over a very small wavelength range [38] about $\lambda_R = Pn$, the light is divided into two beams, one circularly polarized in one sense is transmitted and the other polarized in the opposite sense is totally reflected. No reversal of the sense of rotation occurs on reflection, in contrast to that occurring when a metallic surface reflects a beam of circularly polarized light, i.e., if right-handed circularly polarized light is reflected from a right-handed cholesteric, it remains circularly polarized to the right. Outside the wavelength range about λ_R, perpendicular transmission of light gives the very high optical rotation mentioned above.

If the small wavelength range about λ_R lies in the visible region of the spectrum, the total reflection which separates waves with opposite circular polarizations, gives rise to bright colours. The wavelength (λ_{max}) and the colour of the reflected light depend upon three main factors: (1) the temperature of the film, (2) the angle of incidence of the light illuminating the film and the viewing angle of the reflected light and (3) the nature of the organic material comprising the mesophase. These factors are now discussed.

1. Temperature Dependence

The pitch of the helix, P, is for many substances highly dependent, though in a rather irregular way, on temperature, and therefore the coloured light reflected from the film is also strongly dependent on temperature. For this reason, it is possible to use films of cholesteric material in the Grandjean plane texture to indicate the temperatures of surfaces on which they are coated; this use of cholesterics is commonly called surface thermography. The pitch of the helix for cholesteric materials is usually in the range 0·2 to 20 μ, and the light reflected from the film may change from the ultraviolet to the infrared through the entire visible spectrum over a wide or a narrow temperature range, dependent upon the chemical nature of the cholesterogen or mixture of cholesterogens. Although the change in colour of the reflected light can be interpreted in terms of change in pitch of the helix [8], it is important to note that the colour dependence on temperature may be positive or negative. Thus, as the temperature falls, the first colour to appear may be violet; this may then change to blue, then to green, then to yellow and then to red, the wavelength maximum of the reflected light then passing beyond the visible range into the infrared. This is the case for the Grandjean plane textures of the cholesteric mesophases of the cholesteryl n-alkanoates, although only part of the colour sequence may be observed in practice before crystallization occurs. With other cholesteric materials, the sequence of colours is reversed on cooling. In other cases, the change on cooling may be from red to green and back again to red. The film of cholesteric mesophase does not reflect coloured light when the wavelength maximum is in either the ultraviolet or infrared ranges and remains there throughout the temperature range in which the mesophase exists.

2. Angular Dependence

When the film of cholesteric mesophase is illuminated by incident light directed normal to the plane of the film, the wavelength of the reflected light is a maximum in the direction normal to the film. If a mechanical disturbance is caused in the film, the wavelength maximum changes momentarily and then reverts to the original value when the film recovers. The wavelength maximum of the reflected light also changes, at a constant temperature, with change in the angle at which the film is illuminated and in the angle from which it is observed. The relationship between the wavelength of the reflected light and the angles made by the incident and reflected rays with the direction normal to the film can be explained approximately [9] on the basis of the reflections arising from a number of regularly spaced planes, i.e., in terms of a Bragg reflection [7, 10].

In the use of cholesteric films to measure absolute or real surface temperatures, angular dependence of the wavelength of the reflected light must be taken into account, since the observed colour will change with angle of illumination and angle of viewing. Carefully defined optical conditions are therefore necessary when measuring real temperatures from the wavelength of the reflected light [7, 11].

3. Dependence on the Nature of the Cholesterogen

Any optically active compound whose molecules are lath-like in nature may form a cholesteric mesophase. However, the commonest cholesterogens, and consequently those most widely used for surface thermography, are derivatives of the naturally occurring, optically active alcohol, cholesterol. Examples are provided by cholesteryl chloride, cholesteryl n-alkanoates, cholesteryl n-alkyl carbonates and cholesteryl alkenyl carbonates. Derivatives of this kind are usually crystalline at room temperature, and some may be quite high melting. For many applications of surface thermography, it is desirable to have a cholesterogen that can be applied to the surface in the cholesteric state around or below the room temperature range. For these reasons, cholesterogens are seldom used in the pure state, and normally mixtures of two or more cholesterogens are employed, primarily to reduce the melting points compared with those of pure mesogens. Non-mesogens, including certain dyes, may also be added to alter the physical properties and depress the melting point, although the temperature range in which the cholesteric mesophase exists must not be too adversely affected. A large body of useful, but empirical data has been accumulated with regard to colour dependence (positive or negative over a narrow or a wide temperature range) of pure cholesterogens and mixtures of cholesterogens. This information is important in the choice of the best materials, whether of high, medium or low temperature sensitivity, for measuring thermal gradients and changes over a surface, within a particular temperature range; mixtures of cholesterogens suitable for surface thermography at widely different temperatures and over different temperature ranges are commercially available from several sources which provide information sheets [12].

A particularly useful binary mixture is provided by cholesteryl oleyl carbonate and cholesteryl nonanoate [13]. Pure cholesteryl oleyl carbonate changes to the amorphous liquid at *ca* 40°C and gives a full colour response in a narrow range (*ca* 2°C) just above 20°C. With increasing concentrations of cholesteryl nonanoate, both the actual temperature at which an observed colour change occurs and the temperature range over which the full colour response takes place increase, e.g., with *ca* 80% cholesteryl nonanoate, the change from violet to red occurs over the range 62–52°C. A 1:1 (w/w) mixture of the two esters gives a beautiful and full colour response over *ca* 3°C in the body temperature range and is suitable for skin thermography. Addition of a few per cent of cholesteryl chloride to such a mixture broadens the temperature range, and addition of more cholesteryl chloride makes the mixture still less temperature sensitive, so that with *ca* 20% of the chloride, the colour is mainly blue at body temperature. Another mixture used for skin thermography consists of 80% cholesteryl nonanoate, 15% cholesteryl chloride and 5% cholesteryl oleate [14]. Other materials may be added to these and to similar mixtures to further modify the temperature characteristics. In references [11], [15] and [16] are quoted more examples of mixtures formulated for various purposes and giving different temperature ranges of colour response at different temperatures. A full colour response over as small a temperature range as 0·1°C is claimed in technical literature for certain materials, and other compounds and mixtures of compounds are available which change from violet to red over wide temperature ranges, e.g., from 112–52°C and from 161–74°C.

The observed temperature dependence of the colour of light reflected from a cholesteric film made from a *pure* cholesterogen appears to be a complex function of its chemical constitution and molecular structure. This is not unexpected when it is remembered that some cholesterogens have a positive and some a negative temperature dependence. With the cholesteryl n-alkanoates, the colour changes from violet to red on cooling, and in this series an empirical correlation has been noted [17] to the effect that the reflected light shifts to shorter wavelengths as the homologous series is ascended. The cholesteric mesophase of cholesteryl formate reflects light which is only faintly red, whereas the mesophase of the next homologue, the acetate, appears green at first and then changes to red on cooling, before crystallization occurs. Higher members give cholesteric mesophases which give a rich violet colour at first, the colour changing with greater or lesser rapidity through the spectrum to red, on cooling, provided that crystallization is sufficiently slow to occur. However, still higher homologues such as the laurate, myristate and palmitate give cholesteric mesophases that are coloured only weakly violet, and the cholesteric mesophase of cholesteryl stearate is almost colourless. There is also some evidence that the colour sensitivity increases as the ester alkyl chain of a cholesteryl n-alkanoate is lengthened [7].

Marked variations in the colour of light reflected from a cholesteric mesophase are consequently observed when the structure of the cholesterogen is varied more widely. Thus, cholesteryl *p*-nitrobenzoate [18] gives a high temperature cholesteric mesophase between 186°C and 258–9°C and shows particularly brilliant colours which change from violet to red as the temperature falls from the isotropic–cholesteric transition point. Cholesteryl 3,5-dinitrobenzoate, on the other hand, reflects only red light from its monotropic cholesteric mesophase, resembling in this respect cholesteryl formate and also (-)2-methylbutyl 4-*p*-cyanobenzylideneaminocinnamate.

The effects which composition may have on the optical properties of Grandjean plane textures of cholesteric mesophases are illustrated by an example taken from Friedel's observations [2]. A mixture of cholesteryl benzoate and cholesteryl acetate gives a cholesteric mesophase which reflects coloured light that changes from shorter to longer wavelengths on cooling. However, a mixture of cholesteryl benzoate (*ca* 37%), cholesteryl acetate (*ca* 30%) and *p*-azoxyanisole—a nematogen—(*ca* 33%) gives a mesophase showing a reversed colour sequence (green to blue) on cooling.

From what has been said about the sensitivity of the temperature range of full colour response and the sensitivity of the actual temperature at which a given colour appears to the chemical constitution in the case of a pure cholesterogen and to the composition in the case of a mixture of cholesterogens, it is clear that if surface thermography is used for the measurement of real temperatures, a clean surface must be used and frequent and careful calibration [7] of the cholesteric materials must be made in any continuing series of experiments. The wavelength of light reflected under a defined set of optical conditions from a film of cholesteric mesophase can be related to the real temperature of the film, but the relationship will be affected if the composition of the cholesteric mesophase changes, e.g., through contamination or decomposition [19].

Such difficulties may be overcome by careful control of experimental procedure, but in many applications of surface thermography the aim is only to detect temperature differences, i.e., the contours of temperature changes over a surface, and not to measure real temperatures. Such measurements are less sensitive to slight changes in composition of the cholesterogen through contamination. However the effects of contaminants cannot be ignored, for the performance of a cholesteric liquid crystal—the quality of the colours, the speed of response to temperature change etc.—are adversely affected by over long exposure to the surroundings. Absorption of impurities from the air or the surface on which the liquid crystals are coated, atmospheric oxidation or hydrolysis and ultraviolet light are factors which influence their behaviour. Although unprotected films are marred in a few hours, cholesteric liquid crystals stored in sealed glass containers have a long lifetime.

It is important therefore that the film of cholesteric mesophase applied to the surface (either by painting or by spraying on a solution in a volatile solvent which is allowed to evaporate) should not be contaminated by materials on the surface. The surface must be free from oil, grease etc which would dissolve in the cholesteric and alter its optical properties. In skin thermography, diffusion of lipids into the mesophase must be prevented, and a preliminary defatting using a detergent is recommended [20]. The skin or other surface is then sealed with a black, water based paint which provides a good background against which to view the colours, and one which absorbs the light which is not reflected by the film.

The necessity for a black background may be a problem in some cases where it is undesirable to paint a surface. This may be overcome by using a film of cholesteric between two thin plastic sheets, the lower one with a black back. The sandwich, contained within an O-ring, is then placed over the object [21]. Evacuation of the air pulls the film down over the surface, giving good contact even in the case of uneven areas of surface.

The National Cash Register Company has developed a process for encapsulating the liquid crystal in spheres of gelatin of *ca* 40–50 μ diameter [22]. Aqueous suspensions of the encapsulated cholesteric may be sprayed on black plastic, paper etc. to form films which, when dry, may be used [41] for many of the same applications as unencapsulated materials. Although some disadvantages of encapsulated films have been reported [7], they do have the advantage that the cholesteric is protected within the capsule, and the films have longer lives in the exposed state. The intensity of the light reflected from encapsulated cholesterics is also less angle dependent.

Time Constants for Cholesterics and their Temperature Resolution

In some cases, if a film of cholesteric mesophase is suddenly chilled, the mesophase gives a glass which continues to reflect coloured light of the original wavelength. That is, the cholesteric structure requires time to respond to the temperature change. Fergason [7] defines a time constant, $1/\beta$, for cholesteric liquid crystals, and quotes 0·1 s for cholesteryl nonanoate and 0·2 s for cholesteryl oleyl carbonate. He points out that this does not represent a limitation when viewing by eye and that correction of the observed wavelength for rate of response allows cholesteric liquid crystals to be used when temperature changes occur within a quarter of a time constant. Elsewhere Fergason quotes time constants as low as 10 ms [23] and 30 ms (quoted in reference [31]), but the materials are not specified.

Fergason also describes the high density of information obtained from cholesteric films and quotes a resolution of 1,000 lines per inch with a temperature resolution of $<0{\cdot}1°C$ [7]. Further details on temperature and spatial resolution are given in reference [31].

Examples of the Application of Cholesteric Mesophases in Surface Thermography [43]

Skin Thermography

Areas of skin lying over veins and arteries are slightly warmer than other areas, and the thermal map provided by a film of cholesteric clearly indicates in colour the paths of these veins and arteries. Subcutaneous tumours are also revealed in size and shape by the warm areas of skin overlying them. The applications of skin thermography in the medical field [14, 20, 24] are therefore concerned with the detection and response to treatment of vascular disease and subcutaneous tumours. In the former case, information is obtained about inefficient functioning of the circulatory system through blockage of veins or arteries, the extent of inflammatory disease and the effects of drugs designed to improve circulation. The detection of tumours by this technique is probably most important in connection with breast cancer. References [14] and [20] provide details of experimental methods and photographs of clinical tests. The advantages and disadvantages of using cholesteric films for temperature mapping of skin compared with direct temperature measurement by thermocouples or thermistors or with the use of infrared thermography are discussed.

Surface Thermography in the Electronics Industry

The objective is to obtain temperature distributions for electronic components, to detect points of failure and to identify points of potential failure as indicated by localized hot spots in the circuitry. Considerable technical problems are connected with achieving these objectives, since the sizes of the components may vary from large to microscopic [31], but despite certain difficulties, cholesteric liquid crystals appear [25, 26] to offer a means of achieving them. Cholesteric liquid crystals have the high spatial resolution required and are becoming more widely accepted as valuable materials for such non-destructive testing.

Surface Thermography in Aerodynamic Testing [27, 28, 29]

The use of cholesteric liquid crystals for thermal mapping of the surfaces of aircraft models in wind tunnel tests has been explored. The method appears to be most valuable for detecting the demarcation between areas where laminar flow and turbulent flow are occurring. The problem with using cholesterics under these conditions for measuring real temperatures is that shear on the liquid crystal film affects its temperature response.

Surface Thermography in Non-destructive Testing of Laminates

Laminates with a honeycomb interior, e.g., those used in aircraft structures, owe their mechanical strength to the quality of the bonding between honeycomb and metallic skins. The quality of the bonding can be established by applying a film of cholesteric liquid crystal to one side of the laminate and observing the pattern of heat flow to the other surface. Areas of poor bonding will transfer heat badly and show up as uniformly coloured areas, whereas areas of good bonding will show the pattern of the honeycomb in colour against the cooler background [25, 30].

Miscellaneous Aspects of Surface Thermography

The actual and potential applications of cholesteric liquid crystals in surface thermography are very widespread, and all possibilities cannot be mentioned in a review of this kind. However, attention should be drawn to their applications in evaluating the efficiency of heat exchangers [7, 24], in studies of the conversion of mechanical to thermal energy in specimens undergoing tests for tensile strength or fatigue [24] and in detecting cracks in metal components or voids in weldings by making it possible to observe interruptions in the flow of heat across a surface from a point source [25].

In this general connection, two reports [39, 40] have mentioned the use of cholesteric liquid crystals as temperature sensors in studies of heat flow and convection in weightless liquids and gases. These studies were made during the return flight from the moon of the space ship Apollo 14.

The reader interested in obtaining more information about the technological applications of cholesteric liquid crystals will find useful data in reference [32] in which are gathered together all patents relating to liquid crystals up to 1969. These patents also cover the uses of cholesteric liquid crystals in the detection and measurement of electrical energy and various other forms of electromagnetic radiation. In the latter case, a radiation absorbing film transfers its heat pattern to the cholesteric film. In the case of electrical energy, the voltage may be applied across the thickness of cholesteric film producing changes in optical properties of the mesophase (Chap. 7.2). Alternatively, the electric field may be used to produce heat which is then detected in a cholesteric film; electrical energy may in fact be used to present information which is then displayed in a heat pattern by the cholesteric phase.

Another paper [33] describes developments in the use [34] of a cholesteric film to reproduce a scene by thermal radiography. The effect depends on recording variations in the infrared radiation emitted by objects because of differences in their emissivity and/or temperature. A radiation absorbing membrane in contact with the cholesteric film transfers a heat pattern to the liquid crystal. On illumination of the cholesteric film by monochromatic light of wavelength smaller than the shortest wavelength of light reflected from the film, the observer sees the regions of different temperature, i.e., the scene, the source of the radiation, as regions of different brightness. Temperature differences of 0·2°C were detected, but resolution was limited to about 25 lines per inch. It is thought however that a sensitivity of 0·02°C may be achieved with improved lighting. The technique is certainly interesting, for thermal imaging of this kind would make it possible to carry out the various kinds of surface thermography already described without the need to coat the surfaces with liquid crystals.

Detection of Atmospheric Contaminants

As mentioned before, the diffusion of impurities into a cholesteric film affects the pitch of the helix and changes the wavelength and colour of the reflected light. Fergason [16, 35] first proposed the use of this effect as a means for detecting contaminants in the atmosphere, since vapours of different organic solvents were shown to produce different colours in mixtures of cholesterogens of various compositions [42]. If the vapour is simply dissolved in the cholesteric, the colour change is reversible when the solute is lost by the mesophase by evaporation, but if chemical interaction occurs, the colour change is permanent. The sensitivities of cholesteric films were high (parts per million were detected), but this in itself introduces problems, because the films may be affected by atmospheric contaminants which one may not be concerned with detecting. In addition, they are affected by oxidation, exposure to sunlight etc. Encapsulated films of cholesterics are however more stable, but as a consequence are less sensitive to the contaminants to be detected.

Hysteretic Cholesterics

Attention is drawn finally to hysteretic cholesteric materials which are coloured but change to the amorphous liquid in response to a certain degree of temperature change or applied voltage or to a particular concentration of a contaminant. However, on removal of the stimulus, these materials revert to a colourless cholesteric condition and there is a time lag in the development of the original colour. The materials used are mixtures of cholesterogens, usually containing a cholesteryl or cholestanyl halide [36, 37]. Cholesteric films having this property are of interest for detecting whether a critical temperature, voltage or level of contamination has been exceeded some time after the effect has occurred. For example, a hysteretic film rendered amorphous by a 200 V cm^{-1} field remained colourless for three weeks after reverting to the cholesteric phase, provided that it was not heated to give the amorphous liquid and cooled back to the cholesteric phase. Similarly, a film made colourless by 70–100 p.p.m. of chloroform or trichloroethylene remained colourless for ten minutes after removal from the contaminating atmosphere.

REFERENCES

[1] MATTHIEU, J. *Bull. Soc. franç. Minéral* **62,** 174 (1939).
[2] FRIEDEL, G. *Annls. Phys.* **18,** 273 (1922).
[3] CANO, R. and CHATELAIN, P. *C. r. hebd. Séanc. Acad. Sci., Paris* **251,** 1139 (1960); **253,** 1815 and 2081 (1961); **259,** 352 (1964).
[4] DE VRIES, H. *Acta Crystallogr.* **4,** 219 (1951).
[5] DREHER, R., MEIER, G. and SAUPE, A. *Molec. Crystals Liqu. Crystals* **13,** 17 (1971).
[6] KEATING, P. N. in *Liquid Crystals* **2**, *Part II*, Proc. 2nd Int. Conf. on Liquid Crystals (edited by G. H. Brown), Gordon & Breach, New York (1968), p. 279.
[7] FERGASON, J. L. *Appl. Optics* **7,** 1729 (1968).
[8] ADAMS, J. and HAAS, W. *Molec. Crystals Liqu. Crystals* **16,** 33 (1972).
[9] ADAMS, J. and HAAS, W. *Molec. Crystals Liqu. Crystals* **11,** 229 (1970).
10] FERGASON, J. L. in *Liquid Crystals*, Proc. 1st Int. Conf. on Liquid Crystals (edited by G. H. Brown, G. J. Dienes and M. M. Labes), Gordon & Breach, New York (1967), p. 89.
[11] FERGASON, J. L., GOLDBERG, N. N. and NADALIN, R. J. in *Liquid Crystals*, Proc. 1st Int. Conf. on Liquid Crystals (edited by G. H. Brown, G. J. Dienes and M. M. Labes), Gordon & Breach, New York (1967), p. 105.
[12] *Aldrichimica Acta* **3,** 1 (1970) (Aldrich Chemical Co. Inc., Milwaukee, Winconsin, U.S.A.); *Information Sheet on Liquid Crystals* (Eastman Organic Chemicals, Rochester, New York, U.S.A.); *Information Leaflet on Liquid Crystals* (Liquid Crystal Industries Inc., Turtle Creek, Pennsylvania, U.S.A.); *Licristal* (*Liquid Crystals*) (E. Merck, Darmstadt, Germany); *Information Sheets on Liquid Crystals* (Vari-Light Corporation, Cincinnati, Ohio, U.S.A.).
[13] FERGASON, J. L. *Detection of Liquid Crystals*, August 1965, United States Department of Commerce, AD 620 940.
[14] GAUTHERIE, M. *J. Phys. Radium, Paris* **30,** C4, No. 11–12, 122 (1969).
[15] WOODMANSEE, W. E. *U.S. Patent 3,441,513* (1969)—see also reference [32].
[16] FERGASON, J. L. *U.S. Patent 3,409,404* (1968)—see also reference [32].
[17] GRAY, G. W. *J. chem. Soc.* 3733 (1956).
[18] GRAY, G. W. *Molecular Structure and the Properties of Liquid Crystals*, Academic Press, London and New York (1962), p. 194.
[19] See for example DIXON, G. D. and SCALA, L. C. *Molec. Crystals Liqu. Crystals* **10,** 327 (1970).
[20] SELAWRY, O. S., SELAWRY, H. S. and HOLLAND, J. F. in *Liquid Crystals*, Proc. 1st Int. Conf. on Liquid Crystals (edited by G. H. Brown, G. J. Dienes and M. M. Labes), Gordon & Breach, New York (1967), p. 175.
[21] WATERMAN, G. L. and WOODMANSEE, W. E. *U.S. Patent 3,439,525* (1969)—see also reference [32].

[22] CHURCHILL, D., CARTMELL, J. V. and MILLER, R. E. *British Patent 1,138,590* (1969); CHURCHILL, D., CARTMELL, J. V., MILLER, R. E. and BOUFFARD, P. D. *British Patent 1,161,039* (1969)—see also reference [32]; CHURCHILL, D. in *Liquid Crystals 2, Part I*, Proc. 2nd Int. Conf. on Liquid Crystals (edited by G. H. Brown), Gordon & Breach, New York, (1968), p. 230.
[23] FERGASON, J. L. *Aldrichimica Acta* **3,** 3 (1970).
[24] WOODMANSEE, W. E. in *Liquid Crystals 2, Part I*, Proc. 2nd Int. Conf. on Liquid Crystals (edited by G. H. Brown), Gordon & Breach, New York (1968), p. 228.
[25] WOODMANSEE, W. E. *Materials Evaluation*, October 1966, p. 564.
[26] LUKIANOFF, G. V. in *Liquid Crystals 2, Part I*, Proc. 2nd Int. Conf. on Liquid Crystals (edited by G. H. Brown), Gordon & Breach, New York (1968), p. 219 and 222; LAURIENTE, M. *loc. cit.* p. 224.
[27] KLEIN, E. J. *Astronautics and Aeronautics*, July 1968, p. 70.
[28] KLEIN, E. J. in a paper presented at the American Institute of Aeronautics and Astronautics, Third Aerodynamic Testing Conference, San Francisco, California (1968).
[29] KLEIN, E. J. in *Liquid Crystals 2, Part I*, Proc. 2nd Int. Conf. on Liquid Crystals (edited by G. H. Brown), Gordon & Breach, New York (1968), p. 225.
[30] WOODMANSEE, W. E. in *Liquid Crystals 2, Part I*, Proc. 2nd Int. Conf. on Liquid Crystals (edited by G. H. Brown), Gordon & Breach, New York (1968), p. 228.
[31] LUKIANOFF, G. V. in *Liquid Crystals 2, Part II*, Proc. 2nd Int. Conf. on Liquid Crystals (edited by G. H. Brown), Gordon & Breach, New York (1968), p. 775.
[32] *Liquid Crystals and their Applications* (edited by T. Kallard), Optosonic Press, New York (1970).
[33] ENNULAT, R. D. and FERGASON, J. L. *Molec. Crystals Liqu. Crystals* **13,** 149 (1971).
[34] FERGASON, J. L., VOGL, T. P. and GARSBURY, M. *U.S. Patent 3,114,836* (1963)—see also reference [32].
[35] FERGASON, J. L. *Sci. American* **211,** 77 (1964).
[36] FERGASON, J. L. and GOLDBERG, N. N. *British Patent 1,153,959* (1969)—see also reference [32].
[37] DIXON, G. D. and SCALA, L. C. *Molec. Crystals Liqu. Crystals* **10,** 317 (1970); CILIBERTI, D. F., DIXON, G. D. and SCALA, L. C. *loc. cit.* **20,** 27 (1973).
[38] SAUPE, A. *Angew. Chem. Int. Edn.* **7,** 97 (1968).
[39] GRODZKA, P. G. and BANNISTER, T. C. *Science* **176,** 506 (1972).
[40] *Chem. Eng. News*, January 1971, p. 25.
[41] PICK, P. G., FABIJANIC, J. and STEWART, A. *Molec. Crystals Liqu. Crystals* **20,** 47 (1973); PARKER, R. *loc. cit.* p. 99.
[42] WILLEY, D. G. and MARTIRE, D. E. *Molec. Crystals Liqu. Crystals* **18,** 55 (1972).
[43] CARROLL, P., *Cholesteric Liquid Crystals. Their Technology and Applications*, Ovum Ltd., London (1973).

7.2 LIQUID CRYSTALS IN DISPLAY SYSTEMS

A. Sussman

Introduction

The last few years have been marked by the appearance of a large number of devices utilizing nematic and cholesteric liquids. In particular, those where electric effects can be employed offer the worker in the display field a new assortment of phenomena with operating criteria falling conveniently into the range of voltage and power capabilities of solid state circuits. One feature of liquid crystal devices that sets them apart from other electrically controlled displays is that they are passive, relying on available external illumination for visibility. In display devices with reflecting backs (the controlled reflectance mode of operation), the relative independence of the contrast ratio on illumination brightness and direction is advantageous for outdoor viewing conditions. Conversely, high resolution capability makes it possible to achieve small area displays visible under relatively poor illumination. This review is limited to those device phenomena depending only on voltage applied across a relatively thin film of liquid crystalline material contained between a transparent front electrode, and, depending on mode of operation, a transparent or reflecting rear electrode. The physical processes are quite diverse, depending in some cases on well understood phenomena, in others on effects less clearly understood.

Nematic Liquids in Electric Fields

The response of nematic liquids to electric fields involves a process of orientation of the rod-like molecules, *collectively*, in some direction with respect to that of the field. Devices utilizing nematic liquids may be divided into two broad categories; those in which the orientation by the field is the principal effect, and those in which this orientation is present but is simultaneously disrupted by the hydrodynamic effect of current carriers. Due to the optical anisotropy of nematics, the electric field gives rise to striking optical effects ranging from changes in the direction of the optic axis, in the case of orientation by the field alone, to the production of intense light scattering due to disruption of the order.

In response to an externally applied electric field, a molecule tends to orient with its dipolar axis* parallel to that field. If that axis is more or less in the direction of the major molecular axis, then that molecule is said to have positive dielectric anisotropy; conversely, molecules with negative

* Axis of maximum dielectric constant.

dielectric anisotropy tend to align with their major axes perpendicular to the field. In contrast to amorphous isotropic liquids in which, in the absence of an electric field, there is no preferred molecular orientation, substances in the mesomorphic state show molecular orientation over relatively large volumes. This orientation is strongly dependent on the influence of the walls of the container as will be discussed subsequently. In the case of nematics, the orientation may be made such that the direction of the optic axis is perpendicular to the plane of the electrodes. Optically, thin films so oriented look clear under ordinary light, and also appear isotropic when viewed between crossed polarizers along the direction approximately parallel to the orientation axis, i.e., perpendicular to the electrodes. When the sample is rotated around the optic axis, extinction is maintained. This is the pseudoisotropic or homeotropic orientation. Note however that light entering off axis from the perpendicular direction suffers double refraction. The optic axis can also be oriented parallel to the plane of the electrodes, and in this case, the birefringence is easily observed between crossed polarizers. Again, under ordinary light, the liquid appears clear.

If no special procedures are followed, the orientation of the nematic is likely to vary from point to point, and owing to the strong optical anisotropy of the liquid there will result a cloudy appearance in ordinary light and a patchy bright appearance between crossed polarizers. This is, of course, undesirable for most applications. The "threads" which give the nematic state its name may also appear in thicker samples if the electrodes are not treated to give the homeotropic texture [1], and they, too, are an undesirable feature. Broadly speaking, the optic axis can be considered parallel to the molecular axis, so that reorientation of the dipolar axis by electric fields effectively reorients the optic axis in the same direction as the field for material of positive dielectric anisotropy, and perpendicular to the field for material of negative dielectric anisotropy.

We have thus two major wall orientations and two possible field-produced orientations. If now the effect of ion transport-induced turbulence is added, almost all of the useful phenomena of nematic devices may be classified. They are presented schematically in Fig. 7.2.1 where the cell filled with liquid crystal has been represented by a cube whose left and right faces are the electrodes. The orientation of the bulk material is represented by lines, which may be considered the major (optic) axes of the molecules. Small arrows represent the dipole moment which is generalized to lie directly along the molecular axis for materials with positive dielectric anisotropy and perpendicular to the molecular axis for negative dielectric anisotropy. The right face of the cube represents the orientation of the major part of the liquid as viewed along the direction of the field, i.e., perpendicular to the electrodes. The field-off state is at the left of Fig. 7.2.1, and the field-on state at the right. It is entirely possible that the immediate surface layer adjacent to the electrodes retains the field-off configuration in the field-on state; this has not been shown in order to keep the diagrams relatively simple, but the effect of the possible retention of the original orientation at the surfaces must always be borne in mind.

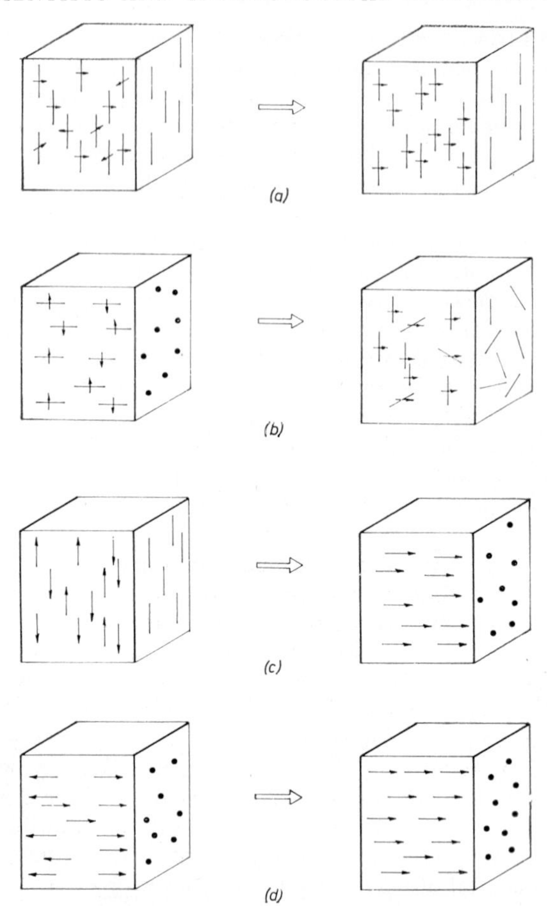

FIG. 7.2.1. Schematic representation of the orientations before and after the application of an electric field—nematic liquids.

The figure attempts to show schematically the effect of a sufficiently high electric field upon the orientation of nematic liquids of negative and positive dielectric anisotropy, whose original orientation shown at the left, is maintained in either the birefringent or homeotropic texture by wall forces. The lines represent the alignment of individual molecules, with arrows representing the dipole moments. The cells have been drawn with electrodes as left and right faces of the cubes, with the right face showing the view perpendicular to the electrodes, i.e., in the viewing direction.

At the right, the orientation caused by the field is represented, and in all cases is the result of the rotation of the dipole moments to be parallel to the field.

If the orientation is initially such that the birefringent texture is formed, i.e., with the long molecular axes parallel to the electrodes, then with molecules of negative dielectric anisotropy, the only effect of the field is to orient the dipoles such that the axis of maximum dielectric constant is in the direction of the field (Fig. 7.2.1(*a*)). This cannot change the molecular orientation, since the dipoles need only rotate about the molecular axis, and one would expect a null result. Experimentally, however, above a threshold voltage the appearance of an ordered pattern of cigar-shaped domains is noted [2, 3, 4], and their origin requires a small conductivity [5, 6].

At still higher voltages, dynamic scattering is observed if the conductivity of the liquid is higher than $0{\cdot}1 \mathrm{n}\Omega^{-1}\mathrm{cm}^{-1}$ [7]. Relaxation to the original birefringent texture follows the removal of the voltage.

With material of negative dielectric anisotropy, an originally homeotropic texture should be changed to one in which the molecular (optic) axis is perpendicular to the field. Note, however, that this need not be a birefringent texture which is uniformly oriented over large areas and this is so indicated in Fig. 7.2.1(*b*). Again, experimentally, domain effects appear and if the necessary conductivity exists, dynamic scattering ensues. Relaxation to the homeotropic texture again occurs on removal of the field.

For material of positive dielectric anisotropy originally in the birefringent nematic texture, application of a field gives the homeotropic texture (Fig. 7.2.1(*c*)), complicated in some instances by circular domains [8] visible with crossed polarizers. By using this effect to alter the orientation of pleochroic dyes, the guest-host interaction phenomenon is obtained [9] (see p. 348).

The fourth case involves a homeotropic texture composed of molecules which have positive dielectric anisotropy (Fig. 7.2.1(*d*)). One might again expect a null result perhaps complicated by domain formation. Experimentally, with the application of sufficient voltage, there is some transient light scattering, with retention of the homeotropic texture [10].

Little or no dynamic scattering is thus observed where the materials have positive dielectric anisotropy, even if they have resistivities in the range where dynamic scattering is found in materials with negative dielectric anisotropy [7].

Influence of the Walls on Orientation

Interaction between the nematic liquid and the surface plays a significent rôle in ordering of the nematic material throughout the volume. Rubbing a surface in a particular direction orients the optic axis in that direction. Orientation can also be obtained by seeding a cooling amorphous isotropic material [11], but the direction is not controllable or uniform. The birefringent texture which results from either procedure is reformed on cooling the amorphous isotropic liquid formed after heating or after the order has been disturbed electrically, so apparently orientation, once obtained, remains in the film next to the surface. The term "pellicle" is sometimes used for this persistent film. While the nature of the forces which give rise to the pellicle have not been clarified, they must depend on strong interactions between the surface and the nematogenic molecules, possibly exerted in certain instances through the presence of some intermediate material rubbed on to the surface. It has been found that flaming the electrodes after rubbing on paper removes the tendency to produce the birefringent texture [12].

Spontaneous homeotropy may be obtained, on the other hand, by careful cleaning of the substrate [13] or may be induced by motion of the cell walls. In this case, the interactions with the wall are considered to be weaker than those responsible for the birefringent texture, but to lead to an orientation of the major molecular axes perpendicular to the surface. Different compounds are considered to have different degrees of attachment to the surface, so orientations may vary in spite of techniques, the commonest case being that in which there is no uniform orientation at all over large areas, the material appearing hazy.

In all cases, the relaxation after disturbance of the bulk orientation is to the original texture with the surface layer supplying the directing influence and/or the nucleating centres. Kinetic studies have shown the relaxed orientation proceeding into the bulk from the walls in all of the nematic orientation processes. Thus the conversion of the initial orientation (controlled by the wall forces) to the eventual orientation (controlled by the field) proceeds through intermediate stages (field and wall forces in combination [5]), each stage having a definite voltage threshold. Rise times are controlled by viscous friction and relaxation times by viscous friction or nucleation kinetics.

Cholesteric Liquids in Electric Fields

Transitions occurring with a cholesteric texture as the starting state are more difficult to categorize because of the inherently more complex nature of the cholesteric structure. Mixtures of cholesteric and nematic liquids are considered in this section because they adopt the cholesteric texture [14]). Although the pitch axis defines the orientation of the cholesteric texture, interaction with the field is through the dielectric anisotropy of the structure due to the anisotropy of the individual molecules, causing torques which compete with the elastic torques of the internal structure [15]; the reorientation of the molecules cannot therefore be a simple reorientation of the pitch axis alone. That is to say, the pitch axis of the cholesteric helix is not analogous to the molecular axis in oriented nematic states.

We shall assume some possible starting configurations and examine the reorientation effects of the field. The simplest cholesteric texture is called planar (Grandjean). In it the orientation of the twist axis is perpendicular to the wall. In Fig. 7.2.2, again the orientation of individual molecules is represented by a line, the length of the line being indicative of the amount of rotation about the twist axis, shown as a dotted line. The arrows represent the dipoles. Note here an important difference between Figs. 7.2.1 and 7.2.2. In Fig. 7.2.2 the lines represent as projections on the plane of the paper the orientations of all the molecules in planes which are perpendicular to the paper and parallel or perpendicular to the electrodes.

With material of negative dielectric anisotropy, the only primary effect of the field would be to rotate the molecules about their long axes giving a

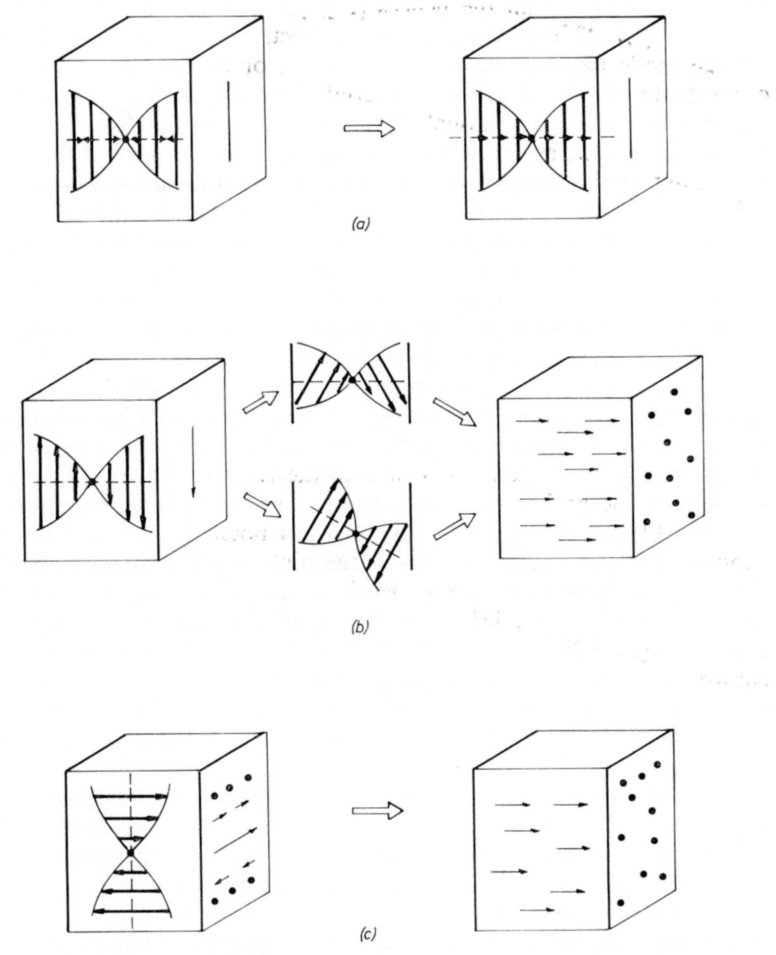

FIG. 7.2.2. Schematic representation of the orientations before and after the application of an electric field—cholesteric liquids.

This figure, which is highly schematic, shows the effect of a sufficiently high field on the cholesteric orientation shown at the left. The twist axis is originally either perpendicular or parallel to the cell walls, and the molecules have either negative or positive dielectric anisotropy. The lines represent the projection on the plane of the paper of individual molecules within planes of uniform orientation in the cholesteric structure whose twist axis is shown dotted. Arrows represent the dipole moments of the individual molecules. For molecules of positive dielectric anisotropy, the dipoles have been shown (*b* and *c*) uniformly oriented in each plane, parallel to the molecular orientation. In fact the dipoles would be arranged in each plane in a manner similar to that shown in the left hand side of Fig. 7.2.1 (*c*) and (*d*) for nematic liquids of positive dielectric anisotropy. Again, the effect of the field (as shown at the right) is to turn the individual molecular dipoles parallel to the field. The result is a quasi-nematic structure.

better alignment within the helices [15], possibly cancelling wall effects (Fig. 7.2.2(*a*)). This is demonstrated indirectly in what is known as the storage mode (see p. 350), in which a d.c. or low frequency a.c. field converts the planar texture of a material of negative dielectric anisotropy into a scattering state through the disruptive effect of ions in transit. However if an a.c. signal of high enough frequency to prevent the effects of ions in transit is then applied, the original planar texture is restored [16].

For a material of positive dielectric anisotropy, the field would be expected to convert the planar orientation to one in which the dipoles are parallel to the field. This may occur in either of two ways [15]; if all the molecules in each plane rotate into the direction of the field (Fig. 7.2.2(*b*), upper path), the resulting texture could be considered a homeotropic nematic texture. If, however, the wall forces are strong enough to restrain the orientation of the planes near the walls (at low fields), the centre regions might rotate as a whole (Fig. 7.2.2(*b*), lower path), followed by an unwinding of the twist axes, with more and more of the wall material becoming oriented parallel to the field as the voltage is raised. This central structure would probably be equivalent to a homeotropic nematic texture, with the film of material next to the walls possibly retaining the birefringent cholesteric texture except at the highest fields. Either or both of these mechanisms might be at work in the field-induced cholesteric-nematic transition [17, 18]. Note that a further simplification has been made in Fig. 7.2.2(*b*) and (*c*) in that the molecules in each plane have been shown as having their dipoles in the same direction; in fact one would expect that in each plane there would be an arrangement similar to that shown on the left hand side of Fig. 7.2.1(*c*) and (*d*) for nematic liquids with positive dielectric anisotropy.

Another common texture of cholesteric material is the so-called "focal-conic" texture, usually obtained by cooling the isotropic liquid without disturbing the electrodes physically; small motions of the electrodes are sufficient to convert it to the planar texture. The "focal-conic" texture is not a uniform texture but consists of many small regions with different orientations of the helix axis [48]. For an analysis of the effect of the electric field on this complicated texture, we may however consider that only two sets of regions are present, one set with the axis perpendicular to the walls (planar) and one with the axis parallel to the walls. This latter orientation of the helix axis is not attainable over large areas but has been observed over microscopic domains or regions [19] in materials of positive dielectric anisotropy when the helix axis has been rotated through 90° (as in the lower path of Fig. 7.2.2(*b*)) by a low enough field so that the helix itself does not unwind. These domain-like regions are stable for long periods, and subsequent application of a sufficiently strong field results in an orientation of the dipoles (by rotation about their centres) in each plane of the structure (the planes in this case are perpendicular to the walls) —see Fig. 7.2.2(*c*). The resulting texture is a homeotropic nematic.

Thus, for material of positive dielectric anisotropy, a sufficiently high field should produce a "nematic" state, independent of the starting texture. Relaxation, however, might be expected to proceed from randomly oriented nucleation centres and to result in a non-uniform texture ("focal-conic"), unless there is a sufficiently strong wall interaction, in which case a planar texture might be favoured.

For material of negative dielectric anisotropy on the other hand, no field induced nematic state is expected, but only a conversion of the "focal-conic" texture to the planar texture by field alone (high frequency) and the reverse conversion by ions in transit (low frequency). This combination of phenomena has been shown to be the basis of the storage effect [20]. Figure 7.2.3 provides a summary of the situation for cholesteric materials of positive or negative dielectric anisotropy.

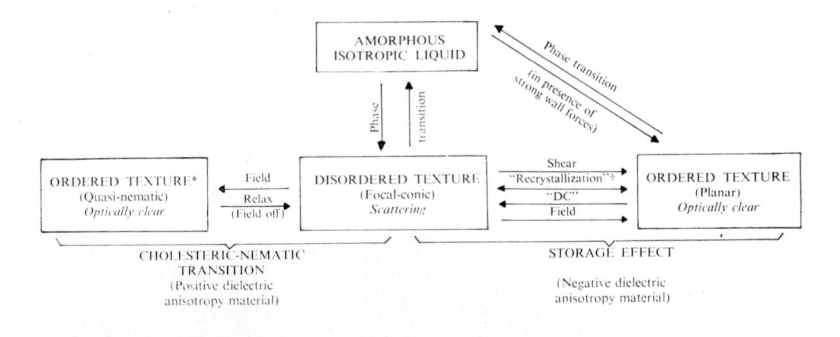

* In field only.
† This refers to the imperfectly understood processes which are involved both in the relaxation of the storage effect and in the sometimes spontaneous reversion of the planar texture to the focal-conic texture.

FIG. 7.2.3. A schematic illustration of various transitions which may occur with cholesteric liquid crystals formed by materials of positive or negative dielectric anisotropy.

All of the observations indicate that the phenomena connected with the cholesteric-nematic conversion and its relaxation are complex [21], partly due to the simultaneous change in orientation of the pitch axis and in the pitch itself [19]. If the pitch dimensions are within the wavelength range of visible light, angle dependent iridescence colours due to Bragg scattering are observed [22]. An applied field causes visible shifts of the coloured light to longer *or shorter* wavelengths depending on the original orientation of the helix axis, on the angle of observation and on whether a reflecting back surface is present; of course the pitch eventually increases with field [19]. These optical phenomena take place at voltages below the threshold for complete pitch unwinding, and may be related to domain formation in nematic phases in that they represent instabilities in the texture induced by the electric field [49]. If the pitch is large enough to result in a colourless planar texture, there will be no chromatic effects due to the field. There is also associated with the plane texture an optical rotatory power [23] which for a given wavelength is inversely proportional to the pitch [24], but there are as yet no related device applications.

Devices: Practical Considerations

A major advantage of flat liquid crystal display devices is that in those devices in which the process controlled is the scattering of light, an increase in illumination will only increase the brightness of the scattering regions and the non-scattering regions in proportion, maintaining the contrast ratio. Thus the power required to maintain a contrast ratio of 20:1 even in bright sunlight is in the range of 10 μWcm^{-2} for the dynamic scattering devices. On the other hand, if the illumination is so dim that artificial light must be supplied, the power for this need only be small because the contrast ratio is still high. Two general classes of cells can be employed: "window" cells having two transparent electrodes, and "mirror" cells having specular reflecting, metallic back electrodes to take advantage of the predominantly forward-scattered light.

Window cells may be employed directly to produce variable density scattering or colour. If either electrode is divided into areas whose electrical activation can be controlled independently, then diverse patterns and even

FIG. 7.2.4. A seven segment numeric display using dynamic scattering and showing the numeral 9. The front and rear electrodes are transparent, with the cell illuminated from the front as well as from behind.

animation can be produced. For instance, the electrode may be divided to give the seven-segment numeric pattern (Fig. 7.2.4), or into other multiple segments to give numbers or letters depending on which segments are activated. Solid state devices which perform this segment selection may be used in some cases to activate the segments directly, because of the low voltage and power required to obtain high contrast. Analogue displays have been fabricated which employ transverse voltage gradients in one or both of the electrodes to produce two-dimensional variation of the voltage between the front and back electrodes [25]. By suitable adjustment of those gradients, moving bands, wedges, spots and apertures can be generated; an all-liquid crystal voltmeter dial face with a moving indicator bar has been demonstrated.

Such a relatively flat display may be viewed by the existing light falling upon its surface or by internally contained illumination. With a window cell, the light impinging on either front or rear (or edge lighting) produces better contrast when the display is seen against a black absorbing background. Because in dynamic scattering the light is mainly scattered forward, illumination from the rear is more desirable, and to this end suitably blackened vane backgrounds can be employed [26]. In the mirror cell, better contrast can be achieved if there is a black surface placed so that the viewer sees the reflection of it in the specular reflecting back of the cell when the cell is in the off state. One may also use the cells to project a display.

Many of the usual difficulties in handling liquids are obviated in the relatively thin cells employed, because capillary forces are sufficient to prevent the liquid from flowing out. Electrode separation may be maintained by spacers at the edges of the electrode sheets provided that the liquid is prevented from getting between the spacer and the electrodes [7, 21]. When electrode plates are specially selected for flatness, cells 25×25 cm with 12 μm spacing can be produced readily. Independently operable areas can be obtained by etching through the conductor or by forming the electrode only where needed using evaporation or sputtering techniques. Contacts may be brought out to the edges of the electrodes in a straightforward manner. A 3 cm square dynamic scattering video (television) display was activated from behind by means of an electron gun through a mosaic of wires imbedded in the glass. Resolution of 150 lines was obtained, limited in this case only by the relative coarseness of the wire mosaic [27].

A photograph can be converted to a liquid crystal picture by the half-tone process with photoresist dots applied directly to either electrode [7].

An image conversion scheme uses a standard cell but a transparent ZnS photoconductor film, *ca.* 1 μm thick, covering one electrode [28]. By projecting a u.v. image onto the photoconductor, a dynamic scattering image is produced, an exposure pulse of 0·1 mJ cm^{-2} at a wavelength of 367 nm giving saturation in 0·1 s. This speed of response is not quite fast enough to show ordinary motion. Using storage material (see p. 350), an image may be held and reprojected using visible light, which does not affect the photoconductor.

Devices: Physical Principles and Parameters

The Guest–Host Interaction

Many crystalline dyes are pleochroic, having an optical absorbance which depends on the orientation of the optic axis of the dye molecule with respect to the electric vector of the light. If the molecular structure of such a dye (the guest) allows it to be "locked into" the nematic structure (the host), the guest-host interaction effects can be obtained [9].

The usual mode of operation is to have the cell prepared so that the host solvent of positive dielectric anisotropy is in the birefringent texture (as in Fig. 7.2.1(*c*)) and illuminated with light polarized parallel to the optic axis. Application of the field produces a perpendicular orientation in the nematic host and thereby the absorption spectrum of the guest molecules is converted from that obtained with the electric vector of the incident light parallel to the optic axis (the characteristic strongly absorbing state of the dye) to that with the electric vector perpendicular to it (usually weakly absorbing) [29]. Operation to saturation, at which nearly all of the guest molecules are realigned, requires a field of *ca.* 50 $kVcm^{-1}$, independent of frequency from d.c. to 50 kHz; above this frequency, higher voltages are required (see p. 341). For a cell 6–25 μm thick, a threshold of 1–2 V is exhibited followed by a gradual shift from the strongly absorbing to the weakly absorbing or colourless state as the field increases. Rise times of 1–5 ms have been achieved at powers of 1 $mWcm^{-2}$, the relaxation occurring in 20–200 ms for the thin and thick cells, respectively. In the system 4-*p*-ethoxybenzylideneaminobenzonitrile (PEBAB) containing 1% indophenol blue, the Gaussian shaped peak at 6,000 Å had an absorbance of 1·9 in the field off condition, and 0·38 with the field on in a 12 μm thick cell. Estimates indicate that the most efficient operation occurs with 100 host molecules per guest molecule. The dye must necessarily be non-reactive and non-ionic to prevent irreversible electrochemical changes. Resistivities in the range 1–10 G Ω cm have been found satisfactory [30].

Another mode of operation has been demonstrated [10] in which a solvent of negative dielectric anisotropy is initially in the homeotropic condition as in Fig. 7.2.1(*d*), and illuminated by polarized light. The guest molecules then have their optic axes perpendicular to the electric vector of the polarized light; hence the cell is initially in its low absorbance state. With d.c. fields of 1–5 $kVcm^{-1}$, a change to the parallel orientation with strong absorbance occurs, while at higher fields the perpendicular orientation with low absorbance reappears. This was first demonstrated in the system 4-*p*-n-butoxybenzylideneaminoacetophenone and 1% indophenol blue. The contrast (absorbance) ratios are not as high as in the opposite effect.

In a window cell, these phenomena could be used for a shutter (no dye being actually necessary, the effect being similar to that in the Kerr cell), for a voltage controlled variable intensity filter, or in either configuration, for a coloured indicator [30].

Another potential device has been demonstrated employing a dye which is both pleochroic and photochromic, i.e., whose configuration and thus optical absorbance can be changed by absorbing light. Colour may be shifted between the two absorptions of the photochromic dye by irradiation with light of suitable wavelengths, and either of the colours may be erased by application of a field [30]. Contrast and absorbance ratios have been found to be low.

Dynamic Scattering

The term dynamic scattering was introduced by its discoverers to describe the phenomenon that ensues when charge in transit through nematic material of negative dielectric anisotropy generates hydrodynamic shear forces leading to turbulent flow, and thus causes spatial variations in the orientation of the material (vortices) of a size requisite to produce light scattering. Details of the physical mechanism are still unresolved, but under a.c. excitation anisotropic conductivity [6] leads to charge separation and subsequent vortical flow [32], while under d.c. net (uncompensated) charge in the bulk, caused by field emission [7] on the electrical double layer gives similar hydrodynamic results. High fields near the electrodes have been observed [33], although uncompensated charge does not account for all the current [34].

Dynamic scattering has a voltage threshold almost independent of cell thickness, chemical composition of nematic material or resistivity [6] (3–5 V in *p*-azoxyanisole). The current is ohmic below threshold, becoming non-linear at fields high enough to cause ion-pair dissociation. Non-linearities, however, are not essential for dynamic scattering to occur. For anisylidene-*p*-aminophenyl acetate (APAPA), with resistivity 50 G Ω cm at 90°, the scattering contrast ratio reaches saturation at a field of 50 kVcm^{-1} in a 12 μm cell. This gives an operating range between 8 and 40 V, the scattering increasing with voltage. Rise times and decay times for scattering are independent of current and directly proportional to viscosity [31]. Rise times and decay times are inversely proportional to the square of the field and the square of thickness, respectively. Such behaviour would be expected if the reorientation were to initiate from an ordered region at each wall propagating towards the centre of the cell. A typical value of rise time for 50 V applied is 1–5 ms; the decay is less than 30 ms for a contrast ratio of better than 20 to 1 [7]. The scattered light of the device is approximately half that of a white standard. Higher contrast ratios have also been reported [35]. With *p*-azoxyphenetole, rise times as short as 0·3 ms have been observed [34], energies to achieve maximum contrast being about 100 $\mu\text{W cm}^{-2}$. No relaxation times were reported.

For applications in matrices where high speed and storage requirements are necessary, special addressing procedures have been developed [36]. In these, the liquid crystal cell may be treated as a circuit element containing capacitance and resistance in parallel.

The relaxation time may be shortened by a quenching process that effectively stiffens the liquid structure, preventing turbulence. This is achieved either by a short high voltage pulse or a burst of audio frequency and reduces the decay time to about 1 ms [37]. Driving the cell into the off state has the advantage of keeping the scattering at a maximum for full on time. Under certain operating conditions, short turn off times of 3 ms have been observed which seem to depend on oscillations of the cell orientation [38].

Improvements in contrast ratio have been obtained by additions of other compounds [39], while widening of the useful temperature range has been obtained by systematic selection of compounds and mixtures [40]. Besides certain proprietary materials [37] and commercially available material [41], various other compounds have been reported [34, 35] which operate in the room temperature range. Operating life in dynamic scattering has been reported to be as high as 3,000 hours [42]; others report 700 hours without substantial reduction in contrast [34]. Individual compounds give different operating lives. When failure is due to electrochemical reactions life may be measured in coulombs/mole of material available at the electrodes. Contrast ratios have been measured as a function of the illumination and viewing angles [36, 43] as well as of applied voltage [27, 36, 34], temperature [44] and electrode material [44].

The dynamic scattering phenomenon has potential usefulness in a variety of devices, some of which have been demonstrated; these include alpha-numeric indicators [42], shutters [42], matrix displays [36], reflective TV [27], controlled reflectance mirrors [45] and a flying spot scanner [25].

Storage Effect

The storage effect appears as the persistence of light scattering in cells containing mixtures of cholesteric and nematic materials of negative dielectric anisotropy when the field, which converted the original transparent material to the scattering state, is removed [16]. A mixture of nematic anisylidene-*p*-aminophenyl acetate with 10% cholesteryl nonanoate can be switched to a maximum contrast of 7:1 by a pulse of energy (0·3 mJcm^{-2}) at a d.c. voltage between 40 and 100 V (or with a.c. up to 100 Hz) in a 12 μm cell. A 100 V pulse achieves maximum contrast in 80 ms, the writing speed or time to achieve full contrast being proportional to the reciprocal of the voltage. During the pulse, the material is turbulent as in dynamic scattering, while with the field off the scattering remains, the cell appearing milky, and the contents resembling the "focal conic" texture [20]. The lower contrast ratios are due to the poorer clarity of the off state rather than to the poorer light scattering in the switched-on condition. Fading of the scattering, after field removal, indicates that the number of light scattering centres decreases at a rate proportional to their number and to a collision cross-section [46]. The reciprocal of the scattering intensity increases linearly with time, reaching in the case quoted one half scattering in 9000 s. By using a 1 kHz a.c. signal, the milkiness may be "erased" in a time proportional to the reciprocal of the voltage. A typical value for erasure with a 12 μm cell is 1 s using 300 V at 1 kHz. The field required for the onset of scattering was found to increase with increasing frequency at fixed composition of material. At fixed frequency, the threshold field increased as the proportion of cholesteric material was increased; the material studied was oleyl cholesteryl carbonate and (nematic) anisylidene-*p*-n-butylaniline (MBBA) [20].

CHOLESTERIC–NEMATIC TRANSITION (FIELD QUENCHED OPALESCENCE)

The application of a high field to an originally opalescent cholesteric mixture, consisting either of cholesteric materials alone or together with nematic materials, can convert the texture to one that is transparent and homeotropic, with the optic axis parallel to the field [17]; this is most probably nematic [18]. The usefulness of this phenomenon is that the scattering state is the "no power" state. Contrast ratios of 10:1 have been achieved.

Two schemes have been advanced for lowering the relatively high operating voltages necessary to cause the phase transformation, both relying on the fact that the critical field for the transformation depends inversely upon the pitch of the cholesteric structure [15].

A mixture of two cholesteric compounds with opposite optical activities gives rise to a material of controllable (but temperature-sensitive) pitch [21]. By this scheme, reductions in the critical field to 2 kV cm^{-1} in a 12 μm cell were made. At 10°C above or below the particular temperature of the minimum in the critical field, the field required was about ten times higher. One drawback of this method is that the rise time for the transition becomes longer as the minimum in the critical field is approached, being hours near the threshold; complete transitions at other voltages were not reported to occur in less than one second. Relaxation times are estimated to be seconds at 26°C, with complete field removal, being slower for field lowering to below threshold. Since the driving force for the relaxation is obtained from the energy of the untwisted pitch, the increase in pitch results in a lower restoring force and consequently a slower process as the pitch is increased. Recent work has demonstrated speeds as high as 50 μs for rise and 60 μs for decay in 6 μm cells.

By adding up to 70% of a nematic mixture of three 4-*p*-alkoxybenzylidenecyanoanilines to a cholesteric compound of similar molecular constitution (active amyl 4-*p*-cyanobenzylideneaminocinnamate), both of positive dielectric anisotropy, the critical field for the removal of opalescence was reduced from 200 kV cm^{-1} for the pure cholesteric phase to 25 kV cm^{-1} the field being inversely proportional to the mole fraction of the nematic compounds [47]. Rise times measured at 25°C were a function of mixture composition, thickness and voltage. In all cases, the rise time was equal to exp $\alpha\, \boldsymbol{E}^{-2}$ with the constant α inversely proportional to the square of the percentage of nematic material. For a cell 12 μm thick, containing 70% nematic material, the rise time ranged between 30 and 100 ms for 75 and 60 V, respectively. The relaxation time was independent of cell thickness, being 25 ms with 70% and 10 ms with 30% nematic material. Measurements over a greater voltage range, with material giving a lower critical field, indicate that the rise times are proportional to the viscosity *times* the inverse square of the field, while the relaxation again depends on the pitch of the final state, the pitch and relaxation time apparently increasing directly with increase in content of nematic material. A hysteresis effect is also demonstrated and is evidence of a nucleation dominated switching process.

Since this section was written*, a paper by M. Schadt and W. Helfrich (*Appl. Phys. Lett.* **18,** 127 (1971)) has been published, and describes the effect of electric fields on a nematic of positive dielectric anisotropy in which the molecules are oriented with their long axes parallel to the electrode surfaces. The orientation at the upper electrode is at 90° to that at the lower electrode plate; a partial helix is induced in the nematic. The applied field orients the molecules normal to the electrode surfaces; on removal of the field, surface orientation effects take over once more and the helical arrangement is reformed. Illumination with plane polarized light gives different optical effects for the "on" and "off" states of the cell and this provides a basis for a further type of display device which is already being marketed by one company.

* See also the additional references following the list of numbered references.

REFERENCES

[1] Gray, G. W. *Molecular Structure and the Properties of Liquid Crystals*, Academic Press, London and New York (1962), p. 34.
[2] Willliams, R. *J. chem. Phys.* **39,** 384 (1963).
[3] Williams, R. and Heilmeier, G. H. *J. chem. Phys.* **44,** 638 (1966).
[4] Heilmeier, G. H. *J. chem. Phys.* **44,** 644 (1966).
[5] Helfrich, W. *J. chem. Phys.* **51,** 2755 (1969).
[6] Helfrich, W. *J. chem. Phys.* **51,** 4092 (1969).
[7] Heilmeier, G. H., Zanoni, L. A. and Barton, L. A. *Proc. IEEE* **56,** 1162 (1968).
[8] Heilmeier, G. H. in *Ordered Fluids and Liquid Crystals*, Advances in Chemistry Series, American Chemical Society, Washington, (1967), p. 68.
[9] Heilmeier, G. H. and Zanoni, L. A. *Appl. Phys. Lett.* **13,** 91 (1968).
[10] Castellano, J. A. and McCaffrey, M. T. in *Liquid Crystals and Ordered Fluids* (edited by J. F. Johnson and R. S. Porter), Plenum Press, New York (1970), p. 293.
[11] Gray, G. W. *Molecular Structure and the Properties of Liquid Crystals*, Academic Press, London and New York (1962), p. 31.
[12] Chatelain, P. *J. Phys. Radium, Paris* **30,** C4, No. 11–12, 3 (1969).
[13] Saupe, A. *Angew. Chem. Int. Edn.* **7,** 97 (1968).
[14] Sackmann, E., Meiboom, S. and Snyder, L. *J. Am. chem. Soc.* **89,** 5981 (1967).
[15] Meyer, R. B. *Appl. Phys. Lett.* **12,** 281 (1968).
[16] Heilmeier, G. H. and Goldmacher, J. E. *Appl. Phys. Lett.* **13,** 132 (1968).
[17] Wysocki, J. J., Adams, J. and Haas, W. *Phys. Rev. Lett.* **20,** 1024 (1968).
[18] Heilmeier, G. H. and Goldmacher, J. E. *J. chem. Phys.* **51,** 1259 (1969).

[19] KAHN, F. J. *Phys. Rev. Lett.* **24,** 209 (1970).
[20] HAAS, W., ADAMS, J. and FLANNERY, J. B. *Phys. Rev. Lett.* **24,** 577 (1970).
[21] WYSOCKI, J. J., ADAMS, J. E. and OLECHNA, D. J. in *Liquid Crystals and Ordered Fluids* (edited by J. F. Johnson and R. S. Porter), Plenum Press, New York (1970), pp. 419 and 463.
[22] FERGASON, J. L. *Appl. Optics* **7,** 1729 (1968).
[23] DAY, G. W. and GADDY, O. L. *Proc. IEEE* **56,** 1113 (1968).
[24] CANO, R. *J. Phys. Radium, Paris* **30**, C4, No. 11–12, 28 (1969).
[25] SOREF, R. A. *Appl. Optics* **9,** 1323 (1970).
[26] SUSSMAN, A. *Illumination Scheme for Liquid Crystal Displays.* U.S. Pat. pending.
[27] VAN RAALTE, J. A. *Proc. IEEE* **56,** 2146 (1968).
[28] MARGERUM, J. D., NIMOY, J. and WONG, S.-Y. *Appl. Phys. Lett.* **17,** 51 (1970).
[29] HEILMEIER, G. H., CASTELLANO, J. A. and ZANONI, L. A. *Molec. Crystals Liqu. Crystals* **8,** 293 (1969).
[30] CASTELLANO, J. A., HEILMEIER, G. H., PASIERB, E. F. and MCCAFFREY, M. T. *Electronically-Tuned Optical Filter*, Contract No. NAS 12-638, Electronics Research Centre, National Aeronautics and Space Administration, Final Report, April 1969.
[31] KOELMANS, H. and VAN BOXTEL, A. M. *Phys. Lett.* **32A,** 32 (1970).
[32] PENZ, P. A. *Phys. Rev. Lett.* **24,** 1405 (1970).
[33] LU, S. and JONES, D. *Appl. Phys. Lett.* **16,** 485 (1970).
[34] ASSOULINE, G. and LEIBA, E. *Revue Tech. Thomson-CSF* **1,** 483 (1969).
[35] JONES, D., CREAGH, L. and LU, S. *Appl. Phys. Lett.* **16,** 61 (1970).
[36] LECHNER, B. J., MARLOWE, F. J., NESTER, E. O. and TULTS, J. *1969 International Solid State Circuits Conference Digest* (University of Pennsylvania, Philadelphia, February 1969).
[37] HEILMEIER, G. H., ZANONI, L. A. and BARTON, L. A. *IEEE Trans. on Electronic Devices* **ED-17,** 22 (1970).
[38] HEILMEIER, G. H. and HELFRICH, W. *Appl. Phys. Lett.* **6,** 155 (1970).
[39] HEILMEIER, G. H. and ZANONI, L. A. *Electro-optical Device*, U.S. Pat. No. 3,499,112 (March 3, 1970); Br. Pat. No. 1,167,486 (Oct. 15, 1969).
[40] GOLDMACHER, J. E. and CASTELLANO, J. A. *Schiff Bases and Mixtures of Schiff Bases for Nematic Behaviors*, Br. Pat. No. 1,170,486 (March 12, 1969).
[41] Room temperature products available from: Liquid Crystal Industries, 460 Brown Avenue, Turtle Creek, Pennsylvania 15145 U.S.A.; Vari-Light Corporation, 9770 Conklin Road, Cincinnati, Ohio 45242, U.S.A.; Eastman Kodak Co., Rochester, New York 14650, U.S.A.
[42] HEILMEIER, G. H. *Appliance Engineer* **2,** 21 (1968).
[43] DEUTSCH, C. and KEATING, P. N. *J. appl. Phys.* **40,** 4049 (1969).
[44] ASSOULINE, G. and LEIBA, E. *J. Phys. Radium, Paris* **30,** C4, No. 11–12, 109 (1969).

[45] Appeared as an advertisement in *Liquid Crystals and Their Applications* (edited by Thomas Kallard), Optosonic Press, New York (1970), p. 212.
[46] HEILMEIER, G. H. and GOLDMACHER, J. E. *Proc. IEEE* **57,** 34 (1969).
[47] HEILMEIER, G. H., ZANONI, L. A. and GOLDMACHER, J. E. in *Liquid Crystals and Ordered Fluids* (edited by J. F. Johnson and R. S. Porter), Plenum Press, New York (1970), p. 215.
[48] CHISTYAKOV, I. G. *Soviet Phys. Usp.* **9,** 551 (1967); *Usp. fiz. Nauk.* **89,** 563 (1966).
[49] HELFRICH, W. *Appl. Phys. Lett.*, **17,** 531 (1970).

The following references relate to some of the device-oriented research during the period up to March, 1973. The literature has become quite voluminous and the list represents only a representative cross section.

Materials and Structures

CASTELLANO, J. *Ferroelectrics* **3,** 1 (1971).
SCHADT, M. *J. chem. Phys.* **56,** 13 (1971).
BERREMAN, D. W. *Phys. Rev. Lett.* **28,** 1683 (1972).
DE JEU, W. H. and WAN DER WEEN, J. *Phil. Res. Rev.* **27,** 172 (1972).
GRAY, G. W. HARRISON, K. J. and NASH, J. A. *Electronic Lett.* **9,** 130 (1973).

Cholesteric-nematic Transition

WYSOCKI, J. *et al. SID Int'l. Symposium Digest II*, 122 (1971).
ASHFORD, A., CONSTANT, J., KIRTON, J. and RAYNES, E. P. *Electronic Lett.* **9,** 118 (1973).

Dynamic Scattering

ORSAY LIQUID CRYSTAL GROUP *Molec. Crystals Liqu. Crystals* **13,** 187 (1971).
KASHNOW, R. A. and COLE, H. S. *J. appl. Phys.* **5,** 2134 (1971).
SUSSMAN, A. *Molec. Crystals Liqu. Crystals* **14,** 182 (1971).
SUSSMAN, A. *Appl. Phys. Lett.* **21,** 269 (1972).

Storage Effect

KENLLENEVICH, B. and COCHE, A. *Appl. Phys.* **42,** 5313 (1971).
HAAS, W., ADAMS, J. and DIR, G. *Chem. Phys. Lett.* **14,** 95 (1972).

Tuneable Birefringence (Controlled Retardation)

KAHN, F. J. *Appl. Phys. Lett.* **20,** 199 (1972).
SCHIEKEL, M. F. and FAHRENSCHON, K. *Appl. Phys. Lett.* **19,** 391 (1971).
SOREF, R. A. and RAFUSE, M. J. *J. appl. Phys.* **43,** 2029 (1972).

Device Performance and Life

RAYNES, E. P. and JAKEMAN, E. *Phys. Lett.* **39A,** 69 (1972).
CREAGH, L. T., KMETZ, A. R. and REYNOLDS, R. A. *IEEE Trans. on Electronic Devices* **18,** 672 (1971).
SUSSMAN, A. *Appl. Phys. Lett.* **21,** 126 (1972).

Photo-Conductor-Controlled Devices

ROSE, A. *IEEE Trans. on Electronic Devices* **19,** 430 (1972).
ASSOULINE, G., HARENG, M. and LEIBA, E. *Proc. IEEE* **59,** 1355 (1971).
MARGARUM, J. D., BEARD, T. D., BLEHA, W. P. and WONG, S.-Y. *Appl. Phys. Lett.* **19,** 216 (1971).

Review Articles

SOREF, R. A. in *The Physics of Opto-Electronic Materials* (edited by W. A. Albers, Jr.), Plenum Press, New York (1971).
SUSSMAN, A. *IEEE Trans. on Parts, Hybrids and Packaging*, Dec. 1972.
ELLIOTT, G. *Chemistry in Britain* **9,** 213 (1973).

7.3 LIQUID CRYSTALS IN GAS-LIQUID CHROMATOGRAPHY

J. P. Schroeder

The use of liquid crystals as stationary phases in gas-liquid chromatography (g.l.c.) provides a convenient experimental method for investigating the solvent and certain other properties of mesophases. It also provides a promising practical g.l.c. technique. The literature on the subject through 1966 has been reviewed by Kelker and Von-Schivizhoffen [1].

The first investigations in this field were made by Kelker [2, 3] and by Dewar and Schroeder [4] in the early 1960's. At the time of writing, twenty four liquid crystalline compounds have been studied as stationary phases. These are listed in Table 7.3.1 with the temperature ranges of their mesophases and literature references to use in g.l.c.

Interest in liquid crystals as solvents stemmed from the orienting effect of the parallel molecular alignment in mesophases on dissolved molecules. Linear solute molecules had been shown [18] to have a smaller disruptive effect on nematics than bulky, nonlinear molecules, presumably because the former fit more readily into the nematic "lattice", being oriented with their long axes parallel to those of the solvent molecules. Such behaviour suggested that nematic solvents might display selectivity toward solutes on the basis of molecular shape, i.e., everything else being equal, a solute consisting of linear molecules should adapt sterically to the parallel alignment in the mesophase more readily than one with relatively bulky molecules and thus should be more soluble. This hypothesis proved to be correct. For a number of pairs of *meta*- and *para*-disubstituted benzenes with very similar boiling points, the more linear *para*-isomer was invariably found to have the longer retention time in g.l.c. on nematic stationary phases [2–4].

Knowledge of liquid crystals as solvents has been given a firmer and more quantitative basis by thermodynamic studies using g.l.c. as the experimental method. Solutes were shown to have higher enthalpies, entropies and free energies of solution in nematics [2, 3, 7], smectics [3, 17] and cholesterics [17] than in the corresponding isotropic liquids. Solution in a mesophase requires more energy (is more endothermic) than solution in the isotropic liquid of the same compound, because energy is required to overcome the steric restraints on entrance of solute molecules into the ordered "lattice". However, there is a nett counteracting entropy increase associated with disordering of the mesophase on dissolution of the solute that is favourable to the process. Accordingly, liquid crystals are reasonably good solvents. They are poorer solvents than the corresponding isotropic liquids for which free energies of solution are always lower, the relatively favourable enthalpy over-riding the entropy effect.

The observed selectivity of nematic and smectic solvents between close-boiling *m*- and *p*-disubstituted benzenes can be rationalized in thermodynamic terms. The enthalpy and entropy of solution of the *para*-isomer in the mesophase are lower, indicating, respectively, stronger solvent–solute interactions and a better ordered solution state. In g.l.c., going from the disordered vapour (similar for both isomers) to the ordered liquid state, the *para*-substituted molecule sacrifices more translational and rotational freedom but, in return, its favourable geometry allows stronger interaction with the mesophase. On balance, the entropy loss is overcome by the enthalpy change and the *para*-isomer is more soluble than the *meta*-isomer [3, 13]. The solubility of long, narrow solute molecules is favoured only if they are also rigid so that there is little sacrifice in entropy on dissolving. Thus, long chain n-alkanes have higher free energies of solution than xylenes because molecules of the former are flexible and must straighten out for maximum interaction with the "lattice", while molecules of the latter are intrinsically rigid and rod-like [17].

The mechanism of solution in cholesterics appears to differ from that in smectics or nematics. For the cholesterogen **24,** the enthalpies and entropies of solution of hydrocarbons are significantly higher and the solvent–solute interactions much weaker relative to the corresponding quantities for the smectic and the amorphous isotropic liquid of this compound [17]. If solution were occurring within the mesomorphous "lattices", one would expect the solvent-solute interactions to be in the order smectic $>$ cholesteric $>$ isotropic. Perhaps instead, solution in the cholesteric is principally between molecular layers where the spatial relationships are less conducive to strong interactions. This would account qualitatively for the high enthalpies and entropies of solution, the fact that cholesteric stationary phases are not selective toward *m*- and *p*-xylene [11, 14, 15, 17] and the lack of sharp, well resolved n.m.r. peaks for benzene dissolved in cholesterics [19] as contrasted with nematic solvents which are highly effective ordering matrices. Solution between molecular layers has also been proposed for smectic solvents [14]. More will be said about this later.

Information concerning transitions between the mesophase and the amorphous isotropic liquid has been obtained from g.l.c. experiments. If the retention time (or volume) of a solute on a nematic [1, 3, 4, 5, 7] or cholesteric [16] g.l.c. column is plotted against temperature, the resulting curve is typically as shown in Fig. 7.3.1. For the mesophase, retention time decreases with temperature in the normal manner, but near the transition point, there is a minimum in the curve, and retention time increases with temperature over a range of several degrees. When the transition to the amorphous isotropic liquid is complete, the curve resumes a normal downward course. The rise at the transition is undoubtedly associated with the increase in solubility deriving from the decrease in free energy of solution on going from ordered mesophase to the isotropic liquid. Thus, for disubstituted benzene isomers on nematic stationary phases, the rise is greater for the *meta*-compound [3, 4] which is less soluble in the mesophase than the *para*-compound but has about the same solubility in the isotropic liquid. That the change begins well below the transition point and

TABLE 7.3.1. *Liquid crystalline compounds that have been studied as stationary phases in g.l.c.*

Compound	Structure	Smectic range (°C)	Nematic range (°C)	References
	$RO-C_6H_4-N{=}NO-C_6H_4-OR$			
1	$R{=}CH_3$	- - - -	118·5–135	2–10
2	$R{=}C_2H_5$	- - - -	138·5–169·5	3, 5, 7, 8, 10–12
3	$R{=}n\text{-}C_4H_9$	- - - -	107–136·5	10
4	$R{=}n\text{-}C_6H_{13}$	(74)[a]	81–129	4, 6, 8–10, 13
5	$R{=}n\text{-}C_7H_{15}$	74–93·5	93·5–124	4
	$RO-C_6H_4-CO{\cdot}O-R'-O{\cdot}OC-C_6H_4-OR$			
6	$R{=}n\text{-}C_7H_{15}$, $R'{=}-C_6H_4-$	83–125	125–206	14
7	$R{=}n\text{-}C_4H_9$, $R'{=}-C_6H_4-C_6H_4-$	171–184	184–358 (decomp.)	14
8	$R{=}n\text{-}C_7H_{15}$, $R'{=}-C_6H_4-C_6H_4-$	150–211	211–316	14
9	$CH_3O-C_6H_4-CH{=}N-C_6H_3(Cl)-C_6H_3(Cl)-N{=}CH-C_6H_4-OCH_3$	- - - -	154·5–363 (decomp.)	8, 14
10	$C_2H_5O-C_6H_4-CH{=}N-C_6H_4-CH{=}CH-CO{\cdot}OC_2H_5$	78–118, 118–157[b]	157–160	14
11	$CH_3O-C_6H_4-CH{=}N-C_6H_4-CN$	- - - -	106–117	5
12	$CH_3O-C_6H_4-CH{=}N-C_6H_4-OCO{\cdot}CH_3$	- - - -	83–107	5

Compound	Structure	Smectic range (°C)	Nematic range (°C)	References
13	$CH_3O-C_6H_4-CH{=}N-C_6H_4-OC_4H_9$	----	48–108	7
14	$C_2H_5O-C_6H_4-N{=}CH-C_6H_4-CH{=}N-C_6H_4-OC_2H_5$	----	200–320	5
15	$C_6H_5-CH{=}N-C_6H_4-C_6H_4-N{=}CH-C_6H_5$	----	239–265	5
16	$CH_3O-C_6H_4-CH{=}N-C_6H_4-C_6H_4-N{=}CH-C_6H_4-OCH_3$	----	266–>390	5
17	$CH_3O-C_6H_4-CH{=}N-C_6H_4-CH{=}CH-C_6H_4-N{=}CH-C_6H_4-OCH_3$	----	274–>340	5
18	$C_2H_5O{\cdot}OC-C_6H_4-N{=}NO-C_6H_4-CO{\cdot}OC_2H_5$	114–120	----	3
19	$C_2H_5O{\cdot}OC-CH{=}CH-C_6H_4-N{=}NO-C_6H_4-CH{=}CH-CO{\cdot}OC_2H_5$	140–251 (decomp.)	----	14
			Cholesteric range (°C)	
20	Cholesteryl benzoate	----	149–180	11, 14, 15
21	Cholesteryl acetate	----	(94·5)[c,d]	16
22	Cholesteryl n-valerate	----	93–101·5[d]	16
23	Cholesteryl n-nonanoate	(77·5)[d,e]	80·5–92[d]	14, 16
24	Cholesteryl myristate	71–81	81–86·5	17

a Monotropic nematic–smectic transition.
b This compound exhibits two distinct smectic mesophases.
c Monotropic isotropic–cholesteric transition.
d Determined by hot stage microscopy. Differential thermal analysis data [16] are not in agreement with these values.
e Monotropic cholesteric–smectic transition.

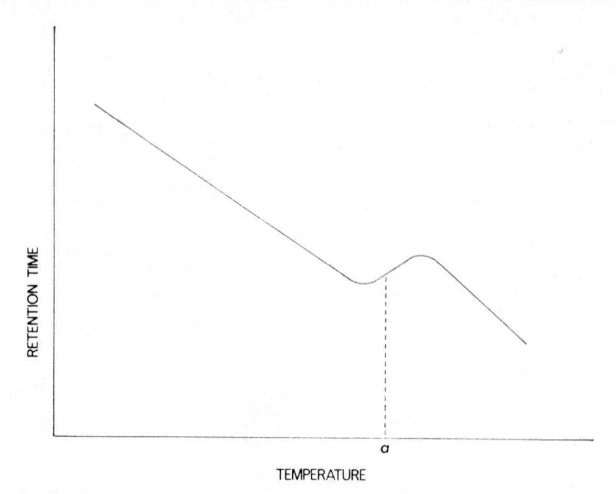

FIG. 7.3.1. Typical plot of retention time *v.* temperature for a solute chromatographed on a nematic or cholesteric stationary liquid phase. (a = mesophase-amorphous isotropic liquid transition temperature.)

continues above it is in agreement with other evidence, e.g., viscosity, thermal expansion and specific heat variations, that nematic- and cholesteric-isotropic transitions involve processes which occur over a significant temperature range [1, 20, 21].

A plot of the type shown in Fig. 7.3.1 would be expected for smectic-amorphous isotropic transitions also, in consideration of the greater difference in molecular order between the two states of matter; this should provide a large enthalpic effect favouring solubility in the amorphous isotropic liquid. No rise in the curve at the transition point was detected for the one compound (**18**) that has been studied in this manner [3]. However, drawing broad conclusions from this result would be premature because the smectic mesophase of compound **18** appears to be poorly ordered. It exists over only a short range (114–120°C) and is not selective toward *m*- and *p*-xylene. A well ordered smectic might behave differently.

Separation of compounds on the basis of molecular shape is the most promising practical application of liquid crystals in g.l.c. Nematic stationary phases have proved to be particularly selective and, of those studied to date, those of nematic compounds **2** and **9** are the best. Examples of their ability to differentiate between linear and non-linear solute molecules are given in Table 7.3.2 along with data for a silicone resin, which discriminates primarily on the basis of boiling point, for comparison.

For all of the disubstituted benzenes in Table 7.3.2, the *meta*- and *para*-isomers have similar volatilities. This is reflected in the relative retention (α) values obtained with the silicone resin, most of which are near unity. On the nematic stationary liquid phases, the more linear *p*-substituted compound is invariably eluted well after the *meta*-isomer. The separation of *m*- and *p*-xylene by g.l.c. is a classically difficult one because of their similar boiling points and polarities. However, complete resolution can be

TABLE 7.3.2. *Relative retentions* (α) *of* p-*disubstituted benzenes* (m-*isomer* = 1·00) *and alkyl benzoates* (*methyl benzoate* = 1·00) *on a silicone and on two nematic stationary liquid phases* [8, 14]

Compound	Stationary Liquid Phase					
	SE-30 Silicone		Nematic **2**		Nematic **9**	
	α	Temp. (°C)	α	Temp. (°C)	α	Temp. (°C)
p-$CH_3C_6H_4CH_3$	0·99	104	1·10 [a]	141	1·16	160
p-$CH_3OC_6H_4OCH_3$	0·99	100	—	—	1·24	158
p-$C_2H_5C_6H_4CH_3$	1·01	100	1·15	141	1·19	160
p-$CH_3OC_6H_4CH_3$	1·02	100	1·18	141	1·18	160
p-BrC_6H_4Br	1·02	100	—	—	1·20	158
p-ClC_6H_4Cl	1·04	100	1·26	141	1·20	160
p-$ClC_6H_4CO{\cdot}CH_3$	1·06	100	—	—	1·28	158
p-$CH_3C_6H_4CO{\cdot}OCH_3$	1·07	100	—	—	1·33	158
$C_6H_5CO{\cdot}OC_2H_5$	1·64	104	1·23	141	1·11	160
$C_6H_5CO{\cdot}OC_3H_7$-i	1·98	104	1·08	141	0·85	160

a Kelker and co-workers [3, 5] report 1·09 at 140°C.

accomplished with compounds **2** or **9** (α values 1·10 and 1·16) using a relatively short column.

The excellent selectivities of compounds **2** and **9** are further exemplified by the data for the alkyl benzoates. For the silicone column, the order of elution is methyl ester, then ethyl and finally isopropyl, with a large interval between the first two. This is in accord with the respective boiling points: 199, 212 and 218·5°C. With compound **2**, the retention times of methyl and ethyl benzoate are closer together and the isopropyl ester is eluted long before the lower boiling ethyl ester. The same trends are even more evident in the case of compound **9** where the isopropyl ester is eluted *first*. The results are consistent with the previous explanation of selectivity by nematic solvents. The methyl benzoate molecule is the most linear of the three and, therefore, fits most readily into a nematic "lattice". For this reason, the methyl ester has a long retention time on columns containing nematic compounds **2** or **9** in relation to its boiling point. Conversely, isopropyl benzoate, with its bulky alkyl group, has anomalously short retention times on the nematic stationary phases.

If the selective affinity of nematic solvents toward linear solute molecules is the result of a steric discrimination by the parallel molecular alignment, the regularity of that alignment should determine the degree of selectivity. Based on these premises, the relative order in several nematic "lattices" has been estimated as a function of temperature, using the difficult separation of *m*- and *p*-xylene by g.l.c. as the experimental criterion [8]. A convenient way to present the data is to plot relative retention (α) of *p*-xylene (*m*-xylene = 1·00) against temperature. Such a graph for three 4,4′-dialkoxyazoxybenzenes (compounds **1**, **2**, and **4**) is shown in Fig. 7.3.2. In the curve for compound **1**, note the gradual increase in α on cooling the amorphous isotropic liquid from 150°C to 135°C, the isotropic-nematic transition point. On further cooling, there is a rapid increase in α at the transition temperature and the order of elution reverses as the nematic "lattice" forms and displays a strong preference for the *para*-isomer. Below the transition point, α continues to increase until 115°C, when crystallization occurs. A similar shape is exhibited by the curves for compounds **2** and **4**, but there are significant differences in details. The

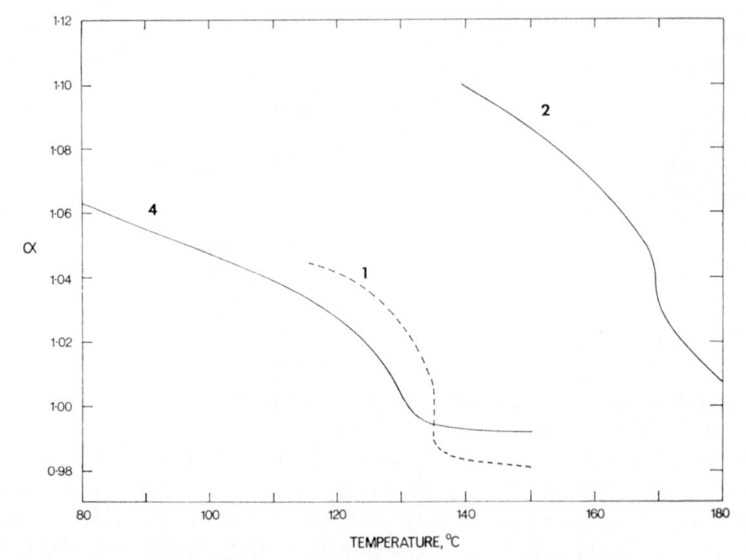

FIG. 7.3.2. Variation of the relative retention (α) of *p*-xylene (*m*-xylene = 1·00) with temperature using compounds **1**, **2** and **4** as stationary liquid phases [8].

slope in the isotropic region is larger for compound **2**, smaller for compound **4**, and α is greater than unity for the isotropic liquid of compound **2** even at 181°C, twelve degrees above the transition point. At a given temperature (it is necessary to extrapolate the curve numbered **2** for this comparison), the selectivities of the nematic mesophases are in the order compound **2** $\gg$ **1** $>$ **4**.

The increases in α on cooling (*a*) the isotropic and (*b*) the nematic liquid indicate development of nematic order above the usually assigned transition point and an increase in "lattice" order with decreasing temperature, respectively. Thus, the data support the evidence presented earlier that the processes that lead to the transition from mesophase to amorphous liquid occur to some extent over a temperature range, only reaching a climax at the transition "point". They also emphasize the conclusion [3–5] that optimum selectivity of a nematic is obtained at the lowest possible operating temperature.

A direct relationship between temperature range of the nematic and selectivity is an attractive concept [4]. The results above and others [1, 5] are consistent with this view. This is reasonable in that a long nematic range indicates strong intermolecular forces in the mesophase which are not easily overcome by thermal agitation and, therefore, a high degree of "lattice" order. However, when available data are examined, exceptions are apparent. In Table 7.3.3, the nematics for which selectivity between *m*- and *p*-xylene has been determined are listed in order of decreasing liquid crystalline range along with selectivity (α) values determined in the lower portion of that range. There is a direct correlation between α and mesophase range except for compounds **4**, **6** and **7**. These are the only compounds in the table with long terminal n-alkyl groups (C_4–C_7), and their

TABLE 7.3.3. *Relative retention* (α) *of* p-*xylene* (m-*xylene* = 1·00) *on nematic stationary phases as a function of nematic temperature range*

Stationary Phase	Nematic Range (°C)	α^a	Reference
9	208·5 (154·5–363 (decomp.))	1·16	8
7	174 (184–358 (decomp.))	1·01[b]	14
6	81 (125–206)	1·05[c]	14
4	48 (81–129)	1·07[d]	8
2	31 (138·5–169·5)	1·10	8
		1·09	3, 5
12	24 (83–107)	1·05	5
1	16·5 (118·5–135)	1·04	5, 8
11	11 (106–117)	1·03	5

a Determined just above the melting point.
b Values of α for *p*-tolualdehyde, methyl *p*-toluate and *p*-chloroacetophenone (*m*-isomer = 1·00) are 1·17, 1·22 and 1·22, respectively [14].
c α for *p*-methylanisole (*m*-isomer = 1·00) is 1·11 [14].
d Values of α for *p*-ethyltoluene, *p*-methylanisole [8], *p*-ethylstyrene and *p*-divinylbenzene [13] (*m*-isomer = 1·00) are 1·09, 1·13, 1·24 and 1·30, respectively.

relatively poor selectivity may reflect comparatively large gaps between parts of the long molecules caused by the bulky end groups. In support of this, greater selectivity is exhibited by compounds **4**, **6** and **7** toward *meta-para*-isomer pairs that have larger substituents (footnotes *b-d*, Table 7.3.3). Selectivity for these three mesophases varies inversely with minimum use temperature. Undoubtedly, the greater thermal disordering of the "lattice" of compound **7** at 184°C accounts for part of its poor performance relative to compounds **4** and **6** which remain nematic down to 81°C and 125°C, respectively. This effect is clearly illustrated in Fig. 7.3.2. Although the nematic of compound **4** is less selective than that of compound **1** at a given temperature, the former has the higher maximum α value because it can be cooled to a lower temperature without crystallizing.

In summary, consideration of three factors appears to be important when predicting the selectivity of a nematic: temperature range of the nematic, molecular rigidity and minimum use temperature. This multiplicity of criteria makes forecasting difficult. For example, compounds **2** and **9** are very effective, but both have relatively high melting points, and compound **2** has only a moderately long nematic range.

G.l.c. data for the smectic and cholesteric mesophases that have been

TABLE 7.3.4. *Relative retention* (α) *of* p-*xylene* (m-*xylene* = 1·00) *on smectic and cholesteric stationary phases*

Stationary Phase	Smectic Range (°C)	Cholesteric Range (°C)	α (temp., °C)[a]	Reference
6	83–125	—	1·10 (68)	14
			1·07 (88)[b]	14
8	150–211	—	1·02 (157)[c]	14
19	140–251 (decomp.)	—	1·01 (160)[d]	14
10	78–118	—	1·00 (85)	14
	118–157[e]		1·02 (122)[f]	14
18	114–120	—	1·00 (115)[g]	3
7	171–184	—	0·99 (175)[h]	14
20	—	149–180	1·01 (155)	14
23	(77·5)[i]	80·5–92	1·00 (70)	14
			1·00 (85)	14
24	71–81	81–86·5	*j*	17

a Values of α (temp., °C) for other *p*-disubstituted benzenes (*m*-isomer = 1·00) are given in subsequent footnotes.
b *p*-Methylanisole 1·33 (88), *p*-chlorotoluene 1·17 (88).
c Methyl *p*-toluate 1·18 (180), *p*-chloroacetophenone 1·14 (180).
d Methyl *p*-toluate 1·18 (145), *p*-dibromobenzene 1·13 (145).
e This compound exhibits two distinct smectic mesophases.
f *p*-Dichlorobenzene, *p*-dibromobenzene and *p*-bromochlorobenzene 1·16 (122).
g *p*-Dichlorobenzene 1·15 (115).
h *p*-Methylanisole and *p*-dichlorobenzene 1·06 (175).
i Monotropic cholesteric–smectic transition.
j Numerical α values are not given in reference 17, but it is stated that the smectic mesophase of compound **24** is selective toward *m*- and *p*-xylene and the cholesteric mesophase is not.

studied are presented in Table 7.3.4. With the exception of compound **6**, these mesophases display poor selectivity between *m*- and *p*-xylene. However, all of the smectic stationary phases are selective toward other *meta-para*-isomer pairs (footnotes *b–d* and *f–h*, Table 7.3.4).

Nematics appear to have a special affinity toward linear solute molecules because the latter fit readily into the solvent "lattice" of parallel, rod-shaped molecules. The molecules of cholesteryl esters are neither as rod-like nor as polar as those of nematogenic compounds and, therefore, the poor selectivity of compounds **20** and **23** is not surprising. However, the smectics of the other compounds in Table 7.3.4 are similar to nematics in the shape, polarity and parallel arrangement of their constituent molecules, and are even more highly ordered. By the argument above, these smectic solvents should be more selective than nematics, but the data show that, in general, they are not. A possible explanation is as follows:

Smectic liquids flow by movement of one molecular layer with respect to the next, showing that the parallel molecules within layers are firmly held together. Perhaps the intralayer "lattice" is so closely packed as to resist strongly the entrance of solute molecules which, as a result, dissolve mainly by insertion between layers [14]. This is a region of relatively weak intermolecular attractive forces and, therefore, should be more accessible, but also poorly selective. There are experimental data to support such a mechanism. Smectic compound **6** exhibits lower solvent power (shorter retention times) than nematic compound **6**. Compound **10** gives two smectic mesophases and the higher temperature modification, which surely has the more "open" molecular arrangement within layers, is more selective (Table 7.3.4). The outstanding selectivity of smectic compound **6** may be the result of dissolution between layers followed by preferential absorption of the *para*-isomer by the parallel n-heptyl chains at the layer surfaces. This does not contradict the earlier proposal that long alkyl end groups disrupt nematic mesophases. In the more tightly packed smectic molecular arrangement, terminal n-alkyl chains may be sterically restrained from coiling and probably assume an extended configuration. If this is correct, the α values for compounds **6** and **8** (the latter also has n-heptyl end groups) indicate that the alkyl chain ordering is highly temperature dependent (Table 7.3.4).

Before leaving the subject of smectic stationary phases, an error in the literature, for which the author is responsible, should be pointed out. This is the claim [4] that smectic compounds **4** and **5** are effective in separating *m*- and *p*-xylene by g.l.c. These liquids were later found to be *nematic* mixtures of 4,4′-dialkoxyazoxybenzenes [6], an observation that led to a systematic study of such blends [8, 10] (see below).

By analogy with the selectivity of nematic and smectic stationary phases toward solutes on the basis of molecular shape, it has been suggested that *d*- and *l*-enantiomorphs might be resolved by g.l.c. using optically active liquid crystals [1, 4, 11, 15]. Experimental success in this endeavour has not been reported.

An inherent practical limitation of mesomorphous stationary phases is their restricted useful temperature ranges. Kelker *et al.* [5] showed that selective nematics of pure compounds with low, medium and high operating temperatures are available. In the same paper, a still more promising means of varying the nematic range was described, i.e., the use of mixtures. At about the same time, as a result of the error referred to earlier, the author recognized the potentials of this technique [6].

Three binary systems have been studied in g.l.c. [5, 6, 8]. As in the case of pure compounds, the best way to present the data is by a plot of α for *p*-xylene (*m*-xylene = 1·00) against temperature (Fig. 7.3.3). System **1–4**

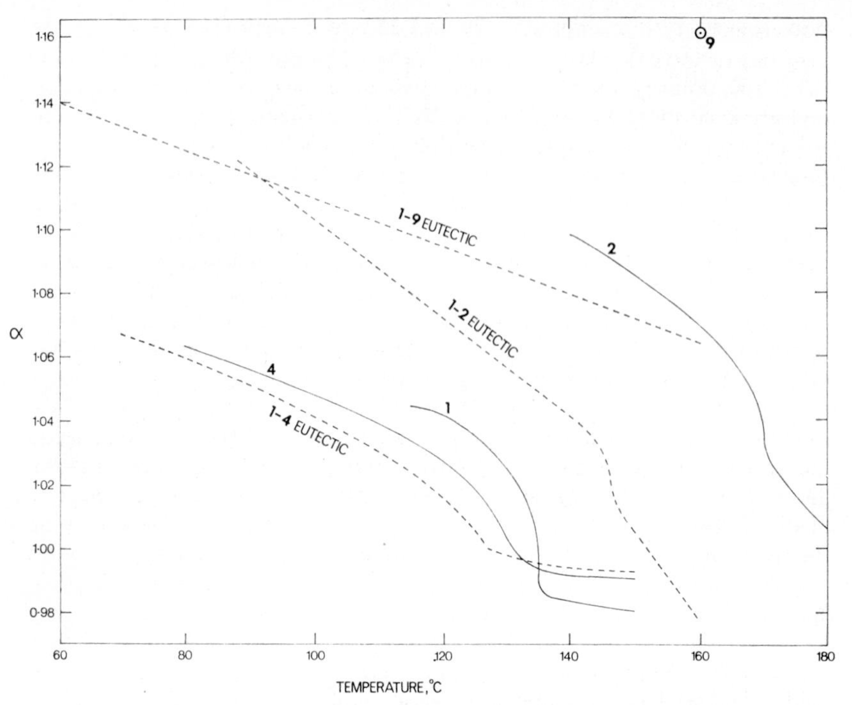

FIG. 7.3.3. Variation of the relative retention (α) of *p*-xylene (*m*-xylene = 1·00) with temperature using compounds **1, 2, 4** and **9** and the eutectic mixtures of compounds **1–2, 1–4*** and **1–9** as stationary liquid phases [8].

illustrates the general principles. Addition of compound **1** to compound **4** lowers the melting point without depressing the nematic–isotropic transition temperature appreciably. The eutectic composition (27 mole % of compound **1**) has an operating nematic range of 70–127°C compared with 81–129°C for pure compound **4.** Nematic mixtures of widely varying composition (27–82 mole % of compound **1**) are very similar in selectivity to one another and to nematic compound **4,** but inferior to nematic compound **1** at a given temperature. Apparently, there is only minor disruption of the "lattice" on adding compound **1** to **4,** and considerably more on addition of compound **4** to **1.** The maximum selectivity for the system (α = 1·07) is shared by compound **4** and the eutectic mixture at 80°C and 70°C respectively. Because of crystallization, compound **1** cannot be used below 115°C (α = 1·045).

The system **1–9** is an especially good example of the practical use of mixed nematic stationary phases. The anil **9** has a very selective mesophase (Table 7.3.2) and a long nematic range, but it is high melting and heat-sensitive, a combination that limits its usefulness severely. *p*-Azoxyanisole (**1**) is lower melting but less selective and has a short nematic range.

* Recent data for the systems **1–3, 2–3, 2–4,** and **3–4** [10] lend support to the results for systems **1–2,** and **1–4.**

Addition of compound **1** to compound **9** depresses the melting point and the mesophase of the eutectic mixture supercools readily to 60°C where its selectivity is almost as high as that of pure compound **9** just above the melting point. At the opposite end of the composition scale, addition of only 3 mole % of compound **9** to compound **1** results in retention of nematic mesomorphism to a higher temperature and improved maximum selectivity. With increasing amounts of compound **9**, these trends are magnified, culminating in the eutectic mixture, for which selectivity over the broad range 60–160°C is markedly better than that attainable with pure compound **1** at any temperature. Thus, relative to heat-sensitive compound **9**, the eutectic mixture permits operation at a lower temperature without a great sacrifice in performance. Relative to compound **1**, it provides superior selectivity over a broader temperature range.

Table 7.3.5 compares the selectivities of compounds **1, 2, 4, 9** and the eutectic mixtures of compounds **1–2, 1–4** and **1–9** toward other *meta-para-* isomer pairs. In each case, the column temperature is in the lower part of the nematic range. All of the nematics are highly selective, but the eutectic mixture is invariably at least as selective as either pure component.

Mesomorphous mixtures of compound **2** and cholesteryl benzoate were found to be less selective than nematic compound **2** [11]. Both these blends and the eutectic mixture of compounds **1–2** [7] exhibit retention time *v.* temperature plots of the type shown in Fig. 7.3.1. As mesophases, both systems are better solvents for *o*-xylene than the pure components, presumably because the "lattice" structure is more open in mixtures.

TABLE 7.3.5. *Relative retentions of* p-*ethyltoluene*, p-*dichlorobenzene and* p-*methylanisole* (meta-*isomer* = 1·00) *on compounds* **1, 2, 4,** *and* **9**, *eutectic mixtures of compounds* **1–2, 1–4** *and* **1–9** *and silicone* [5, 8]

Stationary Phase	Temp. (°C)	p-$C_2H_5C_6H_4CH_3$	p-ClC_6H_4Cl	p-$CH_3OC_6H_4CH_3$
1	120	1·09	1·17	1·12
2	141	1·15	1·26	1·18
1–2 eutectic	100	1·18	1·28	1·23
4	85	1·09	1·11	1·13
1–4 eutectic	75	1·11	1·17	1·16
9	160	1·19	1·20	1·18
1–9 eutectic	100	1·21	1·25	1·23
SE-30 silicone	100	1·01	1·04	1·02

Experimental Considerations

Experimentally, mesomorphous stationary phases are amenable, for the most part, to standard g.l.c. procedures. In the preparation of packed columns, they can be deposited on solid supports from solution in a volatile solvent (dichloromethane is particularly effective). If a suitable solvent cannot be found, dry blending of the solid stationary phase and the support gives satisfactory results [1]. Liquid crystals have also been used in capillary columns [1, 22].

A constant, uniform column temperature is important so that the state of the stationary phase (solid, mesophase, amorphous isotropic liquid) is known and is the same throughout. Zones of uncoated solid support at the ends of the column have been recommended as "temperature buffers" [1]. The coated zone is thus restricted to the thermostated region of the chromatograph and insulated from the injection zone and detector which are usually at different temperatures.

Even when the column temperature is not in question, there may be some uncertainty as to the state of the liquid phase because of substrate effects (the transition temperatures of cholesteryl myristate deposited on Chromosorb W are different from those of a pure sample of the ester [17]) and also the presence of the solute.

Variations in retention volume *v*. temperature data depending on the thermal history of the columns have been observed for smectic [3], nematic [5] and cholesteric [15] stationary phases deposited on solid supports [1]. However, there are no discernible differences in transition temperatures and heats of transition of compound **1** between bulk samples and those deposited on Chromosorb W. In g.l.c., no surface effects are observed at the solid support-nematic liquid interface for compounds **1** and **4**. Slight effects are sometimes detected at the vapour-nematic mesophase interface but these are negligible when the liquid film is at least 1,000 Å thick. For this reason, a relatively high (15 wt %) content of stationary phase in the packing is recommended [9]. These phenomena are not well understood and warrant further research.

The mesomorphous compounds that have been studied in g.l.c. are of three structural types: axoxybenzene derivatives, esters and anils. No stability difficulties are reported for the first two categories, but anil **9** undergoes some thermal decomposition even at 158°C [14], and primary amines give badly tailing peaks when chromatographed on anil **17** as a result of the exchange process in equation (1) [5]. On the other hand, anils **14–17** are recommended

$$\text{—CH=NAr} + \text{RNH}_2 \longrightarrow \text{—CH=NR} + \text{ArNH}_2 \qquad (1)$$

$$\text{—CH=NAr} + \text{RCHO} \longrightarrow \text{—CHO} + \text{RCH=NAr} \qquad (2)$$

for use at elevated temperatures [5] and anil **9** does not appear to undergo exchange with aldehydes (equation (2)) [14]. Liquid crystalline stationary phases, being polar and having high molecular weights, are not very volatile so that bleeding is rarely a problem.

REFERENCES

[1] KELKER, H. and VON-SCHIVIZHOFFEN, E. *Adv. Chromat.* **6,** 247 (1968).
[2] KELKER, H. *Ber. Bunsenges. phys. Chem.* **67,** 698 (1963).
[3] KELKER, H. *Z. analyt. Chem.* **198,** 254 (1963).
[4] DEWAR, M. J. S. and SCHROEDER, J. P. *J. Am. chem. Soc.* **86,** 5235 (1964).

[5] KELKER, H., SCHEURLE, B. and WINTERSCHEIDT, H. *Analytica. chim. Acta* **38,** 17 (1967).
[6] DEWAR, M. J. S., SCHROEDER, J. P. and SCHROEDER, D. C. *J. org. Chem.* **32,** 1692 (1967).
[7] KELKER, H. and VERHELST, A. *J. chromat. Sci.* **7,** 79 (1969).
[8] SCHROEDER, J. P., SCHROEDER, D. C. and KATSIKAS, M. in *Liquid Crystals and Ordered Fluids* (edited by J. F. Johnson and R. S. Porter), Plenum Press, New York and London (1970), p. 169.
[9] CHOW, L. C. and MARTIRE, D. E. *J. phys. Chem.* **73,** 1127 (1969).
[10] ANDREWS, M. A., SCHROEDER, D. C. and SCHROEDER, J. P. *J. Chromat.* **71,** 233 (1972).
[11] KELKER, H. and WINTERSCHEIDT, H. *Z. analyt. Chem.* **220,** 1 (1966).
[12] KELKER, H. *Abh. dt. Akad. Wiss. Berl., Kl. Chem., Geol. Biol.* no. 2, 49 (1966); *Chem. Abstr.* **67,** 76516 (1967).
[13] ZIELINSKI, W. L., Jr., FREEMAN, D. H., MARTIRE, D. E. and CHOW, L. C. *Analyt. Chem.* **42,** 176 (1970).
[14] DEWAR, M. J. S. and SCHROEDER, J. P. *J. org. Chem.* **30,** 3485 (1965).
[15] KELKER, H. in *Gas Chromatographie 1965* (edited by H. G. Struppe), Akademie-Verlag, Berlin (1966), p. B49.
[16] BARRALL, E. M., II, PORTER, R. S. and JOHNSON, J. F. *J. Chromat.* **21,** 392 (1966).
[17] MARTIRE, D. E., BLASCO, P. A., CARONE, P. F., CHOW, L. C. and VICINI, H. *J. phys. Chem.* **72,** 3489 (1968).
[18] DAVE, J. S. and DEWAR, M. J. S. *J. chem. Soc.* 4616 (1954); 4305 (1955).
[19] SACKMANN, E., MEIBOOM, S. and SNYDER, L. C. *J. Am. chem. Soc.* **89,** 5981 (1967).
[20] BARRALL, II, E. M., PORTER, R. S. and JOHNSON, J. F. *J. phys. Chem.* **71,** 895 (1967).
[21] PORTER, R. S. and JOHNSON, J. F. *J. appl. Phys.* **34,** 51, 55 (1963).
[22] DEWAR, M. J. S. and SCHROEDER, J. P., hitherto unpublished results.

Additional References

VOLLERT, U. and RICHTER, H. *Z. Chem.* **10,** 437 (1970).
KRAUS, G., SEIFERT, K., ZASCHKE, H. and SCHUBERT, H. *Z. Chem.* **11,** 22 (1971).
CHOW, L. C. and MARTIRE, D. E. *J. phys. Chem.* **75,** 2005 (1971).
CHOW, L. C. and MARTIRE, D. E. *Molec. Crystals Liqu. Crystals* **14,** 293 (1971).
SCHNUR, J. M. and MARTIRE, D. E. *Analyt. Chem.* **43,** 1201 (1971).
RICHMOND, A. B. *J. chromat. Sci.* **9,** 571 (1971).
KIRK, D. N. and SHAW, P. M. *J. chem. Soc.* (C), 3979 (1971).
TAYLOR, P. J., CULP, R. A., LOCHMUELLER, C. H., ROGERS, L. B. and BARRALL, II, E. M., *Separ. Sci.* **6,** 841 (1971).
ONO, A. *Nippon Kagaku Kaishi*, no. 4, 811 (1972).
PETERSON, H. T., MARTIRE, D. E. and LINDNER, W. *J. phys. Chem.* **76,** 596 (1972).

Index

N